Dychweler y llyfr hwn erbyn y dyddiad isod,
neu pan elwir amdano gan y Llyfrgellydd.
Dirwyir y darllenydd na ddychwel y llyfr erbyn
y dyddiad iawn.

This book must be returned on or before the
date shown below, or when required by the
Librarian. Fines will be imposed for the late
return of this book.

12/3/13

SEPTORIA ON CEREALS

A Study of Pathosystems

Edited by

J.A. Lucas, P. Bowyer and H.M. Anderson

IACR-Long Ashton Research Station
Bristol, UK

CABI *Publishing*

CABI *Publishing* – a division of CAB INTERNATIONAL

CABI *Publishing*
CAB INTERNATIONAL
Wallingford
Oxon OX10 8DE
UK

Tel: +44 (0)1491 832111
Fax: +44 (0)1491 833508
Email: cabi@cabi.org

CABI *Publishing*
10 E. 40th Street
Suite 3203
New York, NY 10016
USA

Tel: +1 212 481 7018
Fax: +1 212 686 7993
Email: cabi-nao@cabi.org

©CAB INTERNATIONAL 1999. All rights reserved. No part of this publication may be reproduced in any form or by any means, electronically, mechanically, by photocopying, recording or otherwise, without the prior permission of the copyright owners.

A catalogue record for this book is available from the British Library, London, UK.

Library of Congress Cataloging-in-Publication Data

Septoria on cereals : a study of pathosystems / edited by J.A. Lucas,
P. Bowyer, and H.M. Anderson.
p. cm.
Includes bibliographical references and index.
ISBN 0-85199-269-2 (alk. paper)
1. Grain- -Diseases and pests. 2. Septoria diseases.
3. Stagonospora diseases. 4. Septoria. 5. Stagonospora.
I. Lucas, John Alexander. II. Bowyer, P. (Paul) III. Anderson, H.
M. (Harry M.)
SB608.G6S46 1999 98-44780
633.1'1945- -dc21 CIP

ISBN 0 85199 269 2

UWB Deiniol Library
SB608.G6 S46 1999
30110006564232

Printed and bound in the UK by the University Press, Cambridge, from copy supplied by the editors.

Contents

Part Three: DISEASE MANAGEMENT AND DECISION SUPPORT

Contributors

Ahmed, H.U.
Plant Pathology Division, Bangladesh Rice Research Institute, Joydepur, Gazipur 1701, Bangladesh

Ballance, G.M.
Department of Plant Science, University of Manitoba, Winnipeg, Manitoba, Canada R3T 2N2

Bartlett, D.W.
Zeneca Agrochemicals, Jealott's Hill Research Station, Bracknell, Berkshire RG42 6ET, UK

Bernier, C.C.
Department of Plant Science, University of Manitoba, Winnipeg, Manitoba, Canada R3T 2N2

Bonants, P.J.M.
DLO-Research Institute for Plant Protection (IPO-DLO), PO Box 9060, 6700 GW Wageningen, The Netherlands

Butters, J.
IACR-Long Ashton Research Station, Department of Agricultural Sciences, University of Bristol, Long Ashton, Bristol BS41 9AF, UK

Carroll, A.M.
DuPont Experimental Station, Wilmington, DE 19880, USA

Caten, C.E.
School of Biological Sciences, The University of Birmingham, Birmingham B15 2TT, UK

Coakley, S.M.
Department of Botany and Plant Pathology, 2082 Cordley Hall, Oregon State University, Corvallis, OR 97331-2902, USA

Coker, R.R.
IACR-Long Ashton Research Station, Department of Agricultural Sciences, University of Bristol, Long Ashton, Bristol BS41 9AF, UK

Cook, R.J.
Morley Research Centre, Wymondham, Norfolk NR18 9DB, UK

Cooley, N.
AgrEvo UK Ltd, Chesterford Park, Saffron Walden, Essex CB10 1XL, UK

Cowger, C.
Department of Botany and Plant Pathology, 2082 Cordley Hall, Oregon State University, Corvallis, OR 97331-2902, USA

Dancer, J.
AgrEvo UK Ltd, Chesterford Park, Saffron Walden, Essex CB10 1XL, UK

Daniels, A.
AgrEvo UK Ltd, Chesterford Park, Saffron Walden, Essex CB10 1XL, UK

Daniels, M.J.
The Sainsbury Laboratory, John Innes Centre, Norwich Research Park, Colney Lane, Norwich NR4 7UH, UK

de Koning, J.R.A.
DLO-Research Institute for Plant Protection (IPO-DLO), PO Box 9060, 6700 GW Wageningen, The Netherlands

DiLeone, J.A.
Department of Botany and Plant Pathology, 2082 Cordley Hall, Oregon State University, Corvallis, OR 97331-2902, USA

Eyal, Z.
Department of Plant Sciences and the Institute for Cereal Crops Improvement, The George S. Wise Faculty of Life Sciences, Tel Aviv University, Tel Aviv 69978, Israel

Fiegen, M.
Department of Biochemistry, Max-Planck-Institut für Züchtungsforschung, Carl-von-Linné-Weg 10, D-50829 Köln, Germany

Flegg, L.M.
University of East Anglia, Norwich NR4 7TJ, UK

Foster, S.
AgrEvo UK Ltd, Chesterford Park, Saffron Walden, Essex CB10 1XL, UK

Fraaije, B.
IACR-Long Ashton Research Station, Department of Agricultural Sciences, University of Bristol, Long Ashton, Bristol BS41 9AF, UK

Garton, J.
Department of Plant Pathology and Microbiology, Texas A&M University, College Station, TX 77843-2132, USA

Gierlich, A.
Department of Biochemistry, Max-Planck-Institut für Züchtungsforschung, Carl-von-Linné-Weg 10, D-50829 Köln, Germany

Godwin, J.R.
Zeneca Agrochemicals, Jealott's Hill Research Station, Bracknell, Berkshire RG42 6ET, UK

Gurr, S.
Department of Plant Sciences, University of Oxford, South Parks Road, Oxford OX1 3RB, UK

Hagenaar-de Weerdt, M.
DLO-Research Institute for Plant Protection (IPO-DLO), PO Box 9060, 6700 GW Wageningen, The Netherlands

Halama, P.
Institut Supérieur d'Agriculture, 41 rue du Port, 59046 Lille cedex, France

Hamza, S.
Institut National Agronomique de Tunisie (INAT), 43 Av. Charles Nicolle, 1082 Tunis El Mahrajène, Tunisia

Hargreaves, J.
IACR-Long Ashton Research Station, Department of Agricultural Sciences, University of Bristol, Long Ashton, Bristol BS41 9AF, UK

Heaney, S.P.
Zeneca Agrochemicals, Jealott's Hill Research Station, Bracknell, Berkshire RG42 6ET, UK

Hoffer, M.E.
Department of Botany and Plant Pathology, 2082 Cordley Hall, Oregon State University, Corvallis, OR 97331-2902, USA

Hogan, K.
Department of Plant Pathology and Microbiology, Texas A&M University, College Station, TX 77843-2132, USA

Hollomon, D.W.
IACR-Long Ashton Research Station, Department of Agricultural Sciences, University of Bristol, Long Ashton, Bristol BS41 9AF, UK

Hossy, H.
Danish Institute of Agricultural Sciences, Flakkebjerg, DK-4200 Slagelse, Denmark

Hunter. T.
IACR-Long Ashton Research Station, Department of Agricultural Sciences, University of Bristol, Long Ashton, Bristol BS41 9AF, UK

Hutcheon, J.A.
IACR-Long Ashton Research Station, Department of Agricultural Sciences, University of Bristol, Long Ashton, Bristol BS41 9AF, UK

Jordan, V.W.L.
IACR-Long Ashton Research Station, Department of Agricultural Sciences, University of Bristol, Long Ashton, Bristol BS41 9AF, UK

Jørgensen, H.J.L.
Plant Pathology Section, Department of Plant Biology, The Royal Veterinary and Agricultural University, Thorvaldsensvej 40, DK-1871 Frederiksberg C, Copenhagen, Denmark

Jørgensen, L.N.
Danish Institute of Agricultural Sciences, Flakkebjerg, DK-4200 Slagelse, Denmark

Kema, G.H.J.
DLO-Research Institute for Plant Protection (IPO-DLO), PO Box 9060, 6700 GW Wageningen, The Netherlands

Kendall, S.
IACR-Long Ashton Research Station, Department of Agricultural Sciences, University of Bristol, Long Ashton, Bristol BS41 9AF, UK

Keon, J.
IACR-Long Ashton Research Station, Department of Agricultural Sciences, University of Bristol, Long Ashton, Bristol BS41 9AF, UK

Knogge, W.
Department of Biochemistry, Max-Planck-Institut für Züchtungsforschung, Carl-von-Linné-Weg 10, D-50829 Köln, Germany

Koeken, J.G.P.
DLO-Research Institute for Plant Protection (IPO-DLO), PO Box 9060, 6700 GW Wageningen, The Netherlands

Kruse, D.
Department of Biochemistry, Max-Planck-Institut für Züchtungsforschung, Carl-von-Linné-Weg 10, D-50829 Köln, Germany

Lamari, L.
Department of Plant Science, University of Manitoba, Winnipeg, Manitoba, Canada R3T 2N2

Li, V.
Department of Biochemistry, Max-Planck-Institut für Züchtungsforschung, Carl-von-Linné-Weg 10, D-50829 Köln, Germany

Liedgens, H.
Department of Biochemistry, Max-Planck-Institut für Züchtungsforschung, Carl-von-Linné-Weg 10, D-50829 Köln, Germany

Lovell, D.J.
IACR-Long Ashton Research Station, Department of Agricultural Sciences, University of Bristol, Long Ashton, Bristol BS41 9AF, UK

McDonald, B.A.
Department of Plant Pathology and Microbiology, Texas A&M University, College Station, TX 77843-2132, USA

Melton, R.E.
The Sainsbury Laboratory, John Innes Centre, Norwich Research Park, Colney Lane, Norwich NR4 7UH, UK

Mundt, C.C.
Department of Botany and Plant Pathology, 2082 Cordley Hall, Oregon State University, Corvallis, OR 97331-2902, USA

Oliver, R.P.
Carlsberg Laboratory, Copenhagen, Denmark

Osbourn, A.E.
The Sainsbury Laboratory, John Innes Centre, Norwich Research Park, Colney Lane, Norwich NR4 7UH, UK

Parker, S.R.
ADAS High Mowthorpe, Duggleby, Malton, North Yorkshire YO17 8BP, UK

Paveley, N.D.
ADAS High Mowthorpe, Duggleby, Malton, North Yorkshire YO17 8BP, UK

Pettway, R.E.
Department of Plant Pathology and Microbiology, Texas A&M University, College Station, TX 77843-2132, USA

Rohel, E.
INRA-Rennes, BP 29, 35650, Le Rheu, France

Royle, D.J.
IACR-Long Ashton Research Station, Department of Agricultural Sciences, University of Bristol, Long Ashton, Bristol BS41 9AF, UK

Secher, B.J.M.
Hardi International, Helgehøj Alle 38, DK 2630 Taastrup, Denmark

Shaw, M.W.
Department of Agricultural Botany, School of Plant Sciences, The University of Reading, 2 Earley Gate, Whiteknights, Reading RG6 6AU, UK

Smedegaard-Petersen, V.
Plant Pathology Section, Department of Plant Biology, The Royal Veterinary and Agricultural University, Thorvaldsensvej 40, DK-1871 Frederiksberg C, Copenhagen, Denmark

Steiner, S.
Department of Biochemistry, Max-Planck-Institut für Züchtungsforschung, Carl-von-Linné-Weg 10, D-50829 Köln, Germany

Stevens, C.
Department of Biological Sciences, Wye College, University of London, Wye, Ashford, Kent TN25 5AH, UK

Sweigard, J.A.
DuPont Experimental Station, Wilmington, DE 19880, USA

Titarenko, E.
Centro Nacional de Biotechnologia, Consejo Superior de Investigaciones Cientificas, Campus Universidad Autonoma, Cantoblanco, Madrid, Spain

Valent, B.
DuPont Experimental Station, Wilmington, DE 19880, USA

van der Lee, T.A.J.
Department of Phytopathology, Wageningen Agricultural University, Wageningen, The Netherlands

van 't Slot, A.
Department of Biochemistry, Max-Planck-Institut für Züchtungsforschung, Carl-von-Linné-Weg 10, D-50829 Köln, Germany

Verstappen, E.C.P.
DLO-Research Institute for Plant Protection (IPO-DLO), PO Box 9060, 6700 GW Wageningen, The Netherlands

Vöcking, R.
Department of Biochemistry, Max-Planck-Institut für Züchtungsforschung, Carl-von-Linné-Weg 10, D-50829 Köln, Germany

Waalwijk, C.
DLO-Research Institute for Plant Protection (IPO-DLO), PO Box 9060, 6700 GW Wageningen, The Netherlands

Wubben, J.P.
Section of Molecular Genetics of Industrial Micro-organisms, Agricultural University Wageningen, Wageningen, The Netherlands

Yarden, O.
Department of Plant Pathology and Microbiology, Texas A&M University, College Station, TX 77843-2132, USA

Zhan, J.
Department of Plant Pathology and Microbiology, Texas A&M University, College Station, TX 77843-2132, USA

Acknowledgement

The editors gratefully acknowledge the substantial time and effort that Mrs C.P.P. Cooke put into the preparation of camera-ready copy for this volume.

Chapter one:

Septoria and *Stagonospora* Diseases of Cereals: A Comparative Perspective

Z. Eyal
Department of Plant Sciences and the Institute for Cereal Crops Improvement, The George S. Wise Faculty of Life Sciences, Tel Aviv University, Tel Aviv 69978, Israel

Introduction

The increased recent interest in *Septoria* and *Stagonospora* species as causative agents of blotch diseases of wheat, and as subjects for biological research, has been stimulated by the concern of growers and scientists in the major wheat production areas. These pathogens now receive much greater attention than in the recent past. For many years *Septoria tritici* and *Stagonospora nodorum* were considered as common and conspicuous, but never important, foliar pathogens of wheat. They have now graduated to the status of major diseases with considerable economic impact on wheat production. Their geographical boundaries have also expanded. Pycnidia-bearing pycnidiospores of *S. tritici* were found on plants of wild emmer (*Triticum turgidum* var. *dicoccoides*) collected in 1902 on hills surrounding Jerusalem, and on plants of durum wheat (*T. durum*) collected in the Jordan Valley in 1924, held in the herbarium of the Hebrew University of Jerusalem (Eyal, 1976). The possible economic impact of Septoria tritici blotch on the yield and quality of spring wheat in Israel attracted attention during changes in cultivar profile in the early 1950s and with the introduction, as in many other countries, of high-yielding, susceptible semi-dwarf spring wheat cultivars in the late 1960s and early 1970s. These cultivar changes were often accompanied by alterations in crop management, such as a higher frequency of wheat in the rotation, less management of crop residues, and increases in nitrogen fertilizers. Since then, *Septoria* and *Stagonospora* species have been reported on wheat in many countries and in most parts of the world. A disturbing lag in genetic resistance to the *Septoria* and *Stagonospora* pathogens in commercial wheat cultivars, by comparison with increased resistance to other foliar pathogens such as rusts and powdery mildew, may also have contributed to the shift in economic importance of these diseases.

In their surveys of diseases of winter wheat in England and Wales from 1976-1988, Polley and Thomas (1991) stated that '*Septoria tritici* probably causes the most serious foliar disease of winter wheat in the UK'. Some of the factors that have been implicated for this reported shift in the early 1980s in the UK included greater cultivar susceptibility, earlier sowing, increased nitrogen fertilizer, high summer rainfall, and development of resistance to MBC fungicides (Bayles, 1991). It is likely that these factors can be associated with shifts in prominence from *S. nodorum* to *S. tritici* in other North European countries (e.g. Germany).

These pathogens have been the subject of several international and national workshops, in particular the International *Septoria* Workshops held in 1976, 1983, 1989 and 1994. Three comprehensive reviews on Septoria/Stagonospora blotch diseases have also been published (Shipton *et al.*, 1971; King *et al.*, 1983; Eyal *et al.*, 1987). This does not preclude the need for a new, updated, comprehensive review; the current chapter will stress the uniqueness of each of the two pathogens, but also identify areas of similarity. In the seven year period from 1990-1996 at least 190 peer reviewed articles (excluding proceedings, reports, etc.) were published on the two pathogens. Research efforts have focused mainly on the epidemiology and chemical control of the diseases, and to a lesser extent on breeding for disease resistance, the genetics of resistance, host-pathogen interactions, and pathogenicity. This review will avoid the most investigated areas, and instead direct attention to less well-known aspects of the two pathogens, with the view that their identification may stimulate scientific activity.

The *Septoria/Stagonospora* Pathogens of Wheat

Classification and Nomenclature

The term 'Septoria diseases of wheat' historically refers to diseases caused by the anamorphic form of fungal pathogens in the genus *Septoria*. The grouping together of several diseases was based on crop management considerations and on the difficulty of taxonomic differentiation, especially under field conditions. For the same reasons they were still referred to as the Septoria diseases of wheat during the four International Septoria Workshops held since 1976, despite the disassociation of numerous biological characters. Revisions in the taxonomy of species placed in the Septoria complex have been ongoing for almost 30 years, yet pathologists and other crop scientists were slow to accept and adopt the proposals submitted by mycologists.

Sprague (1950) stated that the conidia of *Septoria nodorum* are closer to the description of the genus *Stagonospora* than *Septoria*, yet no change in the taxonomy was offered. The conidia of *Septoria*, unlike *Stagonospora,* are ten times as long as wide, but this requires microscopic observation to determine and is not an easy character to measure under field conditions where, in many cases, the distinction has to be made.

The nomenclature of *Septoria* and *Stagonospora* adopted in this book is in accordance with that of Farr *et al.* (1989) as summarized by Cunfer (1994) and passed as a resolution by the participants of the 4th International *Septoria* Workshop held in Poland in 1994. The detailed outline of the taxonomy and nomenclature of *Septoria* and *Stagonospora* species on small grain cereals presented by Cunfer (1994) during the 4th International *Septoria* Workshop was recently published (Cunfer, 1997).

The *Septoria* Fungi

- **teleomorph:** *Mycosphaerella graminicola* (Fückel) J. Schrot. in Cohn
- anamorph: *Septoria tritici* Roberge in Desmaz.
  Disease: Septoria tritici blotch of wheat

Forms and species of *Septoria tritici* with no known teleomorph:
- *Septoria tritici* Roberge in Desmaz. f. *avenae* (host: oats)
- *Septoria tritici* Roberge in Desmaz. f. *holci* (host: *Holcus* spp.)
- *Septoria tritici* Roberge in Desmaz. var. *lolicola* R. Sprague and Alb. G. Johnson (host: *Lolium* spp.)
- *Septoria passerinii* Sacc.
  Disease: speckled leaf blotch of barley
- *Septoria secalis* Prill and Delacr.
  Disease: leafspot of rye

The *Stagonospora* Fungi

In this chapter the taxonomic transfer of the genus *Leptosphaeria* to the genus *Phaeosphaeria* as suggested by Hedjaroude (1968) was adopted:

- **teleomorph:** *Phaeosphaeria nodorum* (E. Muller) Hedjaroude
  syn. *Leptosphaeria nodorum* E. Muller
- anamorph: *Stagonospora nodorum* (Berk.) E. Castellani and E.G. Germano
  syn. *Septoria nodorum* (Berk.) in Berk. and Broome
  Disease: Stagonospora nodorum blotch of wheat

The use of host range in addition to morphology as a criterion for classification introduces some unresolved issues concerning other *S. tritici* and *Stagonospora* spp. especially in the case of *Stagonospora avenae* f.sp. *triticea* which infects both barley and wheat (Cunfer, 1994; 1997).

- **teleomorph**: *Phaeosphaeria avenaria* (G.F. Weber) O.E. Erikss.
  syn. *Leptosphaeria avenaria* G.F. Weber
- anamorph: *Stagonospora avenae* (A.B. Frank) Bisset
  Stagonospora avenae (A.B. Frank) Bissett f.sp. *avenaria*
  syn. *Septoria avenae* A.B. Frank
- **teleomorph**: *Phaeosphaeria avenaria* (G.F. Weber) O.E. Erikss. f.sp. *triticea*
  syn. *Leptosphaeria avenaria* G.F. Weber f.sp. *triticea*

- anamorph: *Stagonospora avenae* (A.B. Frank) Bissett f.sp. *triticea*
  syn. *Septoria avenae* A.B. Frank f.sp. *triticea*

A summary of the classification and nomenclature of the *Septoria* and *Stagonospora* spp. is presented in Table 1.1.

Host Range

Both pathogens can infect a wide range of graminaceous hosts in nature and following artificial inoculation. Krupinsky (1989) obtained 27 isolates of *S. nodorum* from diseased leaves of 11 different grass species including *Agropyron* spp., *Elymus* spp., *Hordeum jubatum* and *Leymus* spp., and all caused symptoms on seedlings of six wheat cultivars. Differences in the ability to cause disease were interpreted as variations in aggressiveness of the isolates. In a later study, Krupinsky (1994) reported that isolates from alternative host grasses that manifested stable, low aggressiveness on wheat 'would not adapt easily to wheat, thus do not pose a threat to wheat'. Cross-infection tests with isolates from barley and wheat have suggested the existence of wheat- and barley-adapted types which have been considered biotypes (Cunfer and Youmans, 1983) or *formae speciales* (Smedegård-Peterson, 1974). Newton and Caten (1991) suggested that the many differences between the wheat- and barley-types of *S. nodorum* may indicate that they are genetically distinct populations within the species.

The appearance of Septoria tritici blotch on wheat sown after non-cereal crops was attributed, among other reasons, to alternative hosts as a source of primary inoculum. The ability to infect and sporulate on leaves of alternative hosts or seeds following artificial inoculation served as a criterion for their possible role in the epidemiology. This fact by itself does not necessarily classify this pathogen as a generalist. Several graminaceous species were traced as possible alternative hosts: *Agropyron* spp., *Agrostis* spp., *Bromus* spp., *Brachypodium* spp., *Dactylis* spp., *Festuca* spp., *Glyceria* spp., *Hordeum* spp., *Poa* spp., *Secale cereale* and *Triticum* spp. The function of grass species in the epidemiology of *S. tritici* on wheat under field conditions needs to be resolved. Viable pycnidia of *S. tritici* are readily detected on wild emmer (*T. turgidum* var. *dicoccoides*) grown indigenously in Israel. However, the role of wild emmer as a source of primary inoculum in causing epidemics in commercial durum or bread wheat fields is not known.

Primary Source of Inoculum

Air-borne ascospores discharged from mature pseudothecia serve as an effective source of primary inoculum for both pathogens. The possible role of ascospores as a continuous source of primary inoculum has stimulated extensive epidemiological research in the UK and will be discussed by Hunter *et al.*, Chapter 7 this volume. However, there are many instances where the sexual state was not found and it is, therefore, assumed that infected wheat debris and rain-splashed pycnidiospores serve as a primary source of inoculum.

With *S. nodorum*, Keller *et al.* (1997) attributed great importance to ascospores as a source of primary inoculum, whereas pycnidiospores probably play an important role in epidemiology on small spatial scales (1-2 m). In contrast, Shah *et al.* (1995) considered asexual spores and infected seeds as the major sources as primary inoculum. Seedling infection from infected wheat seeds has been demonstrated as a primary source of inoculum for *S. nodorum* (Halfon-Meiri and Kulik, 1977; Luke *et al.*, 1986; Shah and Bergstrom, 1993; Shah *et al.*, 1995). Cockerell and Rennie (1996) surveyed seed-borne pathogens in certified and farm-saved cereal seed in Britain between 1992 and 1994, and reported the incidence of seed infected with *S. nodorum* (in 1992-1993 >40% of certified winter wheat samples had at least 20% of seeds infected).

Seeds of *T. dicoccum* infected with *S. tritici* were reported by Brokenshire (1975) after artificial inoculation with spores of *S. tritici,* and also by Hampton (1980) in New Zealand. Brokenshire (1975) stated that 'infection of seedlings from infected untreated seed samples has proved unsuccessful with *T. dicoccum*'. Reports of the possible seed transmission of *S. tritici* suggested this transmission mode as a driving force for gene flow on a global scale (McDonald *et al.*, 1995). The epidemiological implications of naturally-infected grasses and seeds (especially for *S. tritici*) both for short-scale dissemination and long-distance transport, warrants additional investigation.

Symptoms and Diagnosis

Discrimination between Septoria tritici blotch and Stagonospora nodorum blotch on wheat plants under field conditions relies on differences in symptom morphology - lesion size and shape and the distribution of pycnidia within the lesion (dispersed vs. central) or on plant organs (leaves, sheath vs. leaves, sheath, nodes and glumes) and pycnidial colour (dark vs. light brown), respectively.

A lens-shaped necrotic lesion surrounded by a yellow-green border bearing pycnidia within the dead leaf tissue at the centre, together with lesions on nodes, stems, leaf sheaths and glumes, is more common for Stagonospora nodorum blotch. Irregular necrotic lesions which are sunken and grey at first, and which may bear pycnidia scattered within the entire lesion area, are more typical of Septoria tritici blotch of wheat. The distribution of pycnidia within a lesion and over the entire leaf in coalesced lesions is dictated, in part, by differences in the mode of penetration and pycnidial formation of the two pathogens. Pycnidia of *S. tritici* can be found on the same plant organs as *S. nodorum*, except for the nodes. These diagnostic measures do not provide defined criteria for differentiation between the two pathogens, especially on susceptible wheat cultivars where the two pathogens occur simultaneously.

Supplementary and more accurate verification is often needed for field diagnosis and can be provided by microscopic, immunological, and molecular tools. The number of cells of the ascospores and the number of septa of the pycnidiospores, in addition to the size and shape of these spores, clearly

Table 1.1. Classification and nomenclature of the *Septoria* spp. and *Stagonospora* spp. fungi on small grain cereals[a].

Genus	Teleomorph	Anamorph	Common name	Host
Septoria spp.	*Mycosphaerella graminicola*	*Septoria tritici*	Septoria tritici blotch	Wheat
	-[b]			
	-	*Septoria tritici* f. *avenae*		Oats
	-	*Septoria tritici* f. *holci*		Holcus
	-	*Septoria tritici* f. *lolicola*		Lolium
	-	*Septoria passerinii*	Speckled leaf blotch	Barley
	-	*Septoria secalis*	Leaf spot	Rye
Stagonospora spp.	*Phaeosphaeria nodorum*	*Stagonospora nodorum*	Stagonospora nodorum blotch	Wheat
	Phaeosphaeria avenaria	*Stagonospora avenae* f. sp. *Avenae*		Oats
	Phaeosphaeria avenaria f. sp. *Triticea*	*Stagonospora avenae* f. sp. *Triticea*		Oats, wheat and triticale

[a]Cunfer, B.M. (1997). [b]Teleomorphic stages not found.

Table 1.2. Descriptive comparison of the *Septoria/Stagonospora* wheat pathogens.

Sexual state	Pseudothecium (μm)	Ascospore (μm)	Number of cells	Chromosomes	Lesion
Mycosphaerella graminicola	70-100	10-15 x 2-3	2		Irregular to rectangular, elongated between veins
Phaeosphaeria nodorum	120-200	23-32 x 4-6	4		Lens-shaped, with chlorotic borders
Asexual state	**Pycnidium (μm)**	**Pycnidiospore (μm)**	**Number of septa**	**Chromosomes**	**Lesion**
Septoria tritici	60-200	35-98 x 1-3	3-5	14-19	Irregular to rectangular, elongated between veins
Stagonospora nodorum	160-210	15-32 x 2-4	0-3	14-16	Lens-shaped, with chlorotic borders

differentiate between the two wheat pathogens (Table 1.2). The overall karyotype of *S. nodorum* suggests 14-19 chromosomes with sizes ranging from approximately 0.5-3.5 megabase (Mb) pairs (Cooley and Caten, 1991), whereas 14-16 bands, believed to correspond to chromosomes, ranging in size from 0.33-3.5 Mb were reported for *S. tritici* (McDonald and Martinez, 1991). It is possible that some bands may represent two chromosomes of equal length. Serological and molecular assays can greatly improve the accuracy of diagnostic procedures over the morphological characters (pycnidial shape, size, colour and distribution within lesions) especially in epidemiological studies and chemical control schemes where clear distinctions are called for. Mittermeier *et al.* (1990) have developed a rapid monoclonal antibody-based immunoassay which specifically distinguishes between *S. tritici* and *S. nodorum* and can be used under field conditions.

Beck and Ligon (1995) developed species-specific polymerase chain reaction (PCR) primers that can detect and differentiate *S. tritici* and *S. nodorum* in infected wheat tissue. The primers are based on sequence analysis of the internal transcribed spacer (ITS) regions of the two pathogens. The authors suggested that the PCR-based species-specific primers appear to produce few, if any, of the false positive reactions that can be encountered with serological assays due to cross-reactivity of antibodies with antigens of plant or microbial origin. In addition, the cost and time required for PCR-assay development is less than the development of a similar serological assay.

Ueng and Chen (1994) examined the intraspecific genetic variation in *S. nodorum*, *S. avenae* and *S. avenae* f.sp. *triticeae* by the use of restriction fragment length polymorphism (RFLP) with 38 anonymous DNA probes, 20 from *S. nodorum* and 18 from the other two *Stagonospora* species. Clear separation was obtained between the *S. nodorum* and the *S. avenae* species, whereas the separation between the two *S. avenae* species themselves was inconsistent. Using amplified fragment length polymorphism (AFLP) genotyping technology, Arseniuk *et al.* (1997) have assessed the genetic similarities between *S. nodorum* and *S. avenae* f.sp. *triticea* originating from Poland and the USA. Among the 14 *S. nodorum* isolates two similarity groups were recognized. Within each group, Polish and USA isolates formed distinct subgroupings.

The choice of which diagnostic tools to apply will depend on multiple factors, among which objectives, ease of use, accuracy, speed and cost are key elements.

Events Associated with the Infection Process

Penetration and Colonization

The two pathogens have distinctly different pycnidiospores, yet their germination follows a similar course, namely, via elongation of the apical cells or by budding. It should be noted that the number of pycnidiospores of *S. tritici* liberated from a pycnidium was reported to approximate 5-10 x 10^3 (Eyal,

1971). The liberation of pycnidiospores is gradual, with the great majority of spores exuding during the first wetting, while the rest are released on consequent wettings. The multiplication of the number of oozed spores by the number of pycnidia per unit leaf area may suggest a very large number of spores available for the infection process, provided environmental conditions are suitable. However, the number of successful penetrations per release cycle does not match that of the available spores. It is suggested that the large number of available pycnidiospores compensates for a relatively inefficient dispersal mechanism dependent on environmental conditions. The relationship between the number of ascospores incoming to the infection court, and the establishment of lesions, and thereafter the progression of the epidemic from additional showers of ascospores and/or from pycnidiospores within the crop stand needs further analysis (Shaw and Royle, 1989).

The two pathogens vary distinctly in the mode of penetration of the germinating pycnidiospores. The almost exclusive penetration of *S. tritici* through stomata (Cohen and Eyal, 1993; Kema *et al.*, 1996c) distinguishes this pathogen from *S. nodorum* (but see Dancer *et al.*, Chapter 22 this volume), which was shown to penetrate directly through the cuticle (Karjalainen and Lounatmaa, 1986). Formation of pycnidia of *S. tritici* occurred solely in conjunction with stomata, and the distribution of pycnidia, therefore, follows the linear pattern of stomata along the veins. Although pycnidial shape might appear to be determined by the contours of the substomatal chamber where it is formed (Cohen and Eyal, 1993), the correlation is tenuous since similarly-shaped pycnidia were produced on artificial medium (Zelikovitch and Eyal, 1989).

Direct penetration of *S. nodorum* through the periclinal wall into the lumen of the epidermal cell is accomplished by the formation of a penetration peg and possibly by enzymatic degradation of the epidermal cell wall (Karjalainen and Lounatmaa, 1986; Lehtinen, 1993). Polygalacturonase, xylanase and cellulase were produced in minimal medium supplemented with wheat cell walls and in wheat leaves inoculated with *S. nodorum* (Magro, 1984), with apparent lysis of the cutin layer (Zinkernagel *et al.*, 1987). No direct evidence was provided for the presence of cutinase(s) in *S. nodorum.*

Colonization of host tissue, cell collapse, and pycnidial formation were associated with the reaction of wheat cultivars to the two pathogens (Bird and Ride, 1981; Karjalainen and Lounatmaa, 1986; Cohen and Eyal, 1993; Kema *et al.*, 1996c). Unsuccessful penetrations by *S. nodorum* were associated, in part, with papilla formation followed by lignification which consequently reduce infection and colonization, although this is not restricted only to resistant cultivars (Bird and Ride, 1981). Kema *et al.* (1996c) reported that mycelial biomass was a determinant of pycnidial formation. In resistant wheat cultivars inoculated with *S. tritici* which harboured few to no pycnidia, mycelium was mainly restricted to the substomatal cavity. In the sequence of post-penetration events in compatible interactions, necrosis and pycnidial formation follow a timed pattern. The formation of pycnidia usually occurs after cell collapse is well advanced, although the initiation of pycnidia is not necessarily associated with necrosis. The quantification of fungal biomass in leaf tissue of susceptible and resistant cultivars using *S. tritici* antigens (Kema *et al.*, 1996c), and by using the expression *in planta* of the reporter gene *uid*A (GUS) in genetically-transformed *S. tritici* isolates (Pnini-Cohen *et al.*, poster presented at the 15th Long Ashton International Symposium 'Understanding Pathosystems: a Focus on Septoria', Bristol, UK, 13-15 September 1997), indicate that in compatible interactions there is a marked increase in fungal development, whereas little increase in fungal biomass was recorded in the incompatible interactions. The former resulted in prolific pycnidial development whereas few to no pycnidia were formed in the latter. The mechanisms and course of events that trigger synchronous pycnidial formation are not known. Specific mRNA families can be detected during the pre-pycnidial formation phase, which suggests that pycnidial formation is a genetically-regulated phenomenon (Pnini-Cohen *et al.*, unpublished results).

Pathogenesis

The factors responsible for pathogenesis by both fungi are poorly defined. Disruption of chloroplast integrity and compartmentalization within host cells was reported by Karjalainen and Lounatmaa (1986) for *S. nodorum* and by Kema *et al.* (1996c) for *S. tritici*. Disintegration of mesophyll cells and the collapse of host tissue several days after inoculation were ascribed to the production of soluble toxic compounds. The production of phytotoxic compounds in culture filtrates of *S. nodorum* was reported during the 1970s (Bousquet and Skajennikoff, 1974; Kent and Strobel, 1976). The latter authors suggested that the extent of mycelial invasion determines the level of toxins produced in the host and hence the degree of lesion formation. The specificity attributed to the reported phytotoxic compounds was rather variable. In culture, *S. nodorum* produces a wide series of phytotoxic metabolites. Three related metabolites belong to the septorin series: septorin, *N*-methoxyseptorine (biologically the most active compound), and *N*-methoxyseptorinol. Septorin causes decoupling of mitochondria of wheat coleoptiles (Barbier *et al.*, 1994). Several phytotoxic compounds belong to the mellein family: mellein (ochracin), 4-hydroxymellein, 7-hydroxymellein, and 5-hydroxymellein (Devys *et al.*, 1994). Information on the direct involvement of the melleins and septorin phytotoxins in pathogenesis and in host specificity is lacking.

Despite circumstantial evidence (loss of cell compartmentalization, collapse of host tissue and necrosis) which might suggest the involvement of phytotoxin(s) in the pathogenesis of *S. tritici*, no pathogen-produced compound has been isolated and biologically categorized. Harrabi *et al.* (1995) reported an *in vitro* selection for *S. tritici* resistance by treating callus tissue derived from scutella with an ethyl acetate fraction from culture filtrate. The adult plant response and the *in vitro* cell reaction in these studies was not strongly correlated. Inhibition of growth of *S. tritici* isolates *in vitro* was reported for organic extracts of culture filtrate, by the compound methyl-3-indole carboxylate (3-indole carboxylic acid methyl ester [ICA.Me]) (Zelikovitch *et al.*, 1992). The authors suggested that ICA.Me may be involved in the regulation of fungal growth, but could not be classified as a phytotoxin. This lack of information leaves a conspicuous gap in our knowledge of the pathogenesis of *S. tritici*.

Septoria tritici was reported to produce three catalases, peroxidase and superoxide dismutase (SOD) (Levy *et al.*, 1992a). One of the three catalase isozymes was identified as catalase-peroxidase, an enzyme which exhibits both catalase and peroxidase activity. Catalase and peroxidase detoxify hydrogen peroxide, whereas SOD catalyses dismutation of superoxide to hydrogen peroxide and molecular oxygen. The activity of some of these protective enzymes is increased when *S. tritici* cells are challenged with toxic oxygen metabolites (Levy *et al.*, 1992b). *Septoria tritici* is capable of protecting itself against the deleterious effects of certain antibiotics (e.g. 1-hydroxyphenazine) which are involved in the production of toxic, intermediate reduction of oxygen, superoxide and hydrogen peroxide. This mechanism may also provide *S. tritici* with a competitive advantage under natural field conditions where toxic metabolites are produced.

Speciation and Specificity

The nature of specificity in the wheat-*Septoria/Stagonospora* pathosystems has been discussed extensively in the literature due to the biological and genetic implications in breeding for disease resistance. Despite extensive research, specificity remains an unresolved issue, partly because of the difficulties in assigning common pathological/genetic methodologies to pathosystems where quantitative, rather than qualitative, assessment of host response governs the system. The pathogenic variation observed in the *S. tritici*- and *S. nodorum*-wheat pathosystems was classified by Johnson (1992) as characteristic of systems without clear gene-for-gene interactions. He posed the question as to whether distinctive races can be identified with specific virulence for particular cultivars, and whether the interaction indicates a gene-for-gene system. To establish this requires stability and repeatability in different cultivar-isolate interactions. In practice, it would be required to demonstrate that resistant cultivars succumb to new pathogen races.

Septoria tritici

Eyal *et al.* (1983) reported on the existence of physiological specialization in *S. tritici*. The authors emphasized the following issues: (a) host response was assessed by quantification of symptoms (percent coverage of necrosis and pycnidia) with preference to the latter; (b) there was a significant interaction between isolates and cultivars; and (c) a possible differentiation due to *Triticum* species. *Septoria tritici* isolates derived from *T. durum* produced low levels of pycnidia on most *T. aestivum* cultivars but high coverage on the durum wheat cv. Etit 38 and, conversely, isolates derived from *T. aestivum* produced high levels of pycnidia on all *T. aestivum* cultivars except the resistant cv. Racine and low levels of pycnidia on cv. Etit 38. Subsequently, several publications confirmed these findings on specificity in the wheat-*S. tritici* pathosystem on wheat cultivars derived from within the same *Triticum* species (e.g. *T. aestivum* L. and *T. durum* Desf.) or across these species (Yechilevitch-Auster *et al.*, 1983; Eyal *et al.*, 1985; Eyal and Levy, 1987; Saadaoui, 1987; Jlibene *et al.*, 1995; Halperin *et al.*, 1996; Kema *et al.*, 1996a,b; Pnini-Cohen *et al.*, 1997a).

The virulence frequencies of 97 *S. tritici* isolates collected from 22 countries were evaluated on 35 wheat cultivars (28 bread wheats, four durum wheats and three triticales). The *S. tritici* isolates from Syria, all obtained from *T. durum,* were virulent on the four durum wheat cultivars but not on the bread wheat cultivars, whereas the durum wheat isolates from Tunisia were virulent on all durum wheat cultivars and also on some of the *T. aestivum* cultivars (Eyal *et al.*, 1985). It should be noted that the isolates from Brazil, Ecuador, Mexico and Uruguay, the majority of which were derived from bread wheat, were virulent both on durum wheat and on bread wheat cultivars. In countries where durum wheat cultivars predominate, the pathogen population is skewed towards durum wheat-adapted virulence (Eyal *et al.*, 1985). Still, within these populations one can also find *T. aestivum*- or *T. durum*/*T. aestivum*-adapted isolates. There is no evidence as to the proportions of each within the populations (Eyal, 1995). The differentiation between *S. tritici* isolates pathogenic on durum wheat and on bread wheat was also reported by Saadaoui (1987) for 19 Moroccan isolates. Similar observations on possible speciation in *S. tritici* were reported by Kema *et al.* (1996a) in a study that included 63 *S. tritici* isolates from 13 countries tested on 19 bread wheat cultivars, four durum wheat cultivars and one triticale cultivar. The authors suggested designating two varieties in *Mycosphaerella graminicola*, bread wheat- and durum wheat-adapted isolates, each derived from the respective *Triticum* species. In Israel, where both bread wheat and durum wheat cultivars are grown and *S. tritici* can also be found on plants of the indigenous wild emmer (*T. turgidum* var. *dicoccoides*), the division into two distinct pathogenicity groups is less clear and exceptions can be found in any of the species-adapted groups. Genetic crosses between durum wheat-adapted and bread wheat-adapted variants, along with DNA technology, may shed more light on this issue.

Variation in virulence patterns within and between populations of *S. tritici* were independently revealed by assessing host response on a selected set of cultivars, although the genotypes comprising the differentiating sets were not standardized between the various studies (Eyal and Levy, 1987; Eyal *et al.*, 1987; Ahmed *et al.*, 1995; Kema *et al.*, 1996a). Most of the tests were conducted on wheat seedlings, although evidence was provided that specificity was also expressed on adult plants of field-grown wheat genotypes inoculated with specific isolates (Danon and Eyal, 1990; Gilchrist and Velazquez, 1994; Kema and Van Silfhout, 1997).

The possible impact of the sexual state on the virulence spectrum was inferred from regions where pseudothecia were found and ascospore dispersal coincided with wheat planting (Shaw and Royle, 1989). Segregation for virulence has been observed in ascopore progeny (Sanderson *et al.*, 1986). In a recent study on the effect of sexual reproduction on the genetic structure of *S. tritici* populations using nuclear RFLPs and DNA fingerprints, Chen and McDonald (1996) concluded that where sexual reproduction is playing a role in the epidemic, new virulence combinations can be selected by corresponding host resistance genes. They further stressed that new combinations of matching virulence genes will overcome combinations of resistance genes. This conclusion, extrapolated from other pathosystems, requires supporting data for the wheat-*S. tritici* pathosystem. Isolates of *S. tritici* sampled from California and Oregon (where sexual reproduction is operative) differed significantly in their pathogenicity (Ahmed *et al.*, 1996), yet they manifested similarity in allele frequencies as measured by RFLPs (McDonald *et al.*, 1996). Conversely, identical virulences of wheat leaf rust (*Puccinia recondita* f.sp. *tritici*) phenotypes from different countries were not necessarily closely related to molecular phenotype (Kolmer and Liu, 1997).

The expression of virulence in *S. tritici* on specific cultivars is distinct and repeatable, and for certain well-studied isolates was shown to be highly stable despite more than 30 years of subculturing on artificial medium and re-isolation from pycnidia on infected leaves (Eyal, 1992). The assessment of quantitative host response dictates that in any virulence studies constant comparisons are required to isolates of known virulence (rejuvenated from a preserved conidial stock at -70°C) on specific differentiating wheat genotypes under controlled environmental conditions.

The presence of a gene-for-gene interaction in the wheat-*S. tritici* pathosystem was inferred from the statistical significance of the mean square (variance) of the cultivars x isolates interaction terms and not from inheritance studies (Eyal *et al.*, 1985; Kema *et al.*, 1996a,b). Ahmed *et al.* (1996) have reported on pathogen x host interactions that are caused by mechanisms other than cultivar-specific selection. They concluded that the *S. tritici* population adapts to its host, namely, susceptible cultivars select for higher levels of pathogen aggressiveness in the field.

Marked suppression in pycnidial coverage was recorded on seedlings and adult plants of specific genotypes inoculated with a mixture of avirulent and virulent *S. tritici* isolates, or upon sequential inoculation by these isolates (Zelikovitch and Eyal, 1991; Eyal, 1992; Halperin *et al.*, 1996, Ezrati *et al.*, poster presented at the 15th Long Ashton International Symposium 'Understanding Pathosystems: a Focus on Septoria', Bristol, UK, 13-15 September 1997). The divergence from the expected host response upon inoculation with an isolate mixture should be considered in virulence, genetic and epidemiological studies (Table 1.3). The reduction in disease severity following inoculation with a mixture of a virulent and an avirulent isolate was explained by the induction of a resistance mechanism by the avirulent isolate, where host product(s) resulting from the *in planta* interaction affected pycnidial production by the virulent isolate (Halperin *et al.*, 1996). Disease suppression in susceptible cultivars inoculated with an isolate mixture was explained by competition between the isolates during colonization of wheat tissue. The effect of induced resistance in certain resistant cultivars and competition in susceptible cultivars on the structure of *S. tritici* populations under natural field conditions needs verification. Isolates from pycnidia formed on wheat genotypes expressing different host responses following inoculation with a mixture of two *S. tritici* isolates differing in their virulence and DNA fingerprinting have shown that virulence and aggressiveness operated as independent traits (Ezrati *et al.*, poster presented at the 15th Long Ashton International Symposium 'Understanding Pathosystems: a Focus on Septoria', Bristol, UK, 13-15 September 1997). The progressive buildup and final ratio between components of the mixture within an environmental regime was dictated by cultivar x isolate interactions, aggressiveness and isolate-isolate interactions, all operating in unison to determine the structure of the pathogen population.

Stagonospora nodorum

Unlike *S. tritici*, specificity in the wheat-*S. nodorum* pathosystem is much less distinct. King *et al.* (1983) in their review stated that 'present evidence suggests that cultivar resistance to *S. nodorum* is non-specific and that isolates differ in aggressiveness but not in the range of cultivars attacked'. Allingham and Jackson (1981) tested the host-pathogen relationships of 282 isolates of *S. nodorum* originating from northern Florida. Despite the recorded differential interactions, the 253 resistance patterns were not categorized into a conventional race classification. Rufty *et al.* (1981) tested nine isolates from North Carolina (eight) and Montana (one) on seedlings of four winter wheat cultivars and found significant cultivar x isolate interactions that were indicative of specificity in the host and the pathogen. Virulence frequencies of 33 isolates from eight countries were assessed on 38 wheat and triticale cultivars based on percentage leaf necrosis (Scharen *et al.*, 1985). Significant main effects of isolate and cultivars and non-significant interactions were recorded in this study indicating a lack of clear specificity in the wheat-*S. nodorum* pathosystem. It appears that in *S. nodorum* physiological specialization is much less pronounced outside a specific experimental situation. Moderate levels of virulence were reported for populations of *S. nodorum* in Switzerland, where sexual reproduction contributed to high genotypic variation (Keller *et al.*, 1997). Earlier virulence studies had shown limited interactions in the wheat-*S. nodorum* pathosystem in that country (Fried and Meister, 1987). *Stagonospora nodorum* isolates from barley and wheat were highly virulent to their original host but nevertheless weakly virulent to the opposite crop in reciprocal inoculations (Cunfer and Youmans, 1983). Similar results were recorded by Krupinsky (1989; 1994) for *S. nodorum* originating from wheat and a variety of grasses.

Breeding for Disease Resistance

Breeding for resistance has been applied successfully and systematically for

Table 1.3. Specificity in *Septoria tritici* x wheat pathosystem.

Host factor[a]	Pathogen factor[b]		
	Avirulence (+)	Avirulence + virulence (+/-)	Virulence (-)
Host gene (+)	None to few pycnidia, restricted mycelial growth	Low pycnidia coverage, restricted mycelial growth	High pycnidia coverage, profuse mycelial growth
Host gene (-)	High pycnidia coverage, profuse mycelial growth	Moderate to high pycnidia, profuse mycelial growth	High pycnidia coverage, profuse mycelial growth

[a]Wheat cultivars: e.g. (+) Seri 82; (-) Shafir.
[b]*Septoria tritici* isolates: e.g. (+) ISR398; (-) ISR8036.

many cereal pathogens yet, to date, there is relatively little information on the types of host resistance to *S. tritici* and *S. nodorum*, or on the inheritance and durability of difference sources. This important topic is reviewed in more detail by Eyal, Chapter 23 this volume.

Concluding Remarks

The economics of wheat production has dictated that *S. tritici* and *S. nodorum* are still often grouped together. The biology of these two contrasting fungi should not allow this unification. The classification of *M. graminicola* since 1976 (Sanderson) and of *P.* (*Leptosphaeria*) *nodorum* since 1952 (Müller) has outlined the taxonomy. Cunfer's orderly publications (1994; 1997) on the taxonomy and nomenclature of *Septoria* and *Stagonospora* species have greatly added to the clarity, and should be adopted.

The suggestion by P.R. Scott to adopt the anamorph state as the common name of the diseases (Septoria tritici blotch of wheat and Septoria nodorum blotch of wheat) was passed as a motion during the 2nd International *Septoria* Workshop, and has been adopted since then by most workers, although variations on the theme exist. The replacement of Septoria nodorum blotch by Stagonospora nodorum blotch of wheat makes the separation more distinct.

The interpretation of quantitative resistance to Stagonospora nodorum blotch had already prompted breeding for resistance in the early 1980s, resulting from the recognition of the rather low level of specificity in the *S. nodorum* x wheat pathosystem. There are surprisingly few detailed genetic studies related to the contol of this type of resistance. The lengthy debate on specificity in the *S. tritici* x wheat pathosystem arose mainly from unjustified interpretation of relatively small but statistically significant isolate x cultivar interaction terms compared to large main effects of isolates and/or cultivars. It seems that the *S. tritici* x wheat pathosystem has graduated from this phase and is now ready for more detailed studies on the mode of the interaction, the extent of specificity and the implications for breeding for disease resistance.

The introduction of genetic and molecular biology tools has resolved issues which otherwise would have remained uncertain. The greatest impact has been in assessing the genetic diversity within and between populations of the pathogens of both asexual and sexual origin (Newton and Caten, 1985; Ueng *et al.*, 1992; McDonald and McDermott, 1993; Zilberstein *et al.*, 1993; Keller *et al.*, 1997; McDonald, 1997). The two pathogens exhibit high genetic diversity and low levels of clonality. It is expected that studies on the population genetics of these fungi will be linked to pathogenicity and to control strategies. The development of procedures for conducting crosses between isolates of *P. nodorum* (Halama and Lacoste, 1992a,b) and of *M. graminicola* (Kema *et al.*, 1996d) has paved the way to studies related to the sexual cycle, mating, inheritance of virulence and other traits, and possible construction of genetic maps. The use of RAPD, or other markers, in tetrad analyses may greatly contribute to this molecular analysis (Kema *et al.*, 1996d). Progeny from crosses between avirulent and virulent isolates assayed on differential wheat seedlings revealed that avirulence for several cultivars was tightly linked (Kema *et al.*, 1997). The authors suggested the presence of a complex locus of tightly-linked avirulence genes.

Specific PCR primers which distinguish between *S. tritici* isolates were used to study the dynamics of isolate interactions on specific wheat genotypes, each inoculated with a mixture of the two isolates (Ezrati *et al.*, poster presented at the 15th Long Ashton International Symposium 'Understanding Pathosystems: a Focus on Septoria', Bristol, UK, 13-15 September 1997). The progressive contributions of wheat genotypes (resistant and susceptible), different environmental regimes (two locations), the interrelations between the isolates (e.g. competition), and cultivar x isolate interactions (e.g. induced resistance) to the structure of *S. tritici* populations may thus be elucidated.

Genes associated with pathogenicity can be identified and cloned through the complementation of induced mutants. This can be achieved through the introduction and expression of foreign DNA sequences into the fungal genome. The selectable marker gene *hph* (hygromycin phosphotransferase) which confers resistance to hygromycin B was stably introduced into the genome of *S. nodorum* by Cooley *et al.* (1988). The same selectable marker and the reporter gene *uid*A, which encodes the enzyme β-glucuronidase (GUS), were introduced by co-transformation into the genome of *S. tritici* isolates varying in their virulence on spring wheat (Pnini-Cohen *et al.*, 1996). The GUS reporter gene facilitates the analysis and quantification of events associated with infection processes (Pnini-Cohen *et al.*, poster presented at the 15th Long Ashton International Symposium 'Understanding Pathosystems: a Focus on Septoria', Bristol, UK, 13-15 September 1997). Transformation of *S. tritici* using a similar approach was recently reported by Payne (1997). The β-tubulin gene of a carbendazim- resistant isolate of *S. tritici* was isolated by screening a genomic DNA library (Payne, 1997). The cloned sequence was transformed into a carbendazim- sensitive *S. tritici* strain. A homologous transformation system was developed for *S. nodorum* based on selection for a benomyl-(MBC)-resistant β-tubulin gene with the anticipation that this gene (*tub*A^R) will aid in pathogenicity studies (Cooley *et al.*, 1991). Mutants of *S. tritici* altered in virulence have been recovered and a putative pathogenicity gene tagged by insertional mutagenesis (Pnini-Cohen *et al.*, 1997b). Analysis of the 'knock-out' events revealed three open reading frames (ORFs) and stop codons in the region of the insertion (Pnini-Cohen, unpublished results). Complementation studies of such mutants with genomic DNA to recover the wild-type phenotype is currently underway. By employing molecular technologies, a glimpse of the genes associated with both avirulence and resistance and their products and roles in the *S. tritici* x wheat pathosystem should be possible.

These new technologies are enabling us to revive and explore questions about these two economically important plant pathogens which have been neglected for lack of proper experimental tools. The role of toxins in the pathogenicity of *S. tritici* remains a mystery. The sequence of events during infection and the regulation of pycnidial formation *in planta* has not been explored for either pathogen. There is little information on the types and mechanisms of resistance of some of the most important resistant genotypes, their mode of inheritance, and incorporation into modern wheat cultivars. The question of speciation in *S. tritici* on different *Triticum* species, and *S. nodorum* on different grass genera, calls for more detailed investigation.

These two economically important pathogens are now 'mature' for our explorations, and should ignite our curiosity, imagination and sense of responsibility as scientists and agriculturists.

References

Ahmed, H.U., Mundt, C.C. and Coakley, S.M. (1995) Host-pathogen relationship of geographically diverse isolates of *Septoria tritici* and wheat cultivars. *Plant Pathology* 44, 838-847.

Ahmed, H.U., Mundt, C.C., Hoffer, M.E. and Coakley, S.M. (1996) Selective influence of wheat cultivars on pathogenicity of *Mycosphaerella graminicola* (anamorph *Septoria tritici*). *Phytopathology* 86, 454-458.

Allingham, E.A. and Jackson, L.F. (1981) Variation in pathogenicity, virulence, and aggressiveness of *Septoria nodorum* in Florida. *Phytopathology* 71, 1080-1085.

Arseniuk, E., Confer, B.M., Mitchell, S. and Kresovich, S. (1997) Characterization of genetic similarities among isolates of *Stagonospora* spp. and *Septoria tritici* by AFLP analysis. *Phytopathology* 87 (Supplement), S5.

Barbier, M., Devys, M., Bousquet, J.F. and Kollmann, A. (1994) Absolute stereochemistry of *N*-methoxyseptorinol isolated from the fungus *Septoria nodorum*. *Phytochemistry* 35, 955-957.

Bayles, R.A. (1991) Varietal resistance as a factor contributing to the increased importance of *Septoria tritici* Rob. and Desm. in the UK wheat crop. *Plant Varieties and Seeds* 4, 177-183.

Beck, J.J. and Ligon, J.M. (1995) Polymerase chain reaction assays for the detection of *Stagonospora nodorum* and *Septoria tritici* in wheat. *Phytopathology* 85, 319-324.

Bird, P. and Ride, J.P. (1981) The resistance of wheat to *Septoria nodorum:* fungal development in relation to host lignification. *Physiological Plant Pathology* 19, 289-299.

Bousquet, J.F. and Skajennikoff, M. (1974) Isolement et mode d'action d'une phytotoxine prodite en culture par *Septoria nodorum* Berk. *Phytopathologische Zeitschrift* 80, 355-360.

Brokenshire, T. (1975) Wheat seed infection by *Septoria tritici*. *Transactions of the British Mycological Society* 64, 331-334.

Chen, R.S. and McDonald, B.A. (1996) Sexual reproduction plays a major role in the genetic structure of populations of the fungus *Mycosphaerella graminicola*. *Genetics* 142, 1119-1127.

Cockerell, V. and Rennie, W.J. (1996) Survey of seed-borne pathogens in certified and farm-saved cereal seed in Britain between 1992 and 1994. *Project Report No. 124*, Home-Grown Cereals Authority, London.

Cohen, L. and Eyal, Z. (1993) The histology of processes associated with the infection of resistant and susceptible wheat cultivars with *Septoria tritici*. *Plant Pathology* 42, 737-743.

Cooley, R.N. and Caten, C.E. (1991) Variation in electrophoretic karyotype between strains of *Septoria nodorum*. *Molecular and General Genetics* 228, 17-23.

Cooley, R.N., Shaw, R.K., Franklin, F.C.H. and Caten, C.E. (1988) Transformation of the phytopathogenic fungus *Septoria nodorum* to hygromycin B resistance. *Current Genetics* 13, 383-389.

Cooley, R.N., van Gorcom, R.F.M., van den Hondel, C.A.M.J.J. and Caten, C.E. (1991) Isolation of a benomyl-resistant allele of the β-tubulin gene from *Septoria nodorum* and its use as a dominant selectable marker. *Journal of General Microbiology* 137, 2085-2091.

Cunfer, B.M. (1994) Taxonomy and nomenclature of *Septoria* and *Stagonospora* species on cereals. In: Arseniuk, E., Goral, T. and Czembor, P. (eds) *Proceedings of the 4th International Septoria of Cereals Workshop.* Radzikow, Poland, pp. 15-19.

Cunfer, B.M. (1997) Taxonomy and nomenclature of *Septoria* and *Stagonospora* species on small grain cereals. *Plant Disease* 81, 427-428.

Cunfer, B.M. and Youmans, J. (1983) *Septoria nodorum* on barley and relationships among isolates from several hosts. *Phytopathology* 73, 911-914.

Danon, T. and Eyal, Z. (1990) Inheritance of resistance to two *Septoria tritici* isolates in spring and winter bread wheat cultivars. *Euphytica* 47, 203-214.

Devys, M., Barbier, M., Bousquet, J.F. and Kollmann, A. (1994) Isolation of the (-)-(3R)-5-hydroxymellein from the fungus *Septoria nodorum*. *Phytochemistry* 35, 825-826.

Eyal, Z. (1971) The kinetics of pycnidiospore liberation in *Septoria tritici*. *Canadian Journal of Botany* 49, 1095-1099.

Eyal, Z. (1976) Research on Septoria leaf blotch of wheat caused by *Septoria tritici* in Israel. In: Cunfer, B.M. and Nelson, L.R. (eds) *Proceedings of the Septoria Diseases of Wheat Workshop. The University of Georgia Experiment Stations Special Publication No. 4,* pp. 49-53.

Eyal, Z. (1992) The response of field-inoculated wheat cultivars to mixtures of *Septoria tritici* isolates. *Euphytica* 61, 25-35.

Eyal, Z. (1995) Virulence in *Septoria tritici*, the causal agent of septoria tritici blotch of wheat. In: Gilchrist, L., Van Ginkel, M., McNab, A. and Kema, G.H.J. (eds) *Proceedings of a Septoria tritici Workshop.* CIMMYT, Mexico D.F., pp. 27-33.

Eyal, Z. and Levy, E. (1987) Variations in pathogenicity patterns of *Mycosphaerella graminicola* within *Triticum* spp. in Israel. *Euphytica* 36, 237-250.

Eyal, Z., Wahl, I. and Prescott, J.M. (1983) Evaluation of germplasm response to septoria leaf blotch of wheat. *Euphytica* 32, 439-446.

Eyal, Z., Scharen, A.L., Huffman, M.D. and Prescott, J.M. (1985) Global insights into virulence frequencies of *Mycosphaerella graminicola. Phytopathology* 75, 1456-1462.

Eyal, Z., Scharen, A.L., Prescott, J.M. and Van Ginkel, M. (1987) *The Septoria Diseases of Wheat: Concepts and Methods of Disease Management.* CIMMYT, Mexico, D.F., 51pp.

Farr, D.F., Bills, G.F., Chamuris, G.P. and Rossman, A.Y. (1989) *Fungi on Plants and Plant Products in the United States.* American Phytopathological Society Press, St. Paul, Minnesota, 1252pp.

Fried, P.M. and Meister, E. (1987) Inheritance of leaf and head resistance of winter wheat to *Septoria nodorum* in a diallel cross. *Phytopathology* 77, 1371-1375.

Gilchrist, L. and Velazquez, C. (1994) Interaction to *Septoria tritici* isolate-wheat as adult plant under field conditions. In: Arseniuk, E., Goral, T. and Czembor, P. (eds) *Proceedings of the 4th International Septoria of Cereals Workshop.* Radzikow, Poland, pp. 111-114.

Halama, P. and Lacoste, L. (1992a) Déterminisme de la production sexuée du *Phaeosphaeria* (*Leptosphaeria*) *nodorum* agent de la septoriose du blé I. Hétérothallisme et rôle des microspores. *Canadian Journal of Botany* 69, 95-99.

Halama, P. and Lacoste, L. (1992b) Déterminisme de la production sexuée du *Phaeosphaeria* (*Leptosphaeria*) *nodorum* agent de la septoriose du blé II. Action de la température et de la lumiére. *Canadian Journal of Botany* 70, 1563-1569.

Halfon-Meiri, A. and Kulik, M.M. (1977) *Septoria nodorum* infection of wheat seeds produced in Pennsylvania. *Plant Disease Reporter* 61, 867-869.

Halperin, T., Schuster, S., Pnini-Cohen, S., Zilberstein, A. and Eyal, Z. (1996) The suppression of pycnidial production on seedlings following sequential inoculation by isolates of *Septoria tritici. Phytopathology* 86, 728-732.

Hampton, J.G. (1980) Fungal pathogens in New Zealand certified wheat seed. *New Zealand Journal of Experimental Agriculture* 4, 89-92.

Harrabi, M., Cherif, M., Amara, H., Ennaiffer, Z. and Daaloul, A. (1995) *In vitro* selection for resistance to *Septoria tritici* in wheat. In: Gilchrist, L., Van Ginkel, M., McNab, A. and Kema, G.H.J. (eds) *Proceedings of a Septoria tritici Workshop*. CIMMYT, Mexico D.F., pp. 109-117.

Hedjaroude, G.A. (1968) Etudes taxonomiques sur les *Phaesophaeria* Miyake et leurs formes voisines (Ascomycetes). *Sydowia* 22, 57-67.

Jlibene, M., Amri, A., Bouhssini, M.E. and Ferrara, O.G. (1995) Status of breeding wheat resistant to Hessian Fly and Septoria tritici blotch in Morocco. In: Li, Z.S. and Xin, Z.Y. (eds) *Proceedings of the 8th International Wheat Genetics Symposium*. Beijing, China, pp. 949-952.

Johnson, R. (1992) Past, present and future opportunities in breeding for disease resistance, with examples from wheat. *Euphytica* 63, 3-22.

Karjalainen, R. and Lounatmaa, K. (1986) Ultrastructure of penetration and colonization of wheat leaves by *Septoria nodorum. Physiological Molecular Plant Pathology* 29, 263-270.

Keller, S.M., McDermott, J.M., Pettway, R.E., Wolfe, M.S. and McDonald, B.A. (1997) Gene flow and sexual reproduction in the wheat glume blotch pathogen *Phaeosphaeria nodorum* (anamorph *Stagonospora nodorum*). *Phytopathology* 87, 353-358.

Kema, G.H.J. and Van Silfhout, C.H. (1997) Genetic variation for virulence and resistance in the wheat-*Mycosphaerella graminicola* pathosystem III. Comparative seedling and adult plant experiments. *Phytopathology* 87, 266-272.

Kema, G.H.J., Annone, J.G., Sayoud, R., Van Silfhout, C.H., Van Ginkel, M. and de Bree, J. (1996a) Genetic variation for virulence and resistance in the wheat-*Mycosphaerella graminicola* pathosystem I. Interactions between pathogen isolates and host cultivars. *Phytopathology* 86, 200-212.

Kema, G.H.J., Sayoud, R., Annone, J.G. and Van Silfhout, C.H. (1996b) Genetic variation for virulence and resistance in the wheat-*Mycosphaerella graminicola* pathosystem II. Analysis of interactions between pathogen isolates and host cultivars. *Phytopathology* 86, 213-220.

Kema, G.H.J., Yu, D.Z., Rijkenberg, F.H.J., Shaw, M.W. and Baayen, R.P. (1996c) Histology of the pathogenesis of *Mycosphaerella graminicola* in wheat. *Phytopathology* 86, 777-786.

Kema, G.H.J., Verstappen, E.C.P., Todorova, M. and Waalwijk, C. (1996d) Successful crosses and molecular tetrad and progeny analyses demonstrate heterothallism in *Mycosphaerella graminicola. Current Genetics* 30, 251-258.

Kema, G.H.J., Koeken, J.G.P. and Verstappen, E.C.P. (1997) Avirulence in *Mycosphaerella graminicola* (anamorph *Septoria tritici*) is controlled by a complex locus of tightly linked genes. *Proceedings of the 19th Fungal Genetics Conference*. March 18-23, 1997, Pacific Grove, California, p. 33.

Kent, S.S. and Strobel, G.A. (1976) Phytotoxin from *Septoria nodorum*. *Transactions of the British Mycological Society* 67, 354-358.

King, J.E., Cook, R.J. and Melville, S.C. (1983) A review of *Septoria* diseases of wheat and barley. *Annals of Applied Biology* 103, 345-373.

Kolmer, J.A. and Liu, J.Q. (1997) Virulence and molecular polymorphism in an international collection of *Puccinia recondita* from common wheat. *Phytopathology* 87 (Supplement), S54.

Krupinsky, J.M. (1989) Comparison of isolates of *Leptosphaeria nodorum* from alternative hosts. In: Fried, P.M. (ed.) *Proceedings of the 3rd International Septoria of Cereals Workshop.* Zurich, Switzerland, pp. 13-18.

Krupinsky, J.M. (1994) Aggressiveness of *Stagonospora nodorum* isolates from alternative hosts after passage through wheat. In: Arseniuk, E., Goral, T. and Czembor, P. (eds) *Proceedings of the 4th International Septoria of Cereals Workshop*. Radzikow, Poland, pp. 123-126.

Lehtinen, U. (1993) Plant cell wall degrading enzymes of *Septoria nodorum*. *Physiological and Molecular Plant Pathology* 43, 121-134.

Levy, E., Eyal, Z., Chet, I. and Hochman, A. (1992a) Resistance mechanisms of *Septoria tritici* to antifungal products of *Pseudomonas*. *Physiological and Molecular Plant Pathology* 40, 163-171.

Levy, E., Eyal, Z. and Hochman, A. (1992b) Purification and characterization of a catalase-peroxidase from the fungus *Septoria tritici*. *Archives of Biochemistry and Biophysics* 296, 321-327.

Luke, H.H., Barnett, R.D. and Pfahler, P.L. (1986) Development of Septoria nodorum blotch on wheat from infected and treated seed. *Plant Disease* 70, 252-254.

Magro, P. (1984) Production of polysaccharide-degrading enzymes by *Septoria nodurum* in culture and during pathogenesis. *Plant Science Letters* 37, 63-68.

McDonald, B.A. (1997) The population genetics of fungi: Tools and techniques. *Phytopathology* 87, 448-453.

McDonald, B.A. and Martinez, J.P. (1991) Chromosome length polymorphisms in *Septoria tritici* population. *Current Genetics* 19, 265-271.

McDonald, B.A. and McDermott, J.M. (1993) Population genetics of plant pathogenic fungi. *BioScience* 43, 311-319.

McDonald, B.A., Pettway, R.E., Chen, R.S., Boeger, J.M. and Martinez, J.P. (1995) The population genetics of *Septoria tritici* (teleomorph *Mycosphaerella graminicola*). *Canadian Journal of Botany* 73 (Supplement 1), S292-S301.

McDonald, B.A., Mundt, C.C. and Chen, R.S. (1996) The role of selection on the genetic structure of pathogen populations: Evidence from yield experiments with *Mycosphaerella graminicola* on wheat. *Euphytica* 92, 73-80.

Mittermeier, L., Dercks, W., West, S.J.E. and Miller, S.A. (1990) Field results with a diagnostic system for the identification of *Septoria nodorum* and *Septoria tritici*. *Proceedings of the 1990 Brighton Crop Protection Conference - Pests and Diseases* 2, 757-762.

Müller, E. (1952) Pilzliche Erreger der Getreideblattdurre. *Phytopathologische Zeitschrift* 19, 403-416.

Newton, A.C. and Caten, C.E. (1985) Heterokaryosis and heterokaryon incompatibility in *Septoria nodorum*. In: Scharen, A.L. (ed.) *Proceedings of the 2nd International Septoria of* Cereals *Workshop.* Montana State University, Bozeman, USA, pp. 13-15.

Newton, A.C. and Caten, C.E. (1991) Characteristics of strains of *Septoria nodorum* adapted to wheat or to barley. *Plant Pathology* 40, 546-553.

Payne, A. (1997) Isolation of a β-tubulin gene conferring resistance to carbendazim in the phytopathogen *Septoria tritici*. *Proceedings of the 19th Fungal Genetics Conference*. March 18-23, 1997, Pacific Grove, California, p. 69.

Pnini-Cohen, S., Zilberstein, A., Schuster, S., Sharon, A. and Eyal, Z. (1996) Genetic transformation in the wheat pathogen *Septoria tritici*. *Phytopathology* 86 (Supplement), S40.

Pnini-Cohen, S., Ezrati, S., Zilberstein, A., Schuster, S. and Eyal, Z. (1997a) The suppression of pycnidial production on wheat seedlings following inoculation with mixtures of *Septoria tritici* isolates. *Proceedings of the 10th Congress of the Mediterranean Phytopathological Union.* Montpellier, France, pp. 597-601.

Pnini-Cohen, S., Zilberstein, A., Schuster, S. and Eyal, Z. (1997b) The interaction between wheat genotypes and transformants of isolate ISR398 of *Septoria tritici* expressing altered virulence. *Phytopathology* 87 (Supplement), S78.

Polley, R.W. and Thomas, M.R. (1991) Surveys of diseases of winter wheat in England and Wales, 1976-1988. *Annals of Applied Biology* 119, 1-20.

Rufty, R.C., Herbert, T.T. and Murphy, C.C. (1981) Variation in virulence in isolates of *Septoria nodorum*. *Phytopathology* 71, 593-596.

Saadaoui, E.M. (1987) Physiologic specialization of *Septoria tritici* in Morocco. *Plant Disease* 71, 153-155.

Sanderson, F.R. (1976) *Mycosphaerella graminicola* (Fucel) Sanderson, comb. Nov., the acogenenous state of *Septoria tritici* Rob. and Desm. *New Zealand Journal of Botany* 14, 359-360.

Sanderson, F.R., Scharen, A.L., Eyal, Z. and King, A.C. (1986) A study of genetic segregation using single ascospore isolates of *Mycosphaerella graminicola*. *Proceedings of the International Plant Breeding Symposium*. Lincoln, New Zealand. pp. 1-16.

Scharen, A.L., Eyal, Z., Huffman, M.D. and Prescott, J.M. (1985) The distribution and frequency of virulence genes in geographically separated populations of *Leptosphaeria nodorum*. *Phytopathology* 75, 1463-1468.

Shah, D. and Bergstrom, G.C. (1993) Assessment of seedborne *Stagonospora nodorum* in New York soft white winter wheat. *Plant Disease* 77, 468-471.

Shah, D., Bergstrom, G.C. and Ueng, P.P. (1995) Initiation of Septoria nodorum blotch epidemics in winter wheat by seedborne *Stagonospora nodorum*. *Phytopathology* 85, 452-457.

Shaw, M.W. and Royle, D.J. (1989) Airborne inoculum as a major source of *Septoria tritici* (*Mycosphaerella graminicola*) infections in winter wheat crops in the UK. *Plant Pathology* 38, 35-43.

Shipton, W.A., Boyd, W.R.J., Rosielle, A.A. and Shearer, B.I. (1971) The common Septoria diseases of wheat. *Botanical Review* 37, 237-262.

Smedegård-Petersen, V. (1974) *Leptospaheria nodorum* (*Septoria nodorum*), a new pathogen on barley in Denmark, and its physiologic specialization on barley and wheat. *Friesia* 10, 251-264.

Sprague, R. (1950) *Diseases of Cereals and Grasses in North America*. Ronald Press, New York, 538pp.

Ueng, P.P. and Chen, W. (1994) Genetic differentiation between *Phaeosphaeria nodorum* and *P. avenaria* using restriction fragment length polymorphisms. *Phytopathology* 84, 800-806.

Ueng, P.P., Bergstrom, G.C., Slay, R.M., Geiger, E.A., Shaner, G. and Scharen, A.L. (1992) Restriction fragment polymorphisms in wheat glume blotch fungus, *Phaeosphaeria nodorum*. *Phytopathology* 82, 1302-1305.

Yechilevich-Auster, M., Levy, E. and Eyal, Z. (1983) Assessment of interactions between cultivated and wild wheats and *Septoria tritici*. *Phytopathology* 73, 1077-1083.

Zelikovitch, N. and Eyal, Z. (1989) Maintenance of *Septoria tritici* cultures. *Mycological Research* 92, 361-364.

Zelikovitch, N. and Eyal, Z. (1991) Reduction in pycnidial coverage after inoculation of wheat with mixtures of isolates of *Septoria tritici. Plant Disease* 75, 907-910.

Zelikovitch, N., Eyal, Z. and Kashman, Y. (1992) Isolation, purification, and biological activity of an inhibitor from *Septoria tritici. Phytopathology* 82, 275-278.

Zilberstein, A., Pnini-Cohen, S., Eyal, Z., Schuster, S., Hillel, J. and Lavi, U. (1993) Application of DNA fingerprinting for detecting genetical variation among isolates of the wheat pathogen *Mycosphaerella graminicola.* In: Chet, I. (ed.) *Biotechnology in Plant Disease Control.* Wiley-Liss, New York, pp. 341-353.

Zinkernagel, V., Reiss, F., Wiedland, M. and Bartscherer, H.C. (1987) Infection structures of *Septoria nodorum* in leaves of susceptible wheat cultivars. *Zeitschrift für Pflanzenkrankheiten und Pflanzenschutz* 95, 169-175.

Chapter two:

Molecular Genetics of *Stagonospora* and *Septoria*

C.E. Caten
School of Biological Sciences, The University of Birmingham, Birmingham B15 2TT, UK

Introduction

The Septoria disease complex of cereals is caused by three recognized anamorphic species which were formerly known as *Septoria tritici*, *Septoria nodorum* and *Septoria avenae* (Shipton *et al.*, 1971; King *et al.*, 1983). The taxonomy and nomenclature of some of these species is in transition (Cunfer, 1994) and, in this chapter, I will follow what seems to be the emerging consensus and refer to the anamorphic states as *Septoria tritici*, *Stagonospora nodorum* and *Stagonospora avenae.* The perfect states are now designated *Mycosphaerella graminicola*, *Phaeosphaeria nodorum* and *Phaeosphaeria avenaria*, respectively (Cunfer, 1994). For simplicity, I will use only the anamorphic names throughout, abbreviated as defined above and refer collectively to the three species as Septoria.

Although the Septoria species have long been recognized as causing significant yield and quality losses, particularly in wheat, under appropriate conditions (Shipton *et al.*, 1971; King *et al.*, 1983; Eyal *et al.*, 1987), investigations of their classical and molecular genetics started only recently. The beginning of the molecular studies may be traced to the initial transformation of *S. nodorum* in 1987 (Cooley *et al.*, 1988). Classical genetic studies depended upon the domestication of the sexual stage and are even more recent, beginning in 1991 in *S. nodorum* (Halama and Lacoste, 1991) and in 1996 in *S. tritici* (Kema *et al.*, 1996). This chapter will review current knowledge of the molecular genetics of the above three species. Although important, the classical genetics will not be considered as this is discussed in other chapters (see Halama, Chapter 4, and Kema *et al.*, Chapter 10 this volume). One area of Septoria biology where molecular technology has been extensively exploited is in population studies (McDonald *et al.*, 1995). This is discussed by McDonald *et al.* (Chapter 3 this volume) and will not be addressed here.

The Genome

Nuclear Genome

The nuclear genomes of *S. nodorum* and *S. tritici* have been analysed by pulsed field gel electrophoresis (PFGE). *Stagonospora nodorum* has an estimated genome size of 28-32 Mb and a chromosome number of 14-19, depending upon the strain. The chromosomes range in size from 0.4-3.5 Mb (Cooley and Caten, 1991). The genome size of *S. tritici* is estimated to be 31 Mb with a chromosome number of 17 or 18; the chromosomes range in size from 0.3-3.5 Mb (McDonald and Martinez, 1991a). These genome sizes are comparable to those of other filamentous fungi, although the chromosome numbers are above average (Clutterbuck, 1994). As a consequence, the chromosomes are smaller on average than those of many other filamentous fungi. In both *S. nodorum* and *S. tritici,* the ribosomal RNA genes are clustered on a single chromosome (Cooley and Caten, 1991; McDonald and Martinez, 1991a). Cloning of random nuclear sequences for RFLP probes has shown that *S. tritici* contains a significant amount of repetitive DNA that could be minisatellite sequences or relic transposition events (McDonald and Martinez, 1990; 1991a). Some of these repeated sequences have been exploited as fingerprinting probes (McDonald and Martinez, 1991b).

In common with many other phytopathogenic fungi, *S. nodorum* and *S. tritici* exhibit extensive karyotypic polymorphism between strains (Cooley and Caten, 1991; McDonald and Martinez, 1991a). In both species, this variation occurs amongst strains in the same local population, as well as between strains from different geographical locations. Studies in *S. nodorum* of the relationship between electrophoretic karyotype and vegetative compatibility (Cooley and Caten, 1991; and unpublished results) indicate that compatible isolates have the same karyotype, a result consistent wth the idea that vegetative compatibility is indicative of clonal origin (Jinks *et al.*, 1966). The converse is not true however, isolates with the same karyotype may or may not be compatible. Using random genomic clones as probes, McDonald and Martinez (1991a) produced evidence that aneuploidy, deletion, duplication and translocation all contribute to karyotypic polymorphism in *S. tritici.* Examination by PFGE of a complete tetrad of *Leptosphaeria maculans,* a close relative of *S. nodorum*, suggests that much of the karyotypic diversity is generated during meiosis (Plummer and Howlett, 1993).

The chromosome complements of *S. nodorum.* and *S. tritici* contain a number of small chromosomes around 500 kb in size. These minichromosomes are polymorphic in size and, at least in *S. tritici,* appear to differ in presence and absence between strains (Cooley and Caten, 1991; McDonald and Martinez,

1991a). In some other phytopathogenic fungi, similar chromosomes have been shown to be non-essential and to be irregularly transmitted through meiosis (Kistler and Miao, 1992). We have attempted to determine whether or not the minichromosome of *S. nodorum* is essential by cloning sequences of it from both a wheat-adapted and a barley-adapted strain (Cooley *et al.*, 1994). No expression could be detected using these clones as probes in Northern analyses. Furthermore, the minichromosome sequences from the two strains were not homologous, in contrast to sequences from the larger chromosomes. The more rapid divergence of the minichromosome indicated by this observation is consistent with the idea that its evolution is not constrained by a requirement to encode function.

Extranuclear Genome

The mitochondrial genomes of *S. tritici* and *S. nodorum* have been studied by McDonald and associates (unpublished results). Six major haplotypes have been recognized in *S. tritici* ranging in size from 46.0-51.1 kb, with the variation mainly attributable to insertions and deletions in one part of the genome. The mitochondrial genome of *S. nodorum* is more variable and a range of sizes from 47.9-62.3 kb has been found among isolates collected from a single field (B.A. McDonald, College Station, 1997, personal communication).

Plasmids and mycoviruses are other potential extranuclear genetic elements present in fungi (Buck, 1986; Kempken, 1995). I am not aware of any reports of DNA plasmids in Septoria, but since they have been found in many other fungi it would be surprising if a deliberate search did not reveal them in some strains. Double-stranded RNA mycoviruses are common in *S. nodorum* (Newton, 1987) and have been found in some strains of *S. tritici* (Zelikovitch *et al.*, 1990). However, no phenotypic effects have been associated with these viruses.

Transformation

Stagonospora nodorum has been transformed using a range of selectable markers, including hygromycin resistance (Cooley *et al.*, 1988), benomyl resistance (Cooley *et al.*, 1991), and nitrate utilization (Cutler, 1996). With all these selective systems, the efficiency of transformation is low (up to 25 transformants per μg DNA) compared to the model fungal systems, but similar to that in many other plant pathogenic fungi (see references in Lemke and Peng, 1995). Transformant colonies are frequently heterokaryotic when they first arise and require purification by single spore propagation. Purified transformants are mitotically stable during growth in culture and in plant tissue (Cooley *et al.*, 1988). Now that controlled sexual crosses can be made (Halama *et al.*, 1994), it would be interesting to examine the transmission of transforming DNA through meiosis in view of the premeiotic instability of repeated sequences in some fungi (Irelan and Selker, 1996).

In all cases analysed, the transforming DNA is integrated into the host chromosomes. With the hygromycin and β-tubulin vectors, integration events are variable and frequently complex with multiple copies inserted at one or more sites (Cooley *et al.*, 1988; 1991). When the *S. nodorum* nitrate reductase gene was used as a selectable marker, however, all 13 transformants analysed showed homologous integration involving either a single cross over or a gene replacement (Cutler, 1996). Targeted integration has also been observed with cosmid clones carrying a *S. nodorum* minichromosome sequence (R.N. Cooley, Birmingham, 1993, personal communication). These observations suggested that homologous integration occurs readily in this species and encouraged the development of procedures for gene disruption (see section on Gene Technology). Interestingly, PFGE of transformants suggested that transformation was associated with rearrangements involving chromosomes other than those into which integration has occurred. Transformants of *Neurospora crassa* show a high frequency of chromosome rearrangements (Perkins *et al.*, 1993).

Transformation of *S. nodorum* to hygromycin resistance or nitrate utilization generally has no phenotypic effect other than that attributable to markers on the vector, although the occasional aberrant colony that presumably results from integrative gene disruption has been observed. In contrast, many benomyl-resistant transformants showed reduced growth and abnormal colony morphology which was attributed to overproduction of β-tubulin (Cooley *et al.*, 1991).

DNA sequences that cannot be directly selected can be readily transferred into *S. nodorum* by cotransformation with a vector carrying a selectable marker such as hygromycin resistance. Cotransformation frequences up to 50% are obtained with equimolar amounts of the two vectors, with still higher frequencies if excess of the unselected vector is used (Cooley *et al.*, 1990).

While transformation has been extensively investigated in *S. nodorum*, it is not well documented in the other two species. However, both *S. tritici* (Pnini-Cohen *et al.*, 1996) and *S. avenae* (J. Wubben, Norwich, 1997, personal communication) have been transformed to hygromycin resistance. *Septoria tritici* has also been transformed to benomyl resistance (A. Payne, Bristol, 1997, personal communication).

Gene Cloning

A genomic library of *S. nodorum* constructed in the cosmid vector pAN7-2 (Cooley *et al.*, 1991) is now available in an arrayed form (P. Bowyer and A. Bailey, Bristol, 1997, personal communication). A cDNA library in the Stratagene Lamdba ZAP®II vector has also been constructed (K. Howard and R.N . Cooley, AgrEvo, 1995, personal communication). Several genes have been cloned from these libraries using either heterologous probes or homologous probes produced by PCR (Table 2.1a). In addition, partial clones and sequence information are available for a number of other genes (Table 2.1a). Cloning of genes for which heterologous sequence information is available and in which conserved regions can be identified is straightforward, and the list in Table 2.1 is likely to increase significantly over the next few years.

Table 2.1. Genes cloned and sequenced from *Stagonospora nodorum*, *Stagonospora avenae* and *Septoria tritici* as of August 1997.

Gene	Complete (C) or partial (P) sequence	Reference
(a) *Stagonospora nodorum*		
β-tubulin (*tub*A)	C	Cooley *et al.*, 1991; GenEMBL#S56922
Chitin synthase (*CHS2*)	P	K. Howard, Birmingham, 1997, pers. Com.
Cytochrome C	C	LARS[a]
Isocitrate lyase	P	LARS
Nitrate reductase (*NIA1*)	C	Cutler, 1996; GenEMBL#Y13654
Nitrite reductase (*NII1*)	P	Cutler, 1996
Ornithine decarboxylase	C	LARS
Peptide synthase	P	LARS
Protease, trypsin-like	P	A. Carlile, Bath, 1997, pers. com.
rRNA sequences	P	Morales *et al.*, 1995 GenEMBL#U04236, U77361-62
Spermidine synthase	P	LARS
(b) *Stagonospora avenae*		
Avenacosidase	C	J. Wubben and A.E. Osbourn, Norwich, 1997, pers. com.
rRNA sequences	P	GenEMBL#U77357-59
(c) *Septoria tritici*		
ABC transporter (*Stat*1, *Stat*2)	C	L.H. Zwiers, Wageningen, 1997, pers. Com.
Acetyl-CoA-acetyl transferase (*Stacat*1)	C	L.H. Zwiers, Wageningen, 1997, pers. Com.
Hydroxyphenylpyruvate dioxygenase	C	J. Hargreaves, Bristol, 1997, pers. com.
Nitrate reductase	C	A. Payne, Bristol, 1997, pers. com.
SLN1 signal sensing domain	P	B. McDonald, College Station, 1997, pers. com.
RRNA sequences	P	GenEMBL#U77363
α-tubulin	C	D. Hollomon, Bristol, 1997, pers. com.
β-tubulin	C	A. Payne, Bristol, 1997, pers. com.

[a]LARS = P. Bowyer and A. Bailey, IACR-Long Ashton Research Station, 1997, personal communucation.

A gene coding for the glycosyl hydrolase enzyme avenacosidase has been cloned from *S. avenae* and sequenced (J. Wubben and A.E. Osbourn, Norwich, 1997, personal communication; Table 2.1b). Avenacosidase is of interest as a determinant of the pathogenicity of *S. avenae* to oats since removal of sugars from the oat leaf saponins destroys their antifungal activity (see Osbourn *et al.*, Chapter 15 this volume). The only information for *S. tritici* in the GenEMBL database is complete sequences for the 5.8S rRNA gene, ITS1 and ITS2 regions, and partial sequences for the 18S and 26S ribosomal RNA genes (GenEMBL#U77363). However, several other genes have been cloned and sequenced (Table 2.1c) including two genes (*Stat*1, *Stat*2) from the ABC-transporter family which is of interest because of its potential role in multidrug resistance (L.-H. Zwiers, Wageningen, 1997, personal communication), hydroxyphenylpyruvate dioxygenase which is a herbicide target (J. Hargreaves, Bristol, 1997, personal communication), a sequence homologous to the SLN1 signal-sensing domain of yeast (B. McDonald, College Station, 1997, personal communication), and α- and β-tubulin genes (D. Hollomon, Bristol, 1997, personal communication). In addition, various random sequences have been cloned from *S. tritici* for use as probes and some have been sequenced (McDonald and Martinez, 1990; Z. Eyal, Tel Aviv, 1997, personal communication; C. Waalwijk, Wageningen, 1997, personal communication). However, none of these random sequences have been related to a functional gene.

The first gene cloned from *S. nodorum* was the β-tubulin gene (*tub*A) and this was of particular interest because β-tubulin is the target for the benzimidazole (MBC) fungicides (Davidse, 1986). The allele cloned (*tub*A^R) came from an induced MBC-resistant mutant and conferred resistance on transfer to a sensitive strain, thereby confirming that resistance to MBC fungicides in *S. nodorum* can originate through modification of β-tubulin (Cooley *et al.*, 1991). Comparison of the sequence of *tub*A^R with that of the wild-type sensitive allele (*tub*A$^+$) identified the mutation as a single C to T transition in the first position of codon 6, leading to a histidine to tyrosine amino acid substitution (Cooley and Caten, 1993). Mutations in codon 6 are found frequently in induced MBC-resistant mutants of *Aspergillus nidulans* (Osmani and Oakley, 1991), but have not been observed in resistant field isolates of phytopathogenic fungi where substitutions at codons 198 or 200 predominate (Koenraadt *et al.*, 1992; Yarden and Katan, 1993). The latter is also the case in *S. tritici* where MBC-resistant field isolates arise from a substitution of alanine for glutamic acid at residue 198 (D. Hollomon, Bristol, 1997, personal communication). It would be interesting to determine the molecular nature of the mutations in MBC-resistant field isolates of *S. nodorum*.

Gene Structure

Introns and Expression Signals

Introns have been found in all the *S. nodorum* genes for which extensive genomic sequence is available. Comparisons suggest the number of introns is less than in other filamentous fungi, particularly *Aspergillus* spp., but their positions are highly conserved (Cooley and Caten, 1993; Cutler, 1996). Furthermore, the 5' and 3' splice sites and the internal lariat sequence generally conform to the consensus sequences for fungal introns (Table 2.2).

Sequences that resemble characteristic fungal expression signals are present in the 5' region of the *S. nodorum tub*A, *NIA1* and *NII1* genes (Table 2.2). The sequence surrounding the translation start codon shows good homology to the fungal consensus and, with the exception of a TATAAA box in *tub*A, putative promoter elements can be found in appropriate positions. The CT-rich motif has been implicated in the start of transcription in filamentous fungi (Gurr *et al.*, 1987). This is the case with *tub*A where S1 nuclease mapping has located the transcription start points just 5 bp and 8 bp downstream of the CT motif (Cooley and Caten, 1993). Polyadenylation signals and transcription termination sites are not well defined in fungal genes. However, the motif AATAAA is considered a consensus for the polyadenylation signal (Gurr *et al.*, 1987) and three truncated forms (ATAA) are present in the 3' sequence of *tub*A (Cooley and Caten, 1993). No such sequences can be recognized in the corresponding region of *NIA1* (Cutler, 1996).

Codon Usage

Sequence analyses of genes from a variety of organisms have shown that synonymous codons are not all equally used, instead there is a bias towards particular preferred codons that is particularly evident in highly expressed genes (Kurland, 1991). This bias is evident in the *tub*A gene of *S. nodorum* which uses only 48 of the 61 sense codons (Cooley and Caten, 1993). Codon usage is also biased in *NIA1* with a preference for pyrimidines, especially cytosine, in the third position. However, the bias is less than with *tub*A and all 61 sense codons are used (Cutler, 1996).

The Nitrate Assimilation Gene Cluster in *Stagonospora nodorum*

Alerted by the fact that the nitrate and nitrite reductase genes are clustered in some fungi (Johnstone *et al.*, 1990; Gouka *et al.*, 1991; Williams *et al.*, 1995), we sequenced around the cloned *NIA1* gene of *S. nodorum* and found the start of the nitrite reductase open reading frame (*NII1*) 829 bp downstream of the translation stop codon of *NIA1* (Fig. 2.1; Cutler, 1996; Cutler *et al.*, 1997). Both genes are transcribed in the same direction, as is also the case in the related

Table 2.2. Expression sequences found in *Stagonospora nodorum* genes.

Gene	5' region			Translation start	Intron structure					3' region
	Putative CAAT box	Putative TATAAA box	CT-rich sequence		Intron number	Length bp	5' site	Internal lariat	3' site	Putative poly (A) signal
*tub*A[a]	+	-	+	CCAAC<u>ATG</u>CG	1	109	GTACGT	TGCTAAC	TAG	ATAA
					2	51	GTAAGC	ATCTAAC	CAG	
					3	52	GTGCGG	TACTGAC	CAG	
NIA1[b]	+	+	+	CCATG<u>ATG</u>AG	1	54	GTAAGT	GGCTAAC	AAG	-
					2	49	GTAAGT	GACTAAT	TAG	
					3	46	GTAAGT	GTCTTAC	CAG	
					4	56	GTGAGT	AACTAAC	AAG	
NII1[c]	+	+	+	CCATA<u>ATG</u>GC	1	58	GTGCGT	GGCTGAC	TAG	?
					2	52	GTAAGC	AGCTGAT	CAG	
Consensus[d]	CAAT	TATAAA	n/a	CCACC<u>ATG</u>GC			GTANGT	YRCTAAC	YAG	AATAAA

[a]Cooley and Caten, 1993; [b]Cutler, 1996; [c]Cutler, 1996. 5' end only sequenced; [d]Filamentous fungal consensus sequences from Gurr *et al.* (1987); Ballance (1991); Jacobs and Stahl (1995).
R = purine, Y = pyrimidine, N = any nucleotide.

species *Leptosphaeria maculans* (Williams *et al.*, 1995). This contrasts with the *Aspergillus* and *Penicillium* clusters that are divergently transcribed (Johnstone *et al.*, 1990; Gouka *et al.*, 1991).

The regulation of nitrate assimilation has been extensively studied in *A. nidulans* and *N. crassa* (Marzluf, 1993). Induction by nitrate is mediated through a positive-acting regulatory protein termed NIRA and NIT4 in *A. nidulans* and *N. crassa*, respectively. Transcription also depends upon another positive-acting regulatory protein, termed AREA or NIT2, that is repressed in the presence of ammonium or glutamine. These two controls ensure that the nitrate assimilation genes are transcribed only in the presence of nitrate and absence of a preferred nitrogen source. Consensus binding sites for both regulatory proteins have been identified in the 5' regions of the nitrate and nitrite reductase genes of *A. nidulans* and *N. crassa*, 5'WGATAR3' for the AREA/NIT2 protein and 5'CTCCGHGG3'/5'TCCNNGGA3' for the NIRA/NIT4 protein (Punt *et al.*, 1995). Multiple copies of both these regulatory protein binding sites are present in the 5' regions of the *NIA1* and *NII1* genes of *S. nodorum* (Cutler, 1996), suggesting that the regulation of this pathway in this phytopathogenic species is similar to that in the saprophytic model species.

Gene Technology

Gene Disruption

The ability to inactivate a specific gene through disruption is a valuable tool in determining the function of an unknown cloned sequence or in confirming the role of a suspected pathogenicity determinant. Most methods of gene disruption depend upon homologous integration of transforming DNA (May, 1992). The integration events observed when transforming *S. nodorum* with the *NIA1* gene (see section on Transformation) suggested that homologous integration occurred readily and we have used this gene to develop methods of gene disruption (Howard *et al.*, 1997). Three methods have been tested in attempts to disrupt *NIA1*: (1) cotransformation of a selectable vector and a vector carrying an internal *NIA1* fragment; (2) integrative disruption using a vector carrying both a selectable marker and an internal *NIA1* fragment; (3) one-step gene replacement involving a nearly complete copy of *NIA1* and flanking regions inactivated by insertion of the selectable marker into the open reading frame. With all three approaches, hygromycin resistance (Hyg^R) was used as the selectable marker. Transformants containing a disrupted *NIA1* gene were produced by both methods 1 and 3 (Table 2.3). No disruptants were recovered by method 2, but the number of Hyg^R transformants analysed was low. These results indicate that one-step gene replacement is an efficient method of gene disruption in *S. nodorum*, while the cotransformation approach would only be a viable option where a simple phenotypic screen is available.

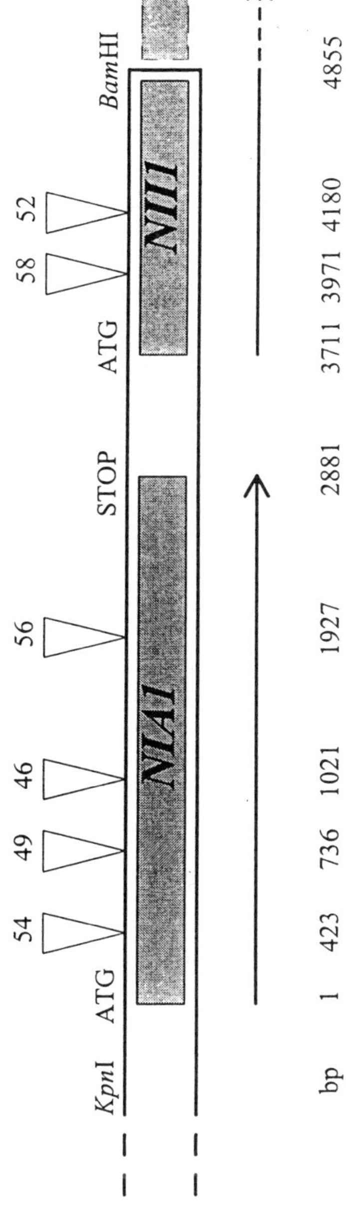

Fig. 2.1. Organization of the nitrate and nitrite reductase gene cluster in *Stagonospora nodorum*. The shaded boxes indicate the *NIA1* and *NII1* open reading frames and the arrows the direction of transcription. Nucleotides are numbered relative to the first base of the *NIA1* ATG start codon; the *NII1* sequence is incomplete. The inverted triangles mark the positions and sizes of introns.

Table 2.3. Targeted disruption of the nitrate reductase (*NIA1*) gene of *Stagonospora nodorum*[a].

Method	Number of Hyg^R transformants	Number of nitrate non-utilizing	Number of *NIA1* disruptants	% *NIA1* disruptants
Cotransformation	139	3	2	1.4
Integrative gene disruption	18	0	0	0
One-step gene replacement	26	14[b]	6	23.0

[a]K. Howard, Birmingham, 1997, personal communication.
[b]5 disrupted in the *NII1* gene part of which was also present on the vector.

An unexpected feature of the results was the recovery of some nitrate non-utilizing transformants that from Southern hybridization still appeared to contain an intact *NIA1* gene (Table 2.2; Fig. 2.2). Eight such strains were recovered among 26 Hyg^R transformants using method 3. Five of these appear to carry a functional *NIA1* gene but an inactivated, and presumably disrupted, *NII1* gene since they excrete nitrite when grown on nitrate and cannot utilize nitrite (K. Howard, Birmingham, 1997, personal communication). These five transformants all contain multiple tandem insertions of vector sequence (Fig. 2.2). Although the vector carried some *NII1* sequence, such integration was not expected as it had been linearized to force the double cross over. The most likely explanation for these *NII1* disruptants is that the vector recircularized and formed multimers *in vivo* which were then integrated by a single cross over in the homologous *NII1* sequence. We were able to detect and interpret these events because of the clustering of the two functionally related genes. These results demonstrate the importance of confirming disruption by molecular analysis rather than relying solely on phenotypic change.

Insertional Mutagenesis

Insertional mutagenesis is receiving increasing attention as a means of defining and cloning genes involved in pathogenicity (Lu *et al.*, 1994). Pnini-Cohen *et al.* (1996; 1997) and Eyal (Chapter 1 this volume) found changes in virulence among transformants of *S. tritici* and are using the vector sequence to recover the flanking regions which should carry genes involved in the pathogenic specificity of this fungus. Other laboratories are using restriction enzyme-mediated integration (REMI) in combination with phenotypic screens to isolate interesting mutants from which the disrupted genes can be cloned and characterized (C. Rushowski and P. Bowyer, Bristol 1997, personal communication).

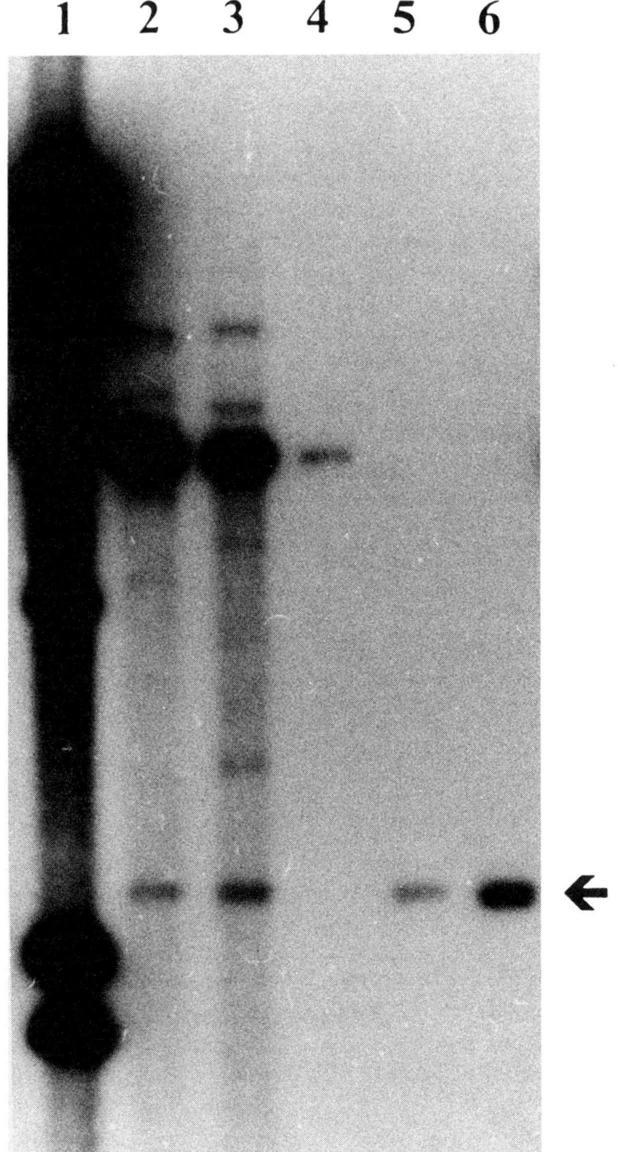

Fig. 2.2. Southern analysis of strains of *Stagonospora nodorum* transformed with the *NIA1* gene replacement vector. Lane 1, λ-*Hin*dIII size standards; lane 2, *nit*4; lane 3, *nit*5; lane 4, *nit*6; lane 5, Nit^+ transformant; lane 6, untransformed wild-type. The DNA was digested with *Cla*I and then probed with the internal 2.5 kb *Cla*I fragment from *NIA1*. The endogenous *NIA1* gene is indicated by an arrow. This is disrupted as expected in *nit*6, but remains intact in *nit*4 and *nit*5 despite the fact these have a nitrate non-utilizing phenotype.

Transposable elements are widely used to mutate and tag genes in prokaryotes and higher eukaryotes (Berg and Howe, 1989). In the last few years, such elements have been discovered in an increasing number of filamentous fungi and their use for insertional mutagenesis proposed (Daboussi, 1996). We have initiated a transposon hunt in *S. nodorum* using the *NIA1* gene as a trap. Spontaneous chlorate-resistant mutants that growth tests indicate carry mutations in *NIA1* are screened by Southern hybridization for disruption of the gene. Preliminary results indicate that gene rearrangements occur at a significant frequency and the origin of these is now being determined (J. Rawson, Birmingham, 1997, personal communication).

Uses of Molecular Genetics

Molecular approaches can be, and are being, applied to the Septoria species to answer a variety of questions. They provide a powerful way to determine the mechanisms these fungi use to parasitize plants, either through the black box approach of using REMI to generate non-pathogenic mutants or as a means of confirming the role of putative pathogenicity factors such as cell wall degrading enzymes or toxins. An extension of this is the use of reverse genetics in the processes of fungicide discovery and target validation (Caten and Hollomon, 1995; Dancer *et al.*, Chapter 22 this volume).

The Septoria species differ in disease potential and fungicide sensitivity and reliable identification is important for control. However, they produce similar symptoms and diagnosis by visual inspection is difficult. To overcome this problem, Beck and Ligon (1995) produced a PCR assay based on the internal transcribed spacer regions of the rDNA that can detect and distinguish *S. tritici* and *S. nodorum* in infected host tissue. Where the molecular basis of fungicide insensitivity is known, such molecular diagnostics can be taken to a sophisticated level allowing the detection of fungicide-resistant strains in a pathogen population (Wheeler *et al.*, 1994; Hollomon *et al.*, Chapter 19 this volume).

An advantage of molecular genetic methods is that they provide large quantities of objective data about a strain, population or species that can be used for taxonomic and phylogenetic purposes. These data may be in the form of presence or absence of RFLP or RAPD bands, or nucleotide or amino acid sequences, all of which can be used for numerical analyses. Phylogenetic analysis of RFLP data has shown that the wheat- and barley-biotypes of *S. nodorum* are genetically as distant from each other as they are from *S. avenae.* In fact, the barley-biotype appears to be more closely related to *S. avenae* than to the wheat-biotype (Ueng and Chen, 1994; Ueng *et al.*, 1995). Comparison of amino acid sequences for β-tubulin and nitrate reductase with those of other Ascomycetes in the database provides little useful information, as *S. nodorum* appears approximately equidistant from all other species. The one exception to this is confirmation of a close relationship to *L. maculans* indicated by the organization of the nitrate assimilation gene cluster and the amino acid sequence of the nitrate and nitrite reductases (Cutler, 1996).

Molecular genetic methods have also been extensively used to develop neutral markers for population studies on *S. tritici* and *S. nodorum* (McDonald *et al.*, 1995; McDonald *et al.*, Chapter 3 this volume). These studies have transformed the common perception of the epidemiology of these pathogens by highlighting the importance of the sexual stage throughout the epidemic cycle (McDonald *et al.*, 1996).

Molecular genetics has already made major contributions to our knowledge of the Septoria pathosystem and this information is being applied in a variety of ways. With the recent increase in interest in Septoria molecular genetics we can look forward to expansion in knowledge and eventually to improved disease control.

Acknowledgements

Much of the work on *S. nodorum* described in this paper has been carried out at The University of Birmingham by a number of postdoctoral fellows and research students, including Dr Neil Cooley, Dr Adrian Newton, Dr Simon Cutler, Miss Kirsty Howard and Mrs Jenny Rawson. Their hard work, enthusiasm and insights are gratefully acknowledged. I would also like to thank those colleagues who willingly provided unpublished information so that I could make this review as up-to-date as possible. The work at Birmingham on *S. nodorum* has been supported over the years by the Agricultural and Food Research Council, The Biotechnology and Biological Sciences Research Council, The Gatsby Charitable Foundation and AgrEvo UK Ltd.

References

Ballance, D.J. (1991) Transformation systems for filamentous fungi and an overview of fungal gene structure. In: Leong, S.A. and Berka, R.A. (eds) *Molecular Industrial Mycology: Systems and Applications for Filamentous Fungi.* Marcel Dekker, New York, pp. 1-29.

Beck, J.J. and Ligon, J.M. (1995) Polymerase chain reaction assays for the detection of *Stagonospora nodorum* and *Septoria tritici* in wheat. *Phytopathology* 85, 319-324.

Berg, D.E. and Howe, M.M. (eds) (1989) *Mobile DNA.* American Society for Microbiology, Washington, DC, 972pp.

Buck, K.W. (ed.) (1986) *Fungal Virology.* CRC Press, Boca Raton, Florida, 305pp.

Caten, C.E. and Hollomon, D.W. (1995) Gene disruption, repeat-induced point mutation and antisense RNA: their application and value in biochemical screening research. In: Dixon, G.K., Copping, L.G. and Hollomon, D.W. (eds) *Antifungal Agents: Discovery and Mode of Action.* BIOS Scientific Publishers, Oxford, pp. 31-47.

Clutterbuck, A.J. (1994) Molecular Biology. In: Gow, N.A.R. and Gadd, G.M. (eds) *The Growing Fungus*. Chapman and Hall, London, pp. 255-274.

Cooley, R.N. and Caten, C.E. (1991) Variation in electrophoretic karyotype between strains of *Septoria nodorum. Molecular and General Genetics* 228, 17-23.

Cooley, R.N. and Caten, C.E. (1993) Molecular analysis of the *Septoria nodorum* β-tubulin gene and characterization of a benomyl-resistance mutation. *Molecular and General Genetics* 237, 58-64.

Cooley, R.N., Shaw, R.K., Franklin, F.C.H. and Caten, C.E. (1988) Transformation of the phytopathogenic fungus *Septoria nodorum* to hygromycin B resistance. *Current Genetics* 13, 383-389.

Cooley, R.N., Franklin, F.C.H. and Caten, C.E. (1990) Cotransformation in the phytopathogenic fungus *Septoria nodorum. Mycological Research* 94, 145-151.

Cooley, R.N., van Gorcom, R.F.M., van den Hondel, C.A.M.J.J. and Caten, C.E. (1991) Isolation of a benomyl-resistant allele of the β-tubulin gene from *Septoria nodorum* and its use as a dominant selectable marker. *Journal of General Microbiology* 137, 2085-2091.

Cooley, R.N., Reynolds, P.A. and Caten, C.E. (1994) Characterisation of the minichromosome of *Septoria nodorum. Abstracts 5th International Mycological Congress*, Vancouver, p. 32.

Cunfer, B.M. (1994) Taxonomy and nomenclature of *Septoria* and *Stagonospora* species on cereals. *Hodowla Roslin Aklimatyzacja i Nasiennictwo (Special Edition)* 38(3-4), 15-19.

Cutler, S.B. (1996) The development of a homologous transformation system for *Septoria nodorum* based on nitrate assimilation. PhD thesis, The University of Birmingham, Birmingham, UK.

Cutler, S.B., Cooley, R.N. and Caten, C.E. (1997) Organization of the nitrate assimilation cluster of *Stagonospora (Septoria) nodorum. Fungal Genetics Newsletter* 44, on-line supplement [http://kumchttp.mc.ukans.edu/research/fgsc/asilomar/genome.html] (poster 19).

Daboussi, M.J. (1996) Fungal transposable elements: generators of diversity and genetic tools. *Journal of Genetics* 75, 325-339.

Davidse, L.C. (1986) Benzimidazole fungicides: mechanism of action and biological impact. *Annual Review of Phytopathology* 24, 43-65.

Eyal, Z., Scharen, A.L., Prescott, J.M. and van Ginkel, M. (1987) *The Septoria Diseases of Wheat: Concepts and Methods of Disease Management.* Centro Internacional de Majoramiento de Maíz y Trigo (CIMMYT), Mexico, D.F., 46pp.

Gouka, R.J., van Hartingsveldt, W., Bovenberg, R.A.L., van den Hondel, C.A.M.J.J. and van Gorcom, R.F.M. (1991) Cloning of the nitrate-nitrite reductase gene cluster of *Penicillium chrysogenum* and use of the *nia*D gene as a homologous selection marker. *Journal of Biotechnology* 20, 189-200.

Gurr, S.J., Unkles, S.E. and Kinghorn, J.R. (1987) The structure and organisation of nuclear genes of filamentous fungi. In: Kinghorn, J.R. (ed.) *Gene Structure in Eukaryotic Microbes*. IRL Press, Oxford, pp. 93-139.

Halama, P. and Lacoste, L. (1991) Déterminisme de la reproduction sexuée de *Phaeosphaeria (Leptosphaeria) nodorum*, agent de la septoriose du blé. I. Hétérothallisme et rôle des microspores. *Canadian Journal of Botany* 69, 95-99.

Halama, P., Skajennikoff, M. and Rapilly, F. (1994) The progeny of two crosses of *Phaeosphaeria nodorum. Hodowla Roslin Aklimatyzacja i Nasiennictwo (Special Edition)* 38(3-4), 115-118.

Howard, K., Foster, S.G., Cooley, R.N. and Caten, C.E. (1997) Gene disruption as a method of fungicide target validation. *Fungal Genetics Newsletter* 44, on-line supplement http://kumchttp.mc.ukans.edu/research/fgsc/asilomar/ biotech. html] (poster 69).

Irelan, J.T. and Selker, E.U. (1996) Gene silencing in filamentous fungi: RIP, MIP and quelling. *Journal of Genetics* 75, 313-324.

Jacobs, M. and Stahl, U. (1995) Gene regulation in mycelial fungi. In: Kück, U. (ed.) *The Mycota II: Genetics and Biotechnology*. Springer-Verlag, Berlin, pp. 155-167.

Jinks, J.L., Caten, C.E., Simchen, G. and Croft, J.H. (1966) Heterokaryon incompatibility and variation in wild populations of *Aspergillus nidulans*. *Heredity* 21, 227-239.

Johnstone, I.L., McCabe, P.C., Greaves, P., Gurr, S.J., Cole, G.E., Brow, M.A.D., Unkles, S.E., Clutterbuck, A.J., Kinghorn, J.R. and Innis, M.A. (1990) Isolation and characterisation of the *crn*A-*nii*A-*nia*D gene cluster for nitrate assimilation in *Aspergillus nidulans*. *Gene* 90, 181-192.

Kema, G.H.J., Verstappen, E.C.P., Todorova, M. and Waalwijk, C. (1996) Successful crosses and molecular tetrad and progeny analyses demonstrate heterothallism in *Mycosphaerella graminicola*. *Current Genetics* 30, 251-258.

Kempken, F. (1995) Plasmid DNA in mycelial fungi. In: Kück, U. (ed.) *The Mycota II: Genetics and Biotechnology*. Springer-Verlag, Berlin, pp. 169-187.

King, J.E., Cook, R.J. and Melville, S.C. (1983) A review of *Septoria* diseases of wheat and barley. *Annals of Applied Biology* 103, 345-373.

Kistler, H.C. and Miao, V.P.W. (1992) New modes of genetic change in filamentous fungi. *Annual Review of Phytopathology* 30, 131-152.

Koenraadt, H., Somerville, S.C. and Jones, A.L. (1992) Characterisation of mutations in the beta-tubulin gene of benomyl-resistant field strains of *Venturia inaequalis* and other plant pathogenic fungi. *Phytopathology* 82, 1348-1354.

Kurland, C.G. (1991) Codon bias and gene expression. *FEBS Letters* 285, 165-169.

Lemke, P.A. and Peng, M. (1995) Genetic manipulation of fungi by DNA-mediated transformation. In: Kück, U. (ed.) *The Mycota II: Genetics and Biotechnology.* Springer-Verlag, Berlin, pp. 109-139.

Lu, S., Lyngholm, L., Yang, G., Bronson, C., Yoder, O.C. and Turgeon, B.G. (1994) Tagged mutations at the *Tox1* locus of *Cochliobolus heterostrophus* by restriction enzyme-mediated integration. *Proceedings of the National Academy of Sciences, USA* 91, 12649-12653.

Marzluf, G.A. (1993) Regulation of sulphur and nitrogen metabolism in filamentous fungi. *Annual Review of Microbiology* 47, 31-55.

May, G. (1992) Fungal technology. In: Kinghorn, J.R. and Turner, G. (eds) *Applied Molecular Genetics of Filamentous Fungi.* Blackie, London, pp. 1-27.

McDonald, B.A. and Martinez, J.P. (1990) DNA restriction fragment length polymorphisms among *Mycosphaerella graminicola* (anamorph *Septoria tritici*) isolates collected from a single wheat field. *Phytopathology* 80, 1368-1373.

McDonald, B.A. and Martinez, J.P. (1991a) Chromosome length polymorphisms in a *Septoria tritici* population. *Current Genetics* 19, 265-271.

McDonald, B.A. and Martinez, J.P. (1991b) DNA fingerprinting of the plant pathogenic fungus *Mycosphaerella graminicola* (anamorph *Septoria tritici*). *Experimental Mycology* 15, 146-158.

McDonald, B.A., Pettway, R.E., Chen, R.S., Boeger, J.M. and Martinez, J.P. (1995) The population genetics of *Septoria tritici* (telemorph *Mycosphaerella graminicola*). *Canadian Journal of Botany* 73 (Supplement 1), S292-S301.

McDonald, B.A., Mundt, C.C. and Chen, R.S. (1996) The role of selection on the genetic structure of pathogen populations: evidence from field experiments with *Mycosphaerella graminicola* on wheat. *Euphytica* 92, 73-80.

Morales, V.M., Jasalavich, C.A., Pelcher, L.E., Petrie, G.A. and Taylor, J.L. (1995) Phylogenetic relationship among several *Leptosphaeria* species based on their ribosomal DNA sequences. *Mycological Research* 99, 593-603.

Newton, A.C. (1987) Occurrence of double-stranded RNA and virus-like particles in *Septoria nodorum. Transactions of the British Mycological Society* 88, 113-141.

Osmani, S.A. and Oakley, B.R. (1991) Cell cycle and tubulin mutations in filamentous fungi. In: Bennett, J.W. and Lasure, L.L. (eds) *More Gene Manipulations in Fungi.* Academic Press, London, pp. 107-125.

Perkins, D.D., Kinsey, J.A., Asch, D.K. and Frederick, G.D. (1993) Chromosome rearrangements recovered following transformation of *Neurospora crassa. Genetics* 134, 729-736.

Plummer, K.M. and Howlett, B.J. (1993) Major chromosomal length polymorphisms are evident after meiosis in the phytopathogenic fungus *Leptosphaeria maculans. Current Genetics* 24, 107-113.

Pnini-Cohen, S., Zilberstein, A., Schuster, S., Sharon, A. and Eyal, Z. (1996) Genetic transformation in the wheat pathogen *Septoria tritici. Phytopathology* 86 (Supplement), S40.

Pnini-Cohen, S., Zilberstein, A., Schuster, S. and Eyal, Z. (1997) The interaction between wheat genotypes and transformants of isolate ISR398 of *Septoria tritici* expressing altered virulence. *Phytopathology* 87 (Supplement), S78.

Punt, P.J., Strauss, J., Smit, R., Kinghorn, J.R., van den Hondel, C.A.M.J.J. and Scazzocchio, C. (1995) The intergenic region between the divergently transcribed *nii*A and *nia*D genes of *Aspergillus nidulans* contains multiple *nir*A binding sites which act bidirectionally. *Molecular and Cellular Biology* 15, 5688-5699.

Shipton, W.A., Boyd, W.R.J., Rosielle, A.A. and Shearer, B.I. (1971) The common Septoria diseases of wheat. *Botanical Review* 37, 231-262.

Ueng, P.A. and Chen, W. (1994) Genetic differentiation between *Phaeosphaeria nodorum* and *P. avenaria* using restriction fragment length polymorphisms. *Phytopathology* 84, 800-806.

Ueng, P.P., Cunfer, B.M., Alano, A.S., Youmans, J.D. and Chen, W. (1995) Correlation between molecular and biological characters in identifying the wheat and barley biotypes of *Stagonospora nodorum. Phytopathology* 85, 44-52.

Wheeler, I., Kendall, S., Butters, J. and Hollomon, D.W. (1994) Rapid detection of benzimidazole resistance in *Rhynchosporium secalis* using allele-specific oligonucleotide probes. In: Heaney, S., Slawson, D., Hollomon, D.W., Smith, M., Russell, P.E. and Parry, D.W. (eds) *Fungicide Resistance.* BCPC, Farnham, UK, pp. 259-263.

Williams, R.S.B., Davis, M.A. and Howlett, B.J. (1995) The nitrate and nitrite reductase-encoding genes of *Leptosphaeria maculans* are closely linked and transcribed in the same direction. *Gene* 158, 153-154.

Yarden, O. and Katan, T. (1993) Mutations leading to substitutions at amino acids 198 and 200 of beta-tubulin that correlate with benomyl-resistance phenotypes of field strains of *Botrytis cinerea. Phytopathology* 83, 1478-1483.

Zelikovitch, N., Eyal, Z., Ben-Zui, B. and Koltin, Y. (1990) Double stranded RNA mycoviruses in *Septoria tritici. Mycological Research* 94, 590-594.

Chapter three:

The Population Genetics of *Mycosphaerella graminicola* and *Stagonospora nodorum*

B.A. McDonald, J. Zhan, O. Yarden, K. Hogan, J. Garton and R.E. Pettway
Department of Plant Pathology and Microbiology, Texas A&M University, College Station, TX 77843-2132, USA

Introduction

In 1989, we began developing DNA-based markers as tools to learn about the population genetics of the wheat leaf blotch pathogen *Mycosphaerella graminicola*. One year later, we began parallel studies using the same genetic tools for the wheat glume blotch pathogen *Stagonospora nodorum*. We began with two elementary questions regarding the population genetics of both fungi. How much genetic diversity is present within populations? How is genetic diversity distributed within and among populations? As our knowledge of the genetic structure of both pathogens increased, we addressed more complex questions. What are the relative contributions of sexual and of asexual reproduction to the genetic structure of populations? How stable are populations over time? Does selection for specific pathogen genotypes occur on particular host genotypes? Is there evidence for host specialization in these pathosystems? To address the latter questions, we have utilized increasingly sophisticated field experiments to differentiate among the various evolutionary forces acting on populations of these fungi. In this chapter, we will present an overview of our understanding of the population genetics of both fungi at this point in time. Some of the results we will present reflect work in progress and our interpretation of these data should be considered preliminary.

Methods

DNA Markers

The RFLP markers utilized in these studies were developed in the same way for both fungi, applying the methods described by McDonald and Martinez (1990b). DNA was extracted from a single, randomly chosen isolate and digested with a restriction enzyme. The resulting DNA fragments were ligated into a plasmid vector and used to transform *Escherichia coli.* The resulting clones containing anonymous segments of fungal DNA were screened against a geographically diverse collection of five to eight fungal strains that had been digested individually with six different restriction enzymes. After screening approximately 50 anonymous clones, we selected a set of probes that gave easy-to-interpret hybridization patterns with one of the restriction enzymes. By 1990, we learned that the probes from each library were specific to the organism from which they originated. The probes from *M. graminicola* did not hybridize to DNA from *S. nodorum*, and the probes from *S. nodorum* did not hybridize to DNA from *M. graminicola.* The probes from both fungi fell into two general categories: those which hybridized to one to three loci in the nuclear genome, and those which hybridized to dispersed, repetitive elements distributed across many chromosomes. We selected from the former category probes which hybridized to single polymorphic loci (Fig. 3.1, panels A and D). These single-locus probes were used to measure gene diversity for individual RFLP loci and to measure population subdivision and genetic similarity among populations (McDonald and Martinez, 1990a; Boeger *et al.*, 1993; McDonald *et al.*, 1994; Keller *et al.*, 1997a,b). The probes which hybridized to repetitive elements were used for DNA fingerprinting (Fig. 3.1, panels B and E) to differentiate clones and to make measures of genotype diversity (McDonald and Martinez, 1991; Keller *et al.*, 1997a,b). The *S. nodorum* library possessed far fewer repetitive DNAs than the *M. graminicola* library (B.A. McDonald, unpublished results).

To measure variation in the mitochondrial (mt) genome, we used caesium chloride ultracentrifugation to separate the mitochondrial and nuclear DNAs (Garber and Yoder, 1983). The purified mtDNA was labelled and used as a probe which hybridized to the entire digested mitochondrial genome for each individual. We discovered, by 1991, that the mtDNA probes were specific to the mtDNA fraction of the total DNA extract, allowing us to visualize the entire mtDNA genome with a single hybridization reaction (Fig. 3.1, panels C and F).

We have been able to generate large data sets for both fungi at relatively low cost by hybridizing different probes to the same nylon membrane many times. We routinely probe the same Southern blot 17 times, allowing us to collect the entire complement of genetic information (DNA fingerprints, 8-12 individual RFLP loci, mtDNA haplotype) from a single digestion of DNA from each individual. All cloned probes are in the public domain and will be made

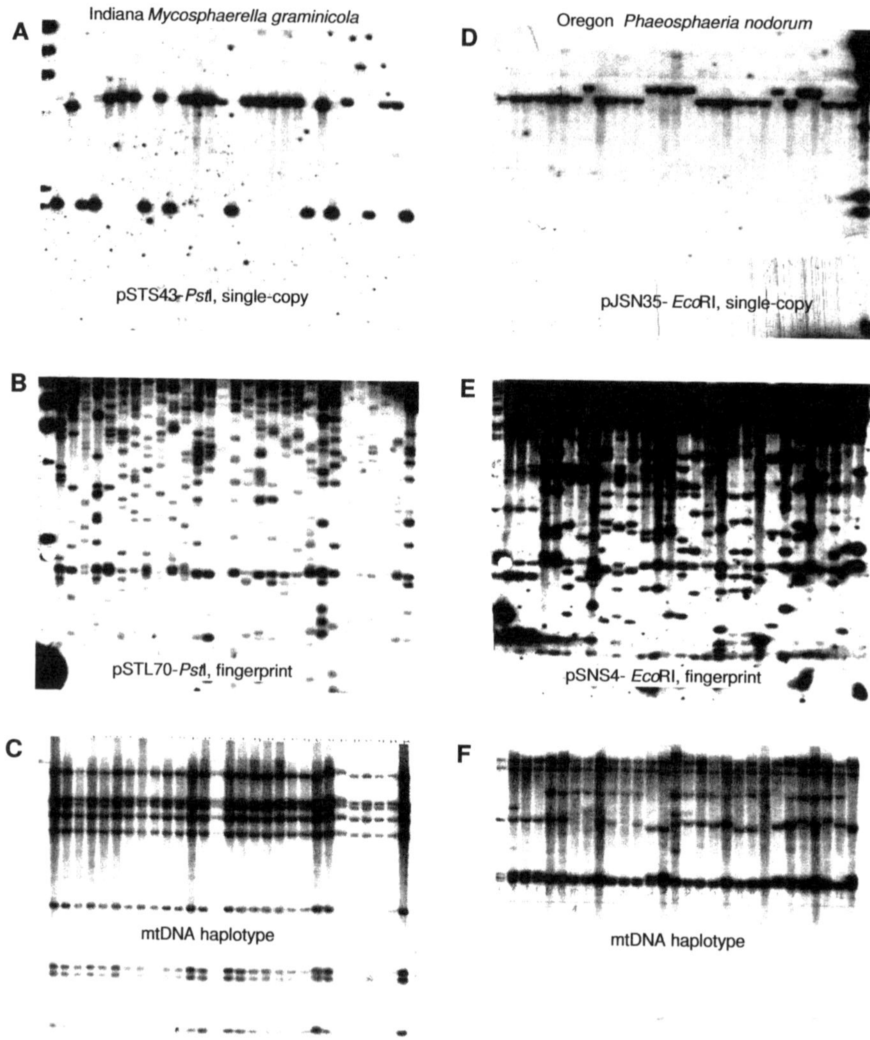

Fig. 3.1. Example of RFLP markers used to characterize populations of *Mycosphaerella graminicola* and *Stagonospora nodorum*. Each set of panels (A-C and D-F) shows the same collection of isolates taken from one field and hybridized with different probes. Panels A-C: *M. graminicola* DNA from a population in Indiana was digested with *Pst*I and hybridized with probes pSTS43, pSTL70 and purified mtDNA respectively. Panels D-F: *S. nodorum* DNA from a population in Texas was digested with *Eco*RI and hybridized with probes pJSN35, pSNS4 and purified mtDNA respectively.

available to interested laboratories upon request.

Sampling Methods

Hierarchical sampling methods have been described previously and now are commonly used in plant pathology (Kohli *et al.*, 1995; McDonald *et al.*, 1995; McDonald, 1997). Hierarchical sampling offers a powerful method to determine the spatial scale on which most of the genetic diversity is distributed for pathogens which have not been studied. For many of the collections described in this chapter, a standardized six-site hierarchy was used to sample field populations (Fig. 3.2). For other field collections, long transects were arranged in a field and leaves were sampled at 1-2 m intervals along each transect. For some collections, leaves were collected at random in a field. All isolates in a collection were taken from a single field at a single time point for the majority of collections. We will refer to these collections as field populations. Four collections of isolates came from the same country or geographical region, but did not originate from a single field. We will refer to these as regional populations to distinguish them from field populations. The regional populations of *M. graminicola* were the Argentina (ARG) population, which comprised 25 isolates from different parts of that country, and the Mediterranean (MED) population, which comprised 99 isolates sampled from countries around the Mediterranean Basin. For *S. nodorum*, the Arkansas (ARK) population comprised isolates collected from infected seed from two different seed lots distributed in Arkansas. The crested wheatgrass (CRW) population comprised isolates made from the weed crested wheatgrass over a two year period in the state of North Dakota. The goal for each collection was to have a large enough field sample to be confident that allele frequencies for individual RFLP loci were representative of the entire field population. Only populations having at least 25 isolates were included in the analysis. A summary of the *M. graminicola* and *S. nodorum* isolates in our collection is presented in Table 3.1.

Data Analysis

Methods used to measure gene and genotype diversity, genetic similarity, population subdivision, and gametic disequilibrium have been published elsewhere (McDonald and Martinez, 1990a; Boeger *et al.*, 1993; Chen *et al.*, 1994; Chen and McDonald, 1996) and the DNA fingerprint probes have been validated for both fungi in previous experiments (McDonald and Martinez, 1991; Keller *et al.*, 1997a). Here we will extend these methods to a larger data set, with special attention placed on the collection of *M. graminicola* isolates from Israel.

Table 3.1. Field populations of *Mycosphaerella graminicola* and *Stagonospora nodorum* that have been assayed for RFLP markers.

Country	Code	Year	Number of fields	Type	Number of isolates	Collector
Mycosphaerella graminicola						
Argentina	ARG	1993	unknown	RG	25	G. Kema
California	CAL	1989	1	H	93	J. McDermott
Canada	CAN	1992	3	H	93	G. Hughes
Chile	CHL	1994	2	R	40	O. Andrade
Denmark	DEN	1994	1	H	123	M. Rasmussen
E. Australia	AUSE	1992	1	H	27	B. Ballantyne
Germany	GER	1992	1	H	48	G. Koch, R. Huang
Indiana	IND	1993	1	R	31	G. Shaner
Israel	ISR	1992	2	H + LT	200	O. Yarden
Mediterranean	MED	1994-5	unknown	RG	99	G. Kema
Mexico	MEX	1993	1	H	312	L. Gilchrist
Missouri	MIS	1994	1	H	37	B. McDonald
Oregon	ORG	1990	1	H	711	J. Boeger, B. McDonald, M. Schmitt
Texas	TEX	1994	2	H	96	B. McDonald, R. Chen
United Kingdom	UK	1992	2	LT	45	M. Shaw, C. Pjils
		1993	2	H	80	

Uruguay	URU	1993	1	H	61	M. Diaz de Ackermann
W. Australia	AUSW	1991	1	R	33	R. Loughman
Stagonospora nodorum						
Arkansas	ARK	1995	seed	RG	190	G. Milus
Denmark	DEN	1994	1	H	16	M. Rasmussen
Mexico	MEX	1993	1	H	21	L. Gilchrist
New York	NY	1991	1	R	56	G. Bergstrom
North Dakota	ND	1993	1	H	25	L. Francl
North Dakota CW	CRW	1990-2	unknown	RG	47	J. Krupinsky
Oregon	ORG	1993	1	H	100	M. Schmitt
South Africa	SA	1995	1	R	68	P. Crous
Switzerland	CH	1994	9	H	432	S. Keller
Texas	TEX	1992	2	H	100	B. McDonald
United Kingdom	UK	1993	2	H	10	M. Shaw, C. Pjils

CW = crested wheatgrass.
H = hierarchical collection of 6-8 sites within a field (Fig. 3.2).
LT = linear transect within a field.
R = random collection, unordered.
RG = regional collection (not single field).

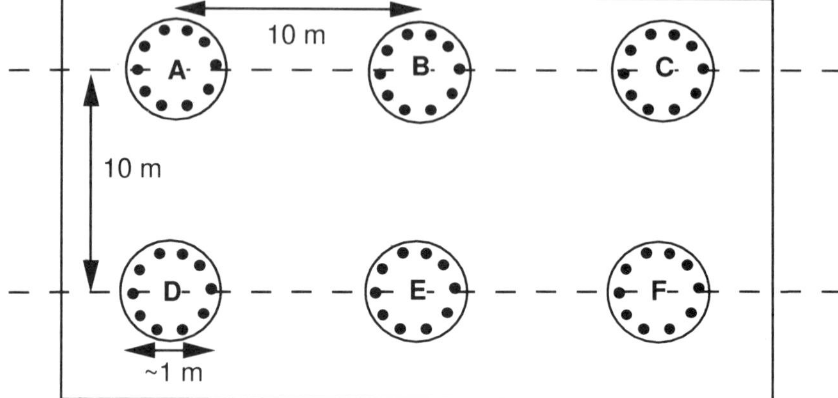

Fig. 3.2. Example of the hierarchical strategy used to sample field populations of *Mycosphaerella graminicola* and *Stagonospora nodorum*. Letters within circles represent six different sites in a field arranged along two transects separated by 10 m. Black circles within sites represent ten infected leaves sampled from ten different tillers by standing in one site and collecting at an arm's radius. This sampling allows gene diversity to be partitioned into two levels in a hierarchy: among leaves within a site and among sites within a field. The hierarchy can be extended by sampling additional fields at different spatial scales.

Results and Discussion

Genetic Diversity, Mating Systems and Genetic Drift

Both gene and genotype diversity were high for both fungi. On average, 18 alleles were present for 11 RFLP loci across approximately 2000 isolates of *M. graminicola* and five alleles were present for seven RFLP loci across approximately 950 isolates of *S. nodorum*. In each population, between three and four alleles were present in around 90% of the isolates (Tables 3.2 and 3.3). The majority of alleles for each locus were present at a low frequency in one or more populations. Different rare or private alleles were found in most populations. Nei's measure of gene diversity across all populations averaged 0.44 across eight RFLP loci for *M. graminicola* and 0.51 across seven RFLP loci for *S. nodorum*. Though both fungi exhibited high gene diversities, *M. graminicola* had a significantly greater number of alleles per locus than *S. nodorum,* which had a more even distribution of allele frequencies across loci. The difference in Nei's diversity between the two fungi is due mainly to the

Table 3.2. Example of common allele frequencies at five RFLP loci for seven field populations of *Mycosphaerella graminicola*. All alleles are based on digestion with *Pst*I. Population abbreviations are given in Table 3.1.

Locus	Allele	ORG N=654	ISR N=160	DEN N=87	UK N=46	URU N=41	MEX N=56	AUSE N=34
pSTS14	1	0.82	0.77	nd	0.89	0.88	1.00	0
	2	0.18	0.21	nd	0	0.05	0	0
pSTL10	1	0.68	0.64	0.66	0.64	0.46	0.79	0.87
	3	0.22	0.30	0.25	0	0.49	0.21	0.13
pSTL53	1	0.50	0.48	0.71	0.72	0.51	0	1.00
	2	0.12	0.06	0.07	0.02	0.32	0.66	0
pSTS43	1	0.66	0.35	0.51	0.50	0.59	0.98	0
	2	0.28	0.09	0.28	0.25	0.24	0.02	0
pSTL31	1	0.70	0.37	0.65	0.51	0.24	0	0.87
	2	0.15	0.04	0.11	0.12	0.22	1.00	0.07

nd = no data.
N = clone-corrected sample size for each field population (McDonald *et al.*, 1995).

Table 3.3. Example of common allele frequencies at seven RFLP loci for eight field populations of *Stagonospora nodorum*. All alleles are based on digestion with *Eco*RI. Population abbreviations are given in Table 3.1.

Locus	Allele	NY N=36	TEX N=50	ARK N=158	ORG N=68	CHG3 N=47	SA N=37	MEX N=18	CW N=17
pJSN121	1	0.50	0.39	0.23	0.25	0.32	nd	0.53	nd
	2	0.19	0.37	0.44	0.38	0.52	nd	0.12	nd
	3	0.31	0.24	0.34	0.37	0.16	nd	0.35	nd
pJSN73	1	0.60	0.85	0.65	0.42	0.59	0.71	0.33	0.29
	2	0.30	0.09	0.22	0.42	0.32	0.24	0.50	0.14
	4	0.	0.02	0	0.10	0	0.06	0	0.57
pSNL15	1	0.16	0.38	0.33	0.12	0.15	0.24	0	0.83
	2	0.57	0.30	0.53	0.28	0.30	0.21	0.82	0.17
	3	0.27	0.32	0.12	0.60	0.54	0.55	0.18	0
pJSN35	1	0.85	0.75	0.60	0.83	0.68	0.70	0.94	1.00
	2	0.15	0.25	0.38	0.17	0.30	0.30	0.06	0
pJSN3	1	0.79	0.82	0.66	0.80	0.75	0.59	1.00	0.15
	2	0.21	0.14	0.32	0.18	0.23	0.41	0	0.85
pJSN27	1	0.91	0.83	0.70	0.80	0.88	0.53	0.76	0.67
	2	0.09	0.13	0.23	0.14	0.10	0.31	0.24	0.17
pSNL13	1	0.74	0.52	0.63	0.86	0.58	0.65	0.35	0.71
	2	0.24	0.31	0.30	0.14	0.42	0.32	0.41	0.29

nd = no data, pJSN121 was not scored for these populations.
N = clone-corrected sample size for each field population (McDonald *et al.*, 1995).

Table 3.4. Nei's measures of gene diversity based on clone-corrected allele frequencies for eight populations of *Mycosphaerella graminicola*. Estimates of total gene diversity for each locus across populations (HT) and population differentiation (GST) were based on analysis of nine populations, excluding Australia and Mexico. Population abbreviations are given in Table 3.1.

Locus	ORG	ISR	DEN	UK	URU	CAN	MEX	AUS	HT	GST
pSTS192A	0.15	0.02	0.22	0.16	0.22	0	0	0.77	0.14	0.04
pSTS192B	0.05	0.48	0.22	0	0	0	0	0	0.15	0.17
pSTS14	0.29	0.36	nd	0.21	0.22	0.53	0	0.26	0.32	0.03
pSTS2	0.53	0.50	0.39	0.48	0.50	nd	0.36	0	0.52	0.02
pSTL10	0.49	0.50	0.48	0.56	0.55	0.61	0.34	0.23	0.50	0.03
pSTL53	0.70	0.64	0.47	0.47	0.62	0.36	0.45	0	0.68	0.03
pSTS43	0.49	0.77	0.65	0.67	0.58	0.72	0.04	0	0.61	0.07
pSTL31	0.48	0.74	0.55	0.68	0.79	0.39	0	0.24	0.63	0.11
Average	0.40	0.50	0.43	0.40	0.44	0.37	0.15	0.19	0.44	0.06

nd = no data.

low gene diversity present for RFLP loci pSTS192A and pSTS192B in *M. graminicola* (Table 3.4). The high gene diversity found in most populations for both fungi suggests that population sizes are very large and the effects of genetic drift are small. This supports our previous finding that populations are stable over time (Chen *et al.*, 1994).

Comparisons of gene diversities across populations revealed some interesting patterns. Perhaps the most significant pattern was shown by the Israeli population of *M. graminicola*. This population had significantly higher gene diversity than other populations around the world (Table 3.4). The Israeli population also had a large number of private alleles that was disproportionate to the number of isolates assayed (Table 3.5). We interpret this finding as evidence that the Middle East is a centre of diversity for *M. graminicola*, and the likely centre of origin for this fungus. It is worth noting that the Australian and Mexican populations exhibited lower gene diversities than other populations. This finding is typical for many pathogens that were introduced since Europeans colonized Australia. The reduced gene diversity probably reflects a founder effect. However, the low gene diversity was unexpected for Mexico. The Mexican collection was made from a CIMMYT disease nursery in Patzcuaro, a location remote from other wheat fields (L. Gilchrist, CIMMYT, Mexico, 1995, personal communication). The Mexican population also exhibited a lower genotype diversity than expected (Table 3.6), having a higher incidence of a few widespread clones than any other population surveyed. These findings are consistent with a small founding population that has not undergone regular sexual reproduction. We speculate that these isolates represent an historic inoculation of the Patzcuaro site with a few isolates that were not sexually compatible. We hope to collect another *M. graminicola* field population from wheat-growing areas of Mexico to obtain a more representative sample of this fungus in Central America.

The gene diversity in the *S. nodorum* population was lowest in the collections from crested wheatgrass and from Mexico (Table 3.7). These were the only two collections that were fixed for one allele at an RFLP locus. The two populations also exhibited substantial differences in allele frequencies for some RFLP loci when compared to other populations (Table 3.3). It is possible that the lower gene diversity and different allele frequencies in the crested wheatgrass population resulted from selection due to host specialization. However, the finding that common alleles are shared between isolates taken from crested wheatgrass and from wheat suggests that there is gene flow between these populations, i.e. crested wheatgrass may serve as a reservoir of inoculum for *S. nodorum* that infects wheat. The differences in the Patzcuaro population may reflect founder events and the geographic isolation of this location.

Table 3.5. Number of private alleles in populations of *Mycosphaerella graminicola* relative to the clone-corrected sample size. The lowest expected allele frequency that could be detected with 95% confidence according to sample size (McDonald and Martinez, 1990a) is shown for each population. For example, the 160 isolates from Israel (ISR) were sufficient to detect all alleles present at a frequency of 2% or greater with 95% confidence. Population abbreviations are given in Table 3.1.

	OR N=654	ISR N=160	DEN N=84	UK N=46	URU N=39	MEX N=55	AUSW N=31
Number of private alleles	31	12	2	2	1	1	1
Lowest expected frequency	0.005	0.02	0.04	0.06	0.07	0.05	0.09

N = clone-corrected sample size for each field population (McDonald *et al.*, 1995).

Table 3.6. Genotype diversity in nine populations of *Stagonospora nodorum* and *Mycosphaerella graminicola*.

	NY	ND	TEX	ARK	ORG	CHG3	SA	MEX	CW
Stagonospora nodorum									
N	40	21	82	195	81	48	73	20	38
(G)	29	18	34	113	58	48	45	17	13
	ORG	IND	CAN	ISR	DEN	UK	URU	MEX	AUSE
Mycosphaerella graminicola									
N	711	29	33	169	115	60	69	122	38
(G)	583	29	28	154	99	47	50	20	17

N was the total number of isolates assayed from each population. Diversity is expressed using Stoddart's measure (G) (Stoddart and Taylor, 1988) which has a maximum possible value of N. Population abbreviations are given in Table 3.1. Genotype diversity of South Africa (SA) and crested wheatgrass (CW) was estimated from multilocus haplotypes instead of DNA fingerprints. Actual diversity was probably higher in these populations.

Table 3.7. Nei's measures of gene diversity based on clone-corrected allele frequencies for eight populations of *Stagonospora nodorum*.

Locus	NY	TEX	ARK	ORG	CHG3	SA	MEX	CW	HT	GST
pJSN121	0.62	0.65	0.64	0.66	0.60	nd	0.58	nd	0.62	0.10
pJSN73	0.54	0.27	0.52	0.63	0.55	0.44	0.62	0.53	0.54	0.06
pSNL15	0.58	0.66	0.59	0.55	0.59	0.61	0.29	0.32	0.65	0.09
pJSN35	0.25	0.38	0.49	0.28	0.45	0.42	0.10	0	0.46	0.07
pJSN3	0.33	0.31	0.46	0.33	0.39	0.48	0	0.22	0.38	0.09
pJSN27	0.16	0.29	0.45	0.33	0.21	0.61	0.36	0.60	0.42	0.06
pSNL13	0.40	0.62	0.51	0.23	0.49	0.47	0.67	0.44	0.51	0.04
Average	0.41	0.45	0.52	0.43	0.47	0.51	0.38	0.35	0.51	0.07

nd = no data.
Estimates of total gene diversity for each locus across populations (HT) and population differentiation (GST) were based on analysis of 11 populations, excluding crested wheatgrass and Mexico. Population abbreviations are given in Table 3.1.

Genotype Diversity

Genotype diversity in the nuclear genome was also high for both fungi (Table 3.6). In the majority of populations surveyed, each leaf, on average, had a unique fungal genotype. When isolates shared the same DNA fingerprint, they usually were sampled from the same leaf. The only evidence for a widespread clone in *M. graminicola* was in the Patzcuaro population and in *S. nodorum* from our first collection in a Texas field (McDonald *et al.*, 1994). In the latter case, we cannot eliminate the possibility that this apparent distribution resulted from a mislabelled tube of DNA.

It appears that the typical population of both fungi exhibits a low degree of clonality at the field level. The genotype diversity values in Table 3.6 should be considered preliminary, because we have not yet corrected for differences in scale of sampling across field populations. In some field populations, it was common to collect three or more isolates from the same leaf while in others, only one isolate was collected from each leaf. Clones were usually distributed across an area of approximately 1 m^2 and did not become widespread. This finding is consistent with limited spread of the splash-dispersed conidia for both fungi.

Our interpretation of the high degree of genotype diversity is that populations of both fungi undergo regular sexual cycles and that the primary inoculum for both fungi is likely to be ascospores (Chen and McDonald, 1996; Keller *et al.*, 1997b). We have used tests for the randomness of associations among RFLP loci to indicate the frequency of sexual recombination for both fungi (Chen and McDonald, 1996). Associations were random for both in cases where sample sizes were adequate to make robust tests for disequilibrium, supporting the hypothesis of random mating. Although the sample size was smaller (N=160), the Israeli population also showed the gametic equilibrium and high genotypic diversity typical of a random mating fungus (Table 3.8), suggesting that the teleomorph is present in the Middle East.

mtDNA Diversity

For both fungi, mtDNA exhibited less diversity than nuclear DNA (nuDNA). The mtDNA genome ranges in size from 48-62 kb for both fungi. Among 482 isolates of *S. nodorum* from Switzerland, Texas and Oregon, we found only 40 different mtDNA haplotypes compared to over 400 nuDNA haplotypes. Twelve of the mtDNA haplotypes were shared among these field populations (S. Keller and B.A. McDonald, unpublished results).

The mtDNA of *M. graminicola* exhibited even lower diversity. We routinely found only two to three mtDNA haplotypes in field populations of *M. graminicola* (Fig. 3.3). Sometimes only one *M. graminicola* mtDNA haplotype was found in a field (Table 3.9). We picked the three most common mtDNA haplotypes from around the world and digested the mtDNA with ten different restriction enzymes that sampled approximately 5% of the entire 48 kb of mtDNA sequence in an attempt to detect cryptic variation. Types 1, 2 and 3 mtDNA haplotypes produced identical digestion patterns with all enzymes in all populations tested, suggesting that these mtDNA haplotypes are identical globally (B.A. McDonald and K. Hogan, unpublished results). We propose that the extremely limited mtDNA diversity is due to a selective sweep that occurred as *M. graminicola* populations became specialized to infect bread wheat. We will present supporting evidence for this hypothesis later in this chapter.

Table 3.8. Test of Brown *et al.* (1980) for multilocus associations in the Israeli population of *Mycosphaerella graminicola*.

N	H	S2k	L	Ia
160	0.503	1.816	1.952	0.129

N is the clone-corrected sample size, H is the mean gene diversity across eight RFLP loci, S2k is the expected variance in the number of differences among individuals under random mating, L is the variance needed to reject the null hypothesis of random mating at $P< 0.05$, and Ia is the index of association.

Population Subdivision and Gene Flow

The distribution of genetic diversity across populations was consistent with a significant level of gene flow for both fungi. Common alleles at individual RFLP loci were shared among nearly all populations (Tables 3.2 and 3.3) and allele frequencies were often quite similar even though populations were separated by thousands of kilometres (e.g. compare Oregon and Israel for *M. graminicola,* and Oregon and Switzerland for *S. nodorum*). We have interpreted this similarity as indicating a significant degree of gene flow among populations of both fungi on an international scale (McDonald *et al.*, 1995; Keller *et al.*, 1997a). If the Australian, Mexican and crested wheatgrass populations are excluded from the analysis, both fungi exhibit nearly identical degrees of population differentiation; G_{ST}=0.06 and 0.07 for *M. graminicola* and *S. nodorum*, respectively. These G_{ST} values are consistent with movement of eleven and seven individuals among populations every generation. The temporal scale of the gene flow cannot, however, be determined from this analysis. Estimates of Nm (where m is the fraction of migrants in a population, and N is the force of genetic drift, which is proportional to the inverse of the population size) based on G_{ST} are an indirect measure that assumes constant gene flow over all generations (McDermott and McDonald, 1993). We have never found the same nuclear genotype in different field populations for either fungus, so we have no direct evidence for gene flow based on nuDNA. However, we have found the same mtDNA haplotypes worldwide for both

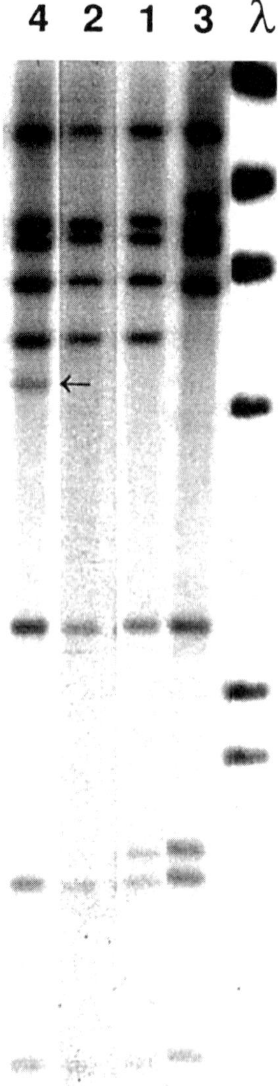

Fig. 3.3. Composite illustration of four common mtDNA haplotypes of *Mycosphaerella graminicola*. Types 1, 2, and 3 are common on bread wheat globally. Type 4 has been found only on durum wheat in the Mediterranean Basin. The 4.7 kb *Pst*I fragment (arrow) in Type 4 is due to a 3.0 kb insertion in the 1.7 kb fragment of Type 1 mtDNA.

Table 3.9. Diversity in nuclear and mitochondrial genomes in 14 populations of *Mycosphaerella graminicola* from around the world.

Population	N	Hnu	Number of nuclear haplotypes	Number of mitochondrial haplotypes	Mitochondrial types detected
California	93	0.36	22	2	Types 1, 2
Canada	33	0.37	27	NA	NA
Denmark	115	0.43	87	2	Types 1, 2
East Australia	38	0.19	15	2	Types 1, 3
Germany	20	NA	NA	2	Types 1, 2
Indiana	29	0.44	29	1	Type 1
Israel	169	0.50	160	3	Types 1, 2, 3
Mexico	122	0.15	56	1	Type 3
Oregon	711	0.39	654	2	Types 1, 2
Russia	19	NA	NA	3	Types 1, 2, 3
Texas	90	0.42	72	1	Type 1
United Kingdom	70	0.40	60	3	Types 1, 2, 3
Uruguay	69	0.44	51	3	Types 1, 2, 3
West Australia	27	0.12	15	2	Types 1, 3

NA = data not yet available.
N is the number of isolates assayed in each field population and Hnu is Nei's average gene diversity across at least eight nuclear RFLP loci. The number of nuclear haplotypes in each population was determined by DNA fingerprints. The number of mitochondrial haplotypes was based on hybridization of purified mtDNA to *Pst*I-digested DNA. The last column summarizes which mtDNA haplotypes were found in each field.

fungi. Unlike nuDNA, mtDNA shows substantial evidence for population subdivision. Different populations may be fixed for different mtDNA haplotypes (Table 3.9).

The present lack of subdivision in the nuclear genome probably reflects historic movement of both fungi around the world. It is possible that gene flow is not a significant evolutionary force at present. However, we consider it more likely that some gene flow continues as a result of global commerce in grain. The obvious mechanism for gene flow on a regional basis is air-dispersed ascospores. In the case of *S. nodorum*, the most likely mechanism for intercontinental dispersal is infected seed (King *et al.*, 1983). Since it has been shown that *M. graminicola* can infect seed (Brokenshire, 1975), we consider it likely that this also is the mechanism for long distance gene flow in *M. graminicola.* Whatever the mechanism, the high degree of similarity in populations of both fungi around the world suggests that they have been transported around by humans.

Evidence for Selection

Our first field experiment to detect selection was begun in 1991 in collaboration with Dr C. Mundt at Oregon State University. In the experiment, four host genotypes that differed for resistance to *M. graminicola* were planted in pure stand and in all possible two-, three- and four-way mixtures in a randomized complete block design with three replications. The results of the experiment are discussed elsewhere (McDonald *et al.*, 1996) but to summarize our findings, among 711 isolates taken from the field, there were 654 nuclear genotypes. No nuclear clones occurred at a high frequency in any host treatment. The genetic structure of the fungal population did not change over the course of the growing season. Frequencies of individual RFLP alleles did not differ between host treatments. In short, there was no evidence for adaptation of the pathogen populations to specific host populations (McDonald *et al.*, 1996). Although this experiment was not successful in detecting selection, it did demonstrate that these nuclear RFLP alleles were selectively neutral; that there was a large amount of genetic diversity within field populations; and that the mtDNA genome had extremely low levels of genetic variation. Out of the 711 isolates collected from the experiment, 705 had one mtDNA haplotype and six had a different mtDNA haplotype that differed from the first by a 1.7 kb deletion. The latter finding was surprising and immediately led us to suspect that mtDNA may be affected by selection.

The strongest evidence for selection operating on *M. graminicola* mtDNA came from a collection of isolates from the Mediterranean Basin supplied by Dr G.H.J. Kema, IPO, The Netherlands in 1994-1995. Kema's group showed that *M. graminicola* isolates originating from bread and durum wheats exhibited host specialization. Isolates collected from durum wheats exhibited greater virulence on durum wheats compared to bread wheats and *vice versa* (Kema *et al.*, 1996). We collected the full complement of RFLP data from 99 isolates in

Kema's collection, including 47 isolates from the bread wheat pathotype and 52 isolates from the durum wheat pathotype. These isolates came from Algeria (N=48), Syria (N=30), Tunisia (N=5), Turkey (N=7), Ethiopia (N=2), Morocco (N=3) and Portugal (N=4). Among them, we found a high frequency of a novel mtDNA haplotype, called Type 4, which has a unique 3.0 kb insertion (Fig. 3.3). The durum wheat pathotypes had a much higher frequency of the Type 4 mtDNA then the bread wheat pathotypes (Table 3.10). For the nuDNA RFLP markers, there was little evidence for differentiation between the bread and durum wheat pathotypes (Table 3.11). We have not found the Type 4 mtDNA among over 1500 isolates sampled from bread wheat around the world (Table 3.9). These data provide indirect evidence that mtDNA may be involved in the host-specialization exhibited by the wheat and durum wheat pathotypes. We hypothesize that the extremely low degree of mtDNA diversity in *M. graminicola* is due to a 'selective sweep' that resulted from selection for a gene (or genes) in the mtDNA that offered a fitness advantage on isolates infecting bread wheat. Upon introduction of this favoured mtDNA haplotype into bread wheat populations around the world through gene flow, its selective advantage was sufficient to allow it to displace previously existing mtDNA haplotypes. It is possible that genetic information in the novel 3.0 kb insertion of the Type 4 mtDNA haplotype confers a selective advantage on durum wheat. The best way to test the selective sweep hypothesis is to make a hierarchical collection of *M. graminicola* isolates from natural populations of wheat ancestors (e.g. *Triticum dicoccoides*) in the Fertile Crescent, which appears to be the centre of origin for *M. graminicola*.

Table 3.10. Frequencies of mtDNA haplotypes in bread and durum wheat pathotypes of *Mycosphaerella graminicola* collected in the Mediterranean Basin.

mtDNA haplotype	Bread pathotype (N=47)	Durum pathotype (N=52)
1	0.49	0.22
2	0.06	0.02
3	0.33	0.27
4	0.02	0.41
5	0.08	0.04
6	0.02	0.04

In collaboration with C.C. Mundt, J. Zhan and B.A. McDonald have conducted a new experiment to measure directly competition among ten genotypes of *M. graminicola* under field conditions. The ten isolates were inoculated on to three host treatments consisting of a moderately resistant wheat cultivar (Madsen) a susceptible wheat cultivar (Stephens) and a 50:50 mixture of these cultivars. A detailed analysis and interpretation of these data will be presented elsewhere. Our most important finding was that intense competition appeared to occur among the different genotypes. Significant changes in the frequencies of specific pathogen genotypes occurred over the season (Fig. 3.4). Some isolates showed evidence for adaptation to particular hosts. The results from this experiment have provided our first direct evidence that selection operates on specific *M. graminicola* pathogen genotypes in a field setting.

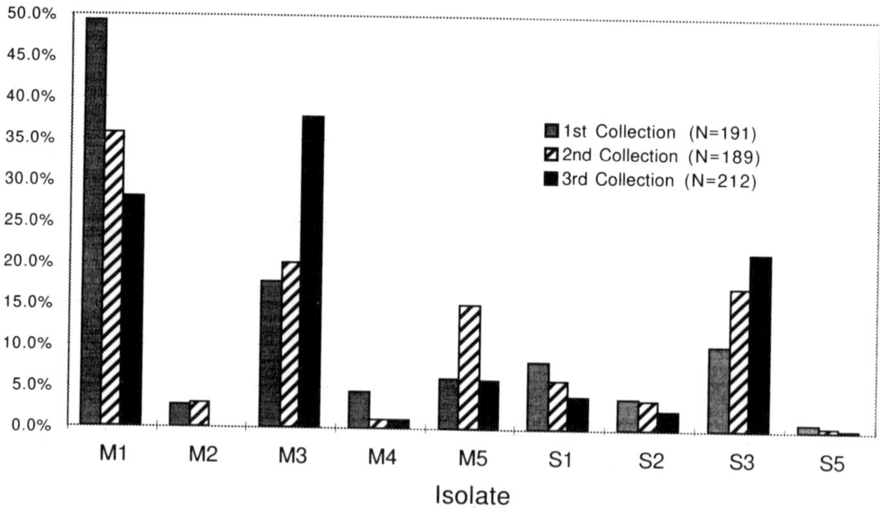

Fig. 3.4. Evidence for competition among strains of *Mycosphaerella graminicola* in a replicated field experiment. Frequencies of nine inoculated isolates of *M. graminicola* sampled from wheat cv. Stephens over a 4 month period. Isolate S4 was not detected on cv. Stephens. Relative frequencies for each isolate were consistent across all three replications of each host treatment.

We have not yet conducted similar replicated field experiments to measure selection in *S. nodorum*. However, we have indirect evidence that selection does not result in widespread clones that are adapted to specific host genotypes. In an experiment conducted in Switzerland in collaboration with M. Wolfe's group, we sampled 50 isolates of *S. nodorum* from each of nine wheat fields near Zurich. Three different wheat cultivars (Arina, Tamaro and Galaxie) were represented three times each among the nine wheat fields. Although fields planted to the same cultivar used the same source of seed, no genotypes were shared among field populations. Only six pairs of clones were found among the 432 isolates that were assayed. Isolates with the same DNA fingerprints always came from the same site within a field (Keller *et al.*, 1997b).

Taken together, all of our experiments suggest that nuclear genotypes do not persist through time for either fungus. Instead, the genes are the units of selection that are carried forward across generations. Selection operates on the

Table 3.11. Nei's measures of gene diversity (Hi), population differentiation (GST), and gene identity (I) for isolates of *Mycosphaerella graminicola* isolated from bread and durum wheat in the Mediterranean Basin. The chi-square test for association is based on contingency table analysis of the associations between nuclear alleles (or mtDNA haplotypes) and host specificity (durum or bread wheat).

RFLP locus	Bread Hi	Durum Hi	Gst	I	γ2
pSTS192 A	0.12	0.16	0.01	1.00	*P*=0.38
pSTS192 B	0	0	NM	1.00	NM
pSTS14	0.21	0.21	0.01	0.99	*P*=0.36
pSTS2	0.27	0.36	<0.01	1.00	*P*=0.74
pSTL10	0.60	0.35	0.05	0.93	*P*=0.18
pSTL53	0.68	0.72	0.02	0.90	*P*=0.04
pSTS43	0.61	0.75	0.01	0.98	*P*=0.38
pSTL31	0.81	0.74	0.05	0.67	*P*=0.07
pSTL2	0.21	0.27	<0.01	1.00	*P*=0.53
mtDNA	0.64	0.70	0.08	0.64	*P*=0.0001

NM = not meaningful: both groups of isolates were fixed for one allele at the pSTS192B locus.

population instead of on the individual. In order to gain a representative spectrum of the diversity for virulence in natural populations, plant breeders should include the widest possible diversity of strains when screening germplasm for resistance to these fungi. Similarly, chemical companies should include at least several hundred strains in their screens for resistance to fungicides.

Conclusions

Given our present data, we have drawn the following conclusions regarding the evolutionary forces that affect the population genetics of *M. graminicola* and *S. nodorum*.

For both fungi, the mating system includes both sexual and asexual reproduction. Asexual reproduction may have an important impact over an area of a few square metres, but sexual reproduction has much greater consequences for the population genetics and biology of both fungi. Genotypes are ephemeral but genes persist in populations through time.

Population sizes are large enough to make genetic drift negligible for both fungi. Large population sizes also ensure that ample mutations are present in every population to allow for a rapid response to selection, e.g. mutations from avirulence to virulence for major resistance genes. Some populations (e.g. Australia) exhibit typical founder effects.

Gene flow is sufficient to unite large geographical areas into a single genetic population. If gene flow is ongoing, then breeders should continue to test their resistant lines over the widest possible geographical area. If gene flow is episodic, continued vigilance is needed to limit the spread of new virulence genes and fungicide resistance genes. Quarantines in areas with low gene diversity, such as Australia, should be enforced to limit the evolutionary potential of these populations. If CIMMYT continues to use Patzcuaro as a field site to screen for resistance to *M. graminicola* and *S. nodorum*, it may want to consider introducing more diverse fungal populations from other parts of Mexico into this disease nursery.

Selection appears to operate on both nuclear and mitochondrial genomes in *M. graminicola*. Although selection may increase the frequency of particular genotypes over the course of a growing season, it appears that particular genotypes are unlikely to reach high frequencies within field populations because of the limited dispersal potential for conidia. However, the genes in the fittest individuals will persist and be recombined to create novel genotypes in the next growing season. Over the course of many growing seasons, selection will change the frequency of genes that affect adaptation to the wheat host, but new genotypes will appear each season. It is too early to say if selection operates in the same way in *S. nodorum*.

In summary, the population genetics of *S. nodorum* and *M. graminicola* are very similar. This probably reflects the similarity in their life histories. Both fungi produce airborne sexual ascospores and splash-dispersed asexual spores.

They both infect above-ground plant parts on the same host and seeds that can be transported globally as part of the world commerce in wheat. Use of multi-allelic, neutral genetic markers combined with hierarchical sampling has allowed us to achieve a much greater understanding of the population biology of both fungi.

Author Contributions

B.A. McDonald is the principal investigator who directed this research; O. Yarden as an Assistant Professor made the collection of infected leaves from Israeli wheat fields; J. Zhan as a PhD student collated and analysed the international *S. nodorum* data; J. Garton as an undergraduate student collated the international *M. graminicola* data; K. Hogan as an undergraduate student collected the majority of the mtDNA data for *M. graminicola;* and R. Pettway as a technician supervised the numerous undergraduate students who collected the majority of the international RFLP data for both fungi.

Acknowledgements

We gratefully acknowledge the many collectors and collaborators around the world who responded to our request for infected leaf material. The majority of these collectors are listed in Table 3.1. This work would not have been possible without the enthusiastic support of a dedicated group of genetics undergraduate students. Funding for this project came from the USDA National Research Initiative Competitive Grants Program (Grant #93-37303-9039), the National Science Foundation (Grant #DEB-9306377), the Texas Agricultural Experiment Station (Hatch project #6928), and the Swiss National Fund (Grant #5002-38966).

References

Boeger, J.M., Chen, R.S. and McDonald, B.A. (1993) Gene flow between geographic populations of *Mycosphaerella graminicola* (anamorph *Septoria tritici*) detected with RFLP markers. *Phytopathology* 83, 1148-1154.

Brokenshire, T. (1975) Wheat seed infection by *Septoria tritici*. *Transactions of the British Mycological Society* 64, 331-335.

Brown, A.H.D., Feldman, M.W. and Nevo, E. (1980) Multilocus structure of natural populations of *Hordeum spontaneum*. *Genetics* 96, 523-536.

Chen, R.S. and McDonald, B.A. (1996) Sexual reproduction plays a major role in the genetic structure of populations of the fungus *Mycosphaerella graminicola*. *Genetics* 142, 1119-1127.

Chen, R.S., Boeger, J.M. and McDonald, B.A. (1994) Genetic stability in a population of a plant pathogenic fungus over time. *Molecular Ecology* 3, 209-218.

Garber, R.C. and Yoder, O.C. (1983) Isolation of DNA from filamentous fungi and separation into nuclear, mitochondrial, ribosomal, and plasmid components. *Analytical Biochemistry* 135, 416-422.

Keller, S.M., McDermott, J.M., Pettway, R.E., Wolfe, M.S. and McDonald, B.A. (1997a) Gene flow and sexual reproduction in the wheat glume blotch pathogen *Phaeosphaeria nodorum* (anamorph *Stagonospora nodorum*). *Phytopathology* 87, 353-358.

Keller, S.M., Wolfe, M.S., McDermott, J.M. and McDonald, B.A. (1997b) High genetic similarity among populations of *Phaeosphaeria nodorum* across wheat cultivars and regions in Switzerland. *Phytopathology* 87, 1134-1139.

Kema, G.H.J., Annone, J.G., Sayoud, R., Van Silfhout, C.H., Van Ginkel, M. and de Bree, J. (1996) Genetic variation for virulence and resistance in the wheat-*Mycosphaerella graminicola* pathosystem I. Interactions between pathogen isolates and host cultivars. *Phytopathology* 86, 200-212.

King, J.E., Cook, R.J. and Melville, S.C. (1983) A review of Septoria diseases of wheat and barley. *Annals of Applied Biology* 103, 345-373.

Kohli, Y., Brunner, L.J., Yoell, H., Milgroom, M.G., Anderson, J.B., Morrall, R.A.A. and Kohn, L.M. (1995) Clonal dispersal and spatial mixing in populations of the plant pathogenic fungus, *Sclerotinia sclerotiorum*. *Molecular Ecology* 4, 69-77.

McDermott, J.M. and McDonald, B.A. (1993) Gene flow in plant pathosystems. *Annual Review of Phytopathology* 31, 353-373.

McDonald, B.A. (1997) The population genetics of fungi: tools and techniques. *Phytopathology* 87, 448-453.

McDonald, B.A. and Martinez, J.P. (1990a) DNA restriction fragment length polymorphisms among *Mycosphaerella graminicola* (anamorph *Septoria tritici*) isolates collected from a single wheat field. *Phytopathology* 80, 1368-1373.

McDonald, B.A. and Martinez, J.P. (1990b) Restriction fragment length polymorphisms in *Septoria tritici* occur at a high frequency. *Current Genetics* 17, 133-138.

McDonald, B.A. and Martinez, J.P. (1991) DNA fingerprinting of the plant pathogenic fungus *Mycosphaerella graminicola* (anamorph *Septoria tritici*). *Experimental Mycology* 15, 146-158.

McDonald, B.A., Miles, J., Nelson, L.R. and Pettway, R.E. (1994) Genetic variability in nuclear DNA in field populations of *Stagonospora nodorum*. *Phytopathology* 84, 250-255.

McDonald, B.A., Pettway, R.E., Chen, R.S., Boeger, J.M. and Martinez, J.P. (1995) The population genetics of *Septoria tritici* (teleomorph *Mycosphaerella graminicola*). *Canadian Journal of Botany* 73 (Supplement), S292-S301.

McDonald, B.A., Mundt, C.C. and Chen, R.S. (1996) The role of selection on the genetic structure of pathogen populations: evidence from field experiments with *Mycosphaerella graminicola* on wheat. *Euphytica* 92, 73-86.

Stoddart, J.A. and Taylor, J.F. (1988) Genotypic diversity: estimation and prediction in samples. *Genetics* 118, 705-711.

Chapter four:

The Sexual Form *Phaeosphaeria nodorum* (Anamorph *Stagonospora nodorum*): Sexual Systems

P. Halama
Institut Supérieur d' Agriculture, 41 rue du Port, 59046 Lille cedex, France

Introduction

Stagonospora nodorum (Berck.) Castellani and Germano (ex *Septoria nodorum* Berk.) is an important leaf spot and glume blotch pathogen with a large distribution. The pathogen may reduce wheat yields and inhibit grain filling. Its negative impact on seed quality and yield (expressed as 1000-kernel weight), especially when conditions are wet, has been recognized worldwide (Shipton *et al.*, 1971).

Weber (1922) observed pseudothecia adjacent to pycnidia of *S. nodorum,* but the first connection between the sexual form, *Phaeosphaeria nodorum* (Müll.) Hedjaroude (ex *Leptosphaeria nodorum* Müll.), and the conidial form was not made until 1952 (Müller, 1952). The sexual stage of the life cycle takes place on wheat stubble during autumn and winter when sexual fruiting bodies are produced. Pseudothecia enable the pathogen to survive difficult environmental conditions and ascospores function as primary inoculum sources. Sexual reproduction increases genetic diversity by creating new combinations for aggressiveness.

The teleomorph has been reported from France (Rapilly *et al.*, 1973), Brazil (Mehta, 1975), New Zealand (Hampton, 1975; Hampton *et al.*, 1978; Sanderson and Hampton, 1978), the UK (Wale and Colhoun, 1979), the USA (Scharen and Sanderson, 1982), Germany (Mittelstädt and Fehrmann, 1987), Ireland (O'Reilly and Bannon, 1988), South Africa (Kemp *et al.*, 1989) and Canada (McFadden and Harding, 1989).

The sexual stage has been induced in the laboratory (Halama and Lacoste, 1989; 1992) and it has been demonstrated that *P. nodorum* exhibits two allele heterothallism with the mating types designated Mat(+) and Mat(-) (Halama and Lacoste, 1991).

The objectives of this study were to bring precision to the heterothallism and to determine whether or not both mating types are present among isolates of different origin.

Materials and Methods

Fungal Strains and Mating Type Determination

The parental strains used in crosses originated from single ascospores isolated from pseudothecia on wheat straw harvested in the field during winter. These wild ascospores, and progeny ascospores produced from pseudothecia *in vitro,* were isolated by micromanipulation using methodology previously described by Halama and Lacoste (1991). The progeny ascospores were not numbered according to their order inside the ascus.

The mating type (MAT) of our collection and of isolates originating from different countries was determined in crosses with two mating type testers, A/5 [MAT(+)] and 6/T [MAT(-)], used in a previous study (Rapilly *et al.*, 1992). To determine the mating type of isolates belonging to a tetrad, all isolates within the tetrad were paired in all combinations. For each cross, pseudothecia were crushed and observed microscopically to confirm their fertility with the presence of mature asci and ascospores.

Cytological Investigations

Host material bearing pseudothecia was fixed in Westbrook solution (Westbrook, 1955) for 24 hours, and embedded in paraffin wax. Five μm thick serial sections were cut and stained with haematoxylin and eosin.

Culture Media and Conditions

A medium made up of sterilized fragments of wheat straw, which had not received any antifungal treatment, was used to obtain pseudothecia. Approximately six 12 cm long pieces of straw and 15 ml of distilled water were put into each tube of 25 200 mm Pyrex tubes. The tubes were cultured under 12 hours of near ultra-violet light (NUV) per day and a temperature of 10°C. The NUV (300 nm $< \lambda <$ 400 nm) was supplied by Sylvania F36 fluorescent tubes with a light energy of approximately 600 $\mu W\ cm^{-2}$.

Results

Precision on Heterothallism

All possible crosses between two series of eight single-ascospore strains (A11-A18 and A31-A38), originating from two asci sampled from pseudothecia harvested in the field, were made. Out of the possible 28 crosses made between the strains of asci A1 and A3, only nine in series A1 and seven in series A3 proved to be fertile, the pseudothecia obtained containing both asci and ascopores (Figs 4.1 and 4.2).

A complete series of eight ascospores (B11-B18) derived from a pseudothecium produced *in vitro* following the fertile cross A32 x A35 was obtained. Of the crosses made between these ascospores, only 11 proved to be fertile (Fig. 4.3). Although this number was higher than the previous results, it was clear that, irrespective of the complete series, the number of fertile crosses observed was lower than the 16 expected for bipolar heterothallism.

The results showed that single-ascospore strains originating from the same ascus are divided into two mating types, MAT(+) and MAT(-). Crosses between monoascosporous strains originating from different asci were made in order to determine whether or not they belonged to MAT(+) or MAT(-). If single-ascospore strains are associated by mating types and within each mating type by twin ascospores on the basis of *in vitro* cultural characteristics, I observed that only certain intergroup crosses were fertile (Figs 4.4, 4.5 and 4.6).

When the ascospores were isolated, their original position in the ascus was not known. Hence, no information on the way compatibility factors are segregated at meiosis was gathered.

In the single-ascospore cultures raised under optimum conditions for pseudothecia production, no fertile pseudothecia were observed although sterile pseudothecia were (Fig. 4.7). In these structures, locule size is reduced and the tissue is predominantly plectenchyma. Typical pseudoparaphyses (pp) originate near the apical region of locules (cl). When pseudoparaphyses grow downward, their tips do not press against the base of the pseudothecial cavity as was observed in an ontogenic study of fertile pseudothecia (Halama *et al.*, 1992). In the present study, no ascogenous development was observed. Sometimes, in fertile crosses of single-ascospore strains, sterile structures are observed among fertile pseudothecia. The sizes of these sterile pseudothecia are similar to fertile pseudothecia locules (Fig. 4.8).

Crossing Between Mating Type Testers and Isolates from Different Countries

Thirty-eight isolates from different countries were sexually compatible with one of the two mating type testers, MAT(+) or MAT(-) (Fig. 4.9). The time required for pseudothecia to form was the same for all isolates, but there were differences in the final densities of the fertile pseudothecia. They produced, with mating type testers, pseudothecia with abundant asci containing eight ascopores. No pseudothecia were produced by isolates grown alone. Only one isolate, originating from Israel, produced pseudothecia which differed from normal mature perithecia. Their size was reduced and few asci were observed with less than the eight ascospores normally found. The degree of interaction

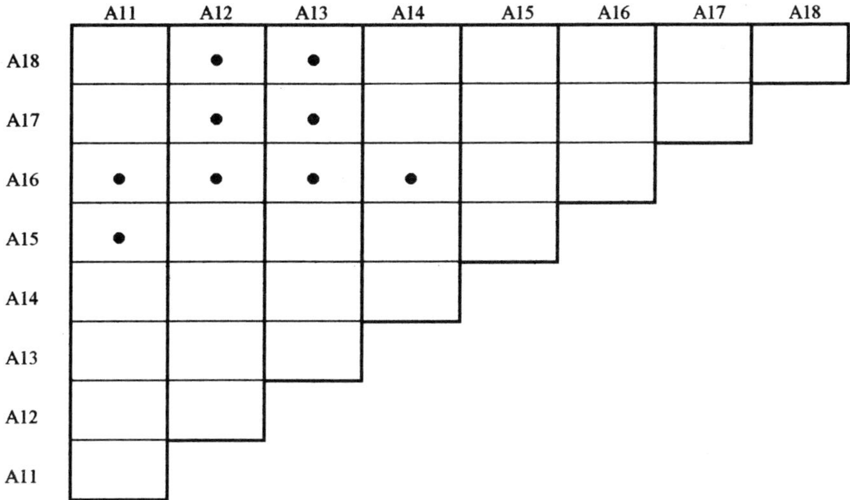

Fig. 4.1. Results of pair-wise crosses between eight single-ascospore strains (A11 to A18) of ascus A1. Presence of fertile perithecia with asci and ascospores is indicated by dots.

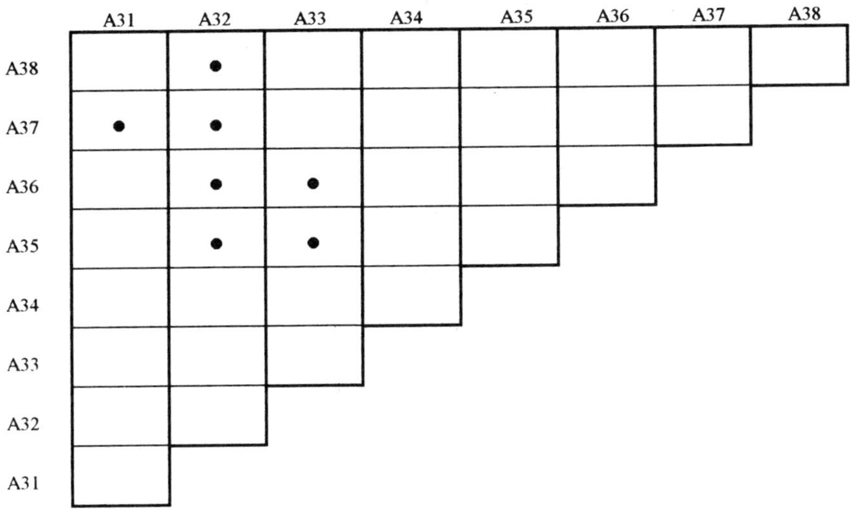

Fig. 4.2. Results of pair-wise crosses between eight single-ascospore strains (A31 to A38) of ascus A3. Presence of fertile perithecia with asci and ascospores is indicated by dots.

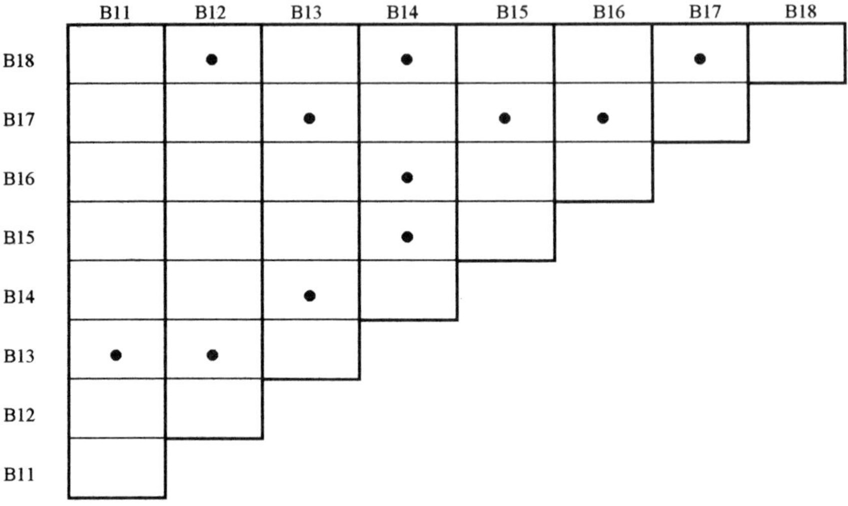

Fig. 4.3. Results of pair-wise crosses between eight single-ascospore strains (B11 to B18) of ascus B. Presence of fertile perithecia with asci and ascospores is indicated by dots.

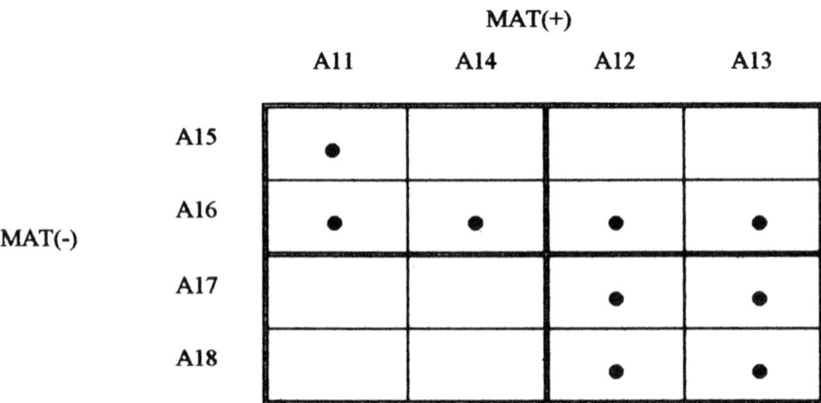

Fig. 4.4. Results of pair-wise crosses between eight single-ascospore strains (A11 to A18) of the tetrad from ascus A1. Twin ascospores for cultural characteristics: (A11-A14), (A12-A13), (A15-A16), (A17-A18). Presence of fertile perithecia with asci and ascospores is indicated by dots.

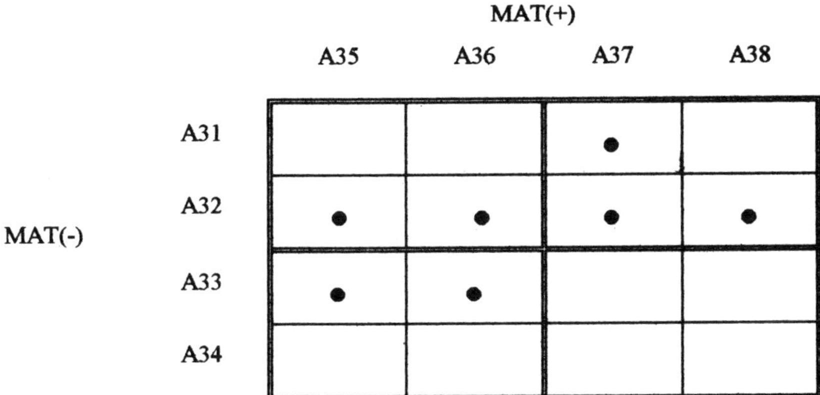

Fig. 4.5. Results of pair-wise crosses between eight single-ascospore strains (A31 to A38) of the tetrad from ascus A3. Twin ascospores for cultural characteristics: (A35-A36), (A37-A38), (A31-A32), (A33-A34). Presence of fertile perithecia with asci and ascospores is indicated by dots.

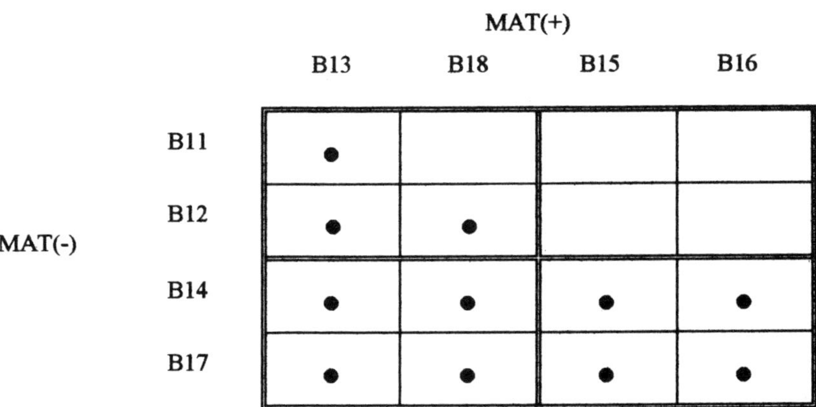

Fig. 4.6. Results of pair-wise crosses between eight single-ascospore strains (B11 to B18) of the tetrad from ascus B. Twin ascospores for cultural characteristics: (B13-B18), (B15-B16), (B11-B12), (B14-B17). Presence of fertile perithecia with asci and ascospores is indicated by dots.

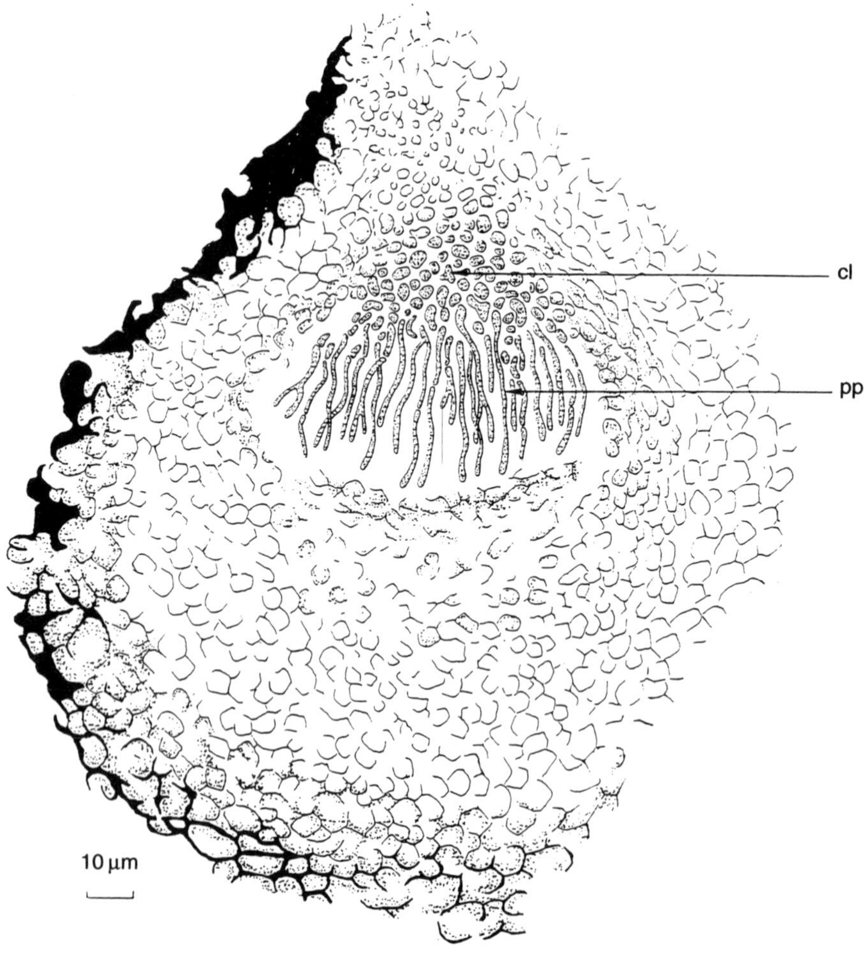

Fig. 4.7. Sterile pseudothecium observed in a culture originated from one single-ascospore strain. pp: pseudoparaphyses; cl: apical region of the locule.

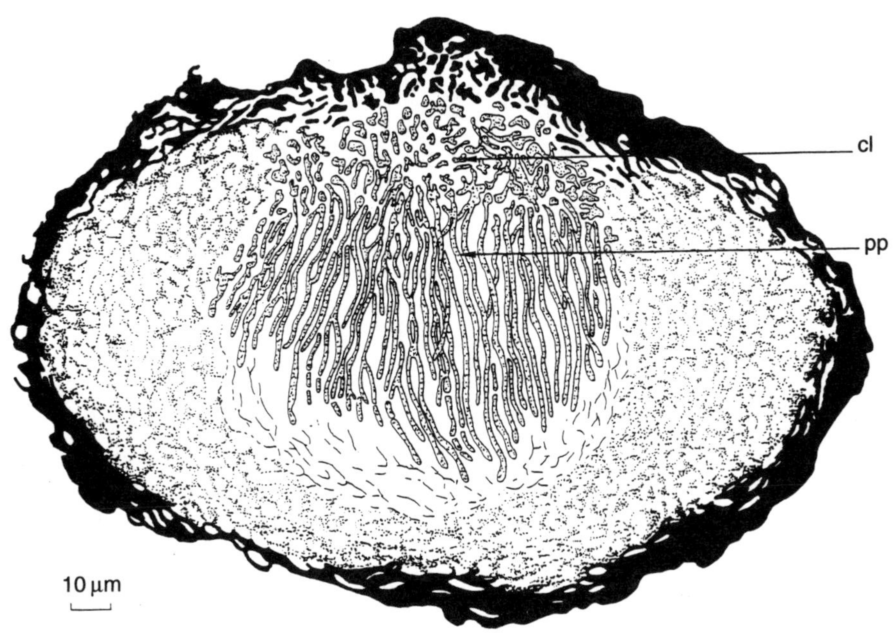

Fig. 4.8. Sterile pseudothecium observed in a fertile culture between two isolates of mating types MAT(+) and MAT(-). pp: pseudoparaphyses; cl: apical region of the locule.

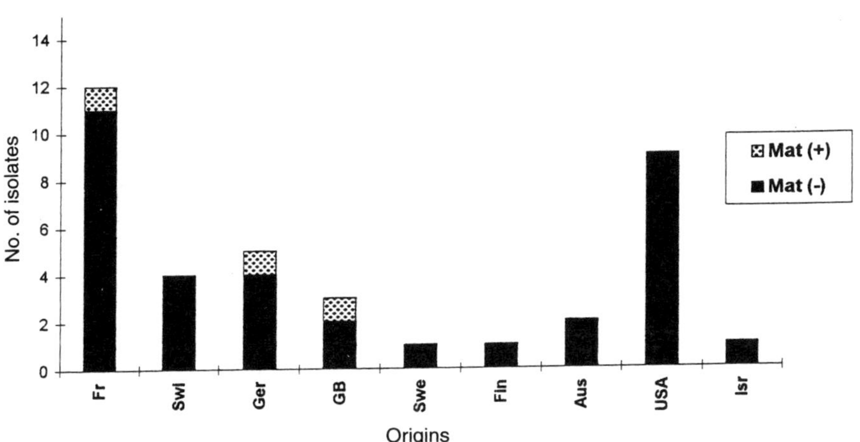

Fig. 4.9. Mating type of *Stagonospora nodorum* isolates from different origins (Fr: France; Swi: Switzerland; Ger: Germany; GB: Great Britain; Swe: Sweden; Fin: Finland; Aus: Australia; USA: United States; Isr: Israel).

between the mating type tester and this isolate was reduced. Variability in the degree of compatibility was also observed in genetic studies on *Cochliobulus heterostrophus* (Nelson, 1959a,b).

Mating type did not segregate evenly between MAT(+) and MAT(-). Mating type (-) occurred more frequently than mating type (+) based on the results of the different crosses (92.1% of the 38 isolates tested). Only three of the isolates, collected from France, the UK and Germany, were MAT(+). For these crosses, no relationship was observed between the level of conidiogenesis and the mating type. Occurrence of *P. nodorum* pseudothecia in nature may, therefore, be affected both by the distribution and by the frequency of mating types.

Discussion

The reported results confirm the existence of a physiological heterothallism as defined by Whitehouse (1949). This heterothallism is bipolar unlike the tetra-polar mechanism in Basidiomycetes. Esser and Kuenen (1967) referred to this type of incompatibility as homogenic, as compared to heterogenic incompatibily when the two partners bear different alleles on all the incompatibility loci. This latter type of incompatibility can be vegetative or sexual. For *Podospora anserina*, Esser (1971) has shown that vegetative and sexual incompatibility can occur together.

The total number of fertile crosses, observed for each of the three complete series of single-ascospore strains originating from the same ascus, was lower than the theoretical value expected in usual bipolar heterothallism. Some of the intergroup crosses did not form fertile pseudothecia. This phenomenon has been observed in other species, in particular *Cochliobolus carbonum* by Nelson (1959b; 1960; 1964; 1970). In that species, I have observed different degrees of pseudothecia, ascal and ascosporal production in fertile crosses. These results suggest that additional genes influenced the expression of the heterothallic locus, the sexual process being controlled by several genes.

Two hypotheses can be put forward to explain the sterility of the intergroup crosses in *P. nodorum.* First, the cultural conditions did not allow the differentiation of pseudothecia for all crosses; in particular where different responses were observed for the two twin single-ascospore strains which should have the same nuclear genome following post-meiotic mitosis (ex: B_{13} x B_{11}:+ and B_{18} x B_{11}:0). Second, another incompatibility mechanism may exist in addition to bipolar heterothallism.

A system of vegetative incompatibility has been suggested by Newton and Caten (1985) following a study of mutant strains of *S. nodorum.* These authors obtained a complementation between certain mutants. However, for other mutant strains, incompatibility appeared to prevent heterokaryosis.

Both mating types of *P. nodorum* were found, but one mating type, MAT(-), occurred more frequently. Unequal frequencies of mating types of heterothallic species have also been obverved in *Sphaerotheca fuliginea* (McGrath, 1994; McGrath *et al.*, 1996) and in *Gibberella fujikuroi* (Elmer, 1995).

In the reported study, where sexual compatibility pairings with standard tester strains revealed unequal distribution of mating types among the isolates, the sexual forms of *P. nodorum* originated from different countries. Only MAT(-) was found exclusively among the USA isolates although genetic studies of the nuclear DNA of *S. nodorum* suggest that sexual ascospores are a significant source of inoculum in the field, as many different multilocus genotypes occur (McDonald *et al.*, 1994). Current isolates originating from different countries may not be representative of natural populations of *S. nodorum*. They are kept in collections to maintain a good level of conidiogenesis *in vitro*. Indeed, difficulties in keeping strains of *S. nodorum* with a high level of sporulation are mentioned by several authors. In a previous study, differences in sporulation between the two mating types *in vitro* were observed (Halama and Lacoste, 1991). It is perhaps an explanation for the most frequent occurrence of one mating type among isolates.

In other respects, a study on aggressiveness of *P. nodorum* after sexual reproduction showed a stronger pathogen strain x cultivar interactions for mating type (+) as compared to mating type (-) (Rapilly *et al.*, 1992). Hence, if plant breeders use isolates of one mating type to make artificial inoculations, different conclusions on the existence or non-existence of pathogen isolate x cultivar interactions may be made and plant breeding programmes influenced differently.

Acknowledgements

I thank C. Caten, G.M. Hoffman, T. Hunter, R. Loughman, B. McDonald, A. Obst, F. Rapilly and M. Skajennikoff for providing isolates.

References

Elmer, W.H. (1995) A single population of *Gibberella fujikuroi* (*Fusarium proliferatum*) prodominates in asparagus fields in Connecticut, Massachusetts, and Michigan. *Mycologia* 87, 68-71.

Esser, K. (1971) Breeding systems in fungi and their significance for genetic recombination. *Molecular and General Genetics* 110, 86-100.

Esser, K. and Kuenen, R. (1967) *Genetics of Fungi.* Springer-Verlag, Berlin, 499pp.

Halama, P. and Lacoste, L. (1989) *Leptosphaeria nodorum* Müller, première obtention contrôlée des périthèces *in vitro*. In: *Compte-Rendu du 1er Congrès S.F.P., Rennes, 19 et 20 Novembre 1987. Sciences Agronomiques Rennes*, pp. 107-108.

Halama, P. and Lacoste, L. (1991) Déterminisme de la reproduction sexuée de *Phaeosphaeria* (*Leptosphaeria*) *nodorum*, agent de la septoriose du blé. I Hétérothallisme et rôle des microspores. *Canadian Journal of Botany* 69, 95-99.

Halama, P. and Lacoste, L. (1992) Déterminisme de la reproduction sexuée du *Phaeosphaeria nodorum*, agent de la septoriose du blé. II. Action de la température et de la lumière. *Canadian Journal of Botany* 70, 1563-1569.

Halama, P., Parguey Leduc, A. and Lacoste, L. (1992) Les organes de reproduction de l'Ascomycète *Phaeosphaeria (Leptosphaeria) nodorum* (E. Müller) Hedjaroude et de sa forme asexuée *Septoria nodorum* Berk. *Canadian Journal of Botany* 70, 1401-1408.

Hampton, J.W. (1975) *Septoria nodorum* infection of wheat in New Zealand. *Australian Plant Pathological Society Newsletter* 4, 25-26.

Hampton, J.G., Close, R.C. and Jones, D.G. (1978) *Leptosphaeria nodorum* infection of wheat in New Zealand. *New Zealand Journal of Agricultural Research* 21, 691-695.

Kemp, G.H.J., Pretorius, Z.A. and Le Roux, J. (1989) First report of the teleomorph of *Leptosphaeria nodorum* on wheat in South Africa. *Mycological Research* 93, 114-115.

McDonald, B.A., Miles, J., Nelson, L.R. and Pettway, R.E. (1994) Genetic variability in nuclear DNA in field populations of *Stagonospora nodorum*. *Phytopathology* 84, 250-255.

McFadden, W. and Harding, H. (1989) Cereals stubble as a source of primary inoculum of leaf spotting pathogens of winter wheat. *Canadian Journal of Plant Pathology* 11, 195.

McGrath, M.T. (1994) Heterothallism in *Sphaerotheca fuliginea*. *Mycologia* 86, 517-523.

McGrath, M.T., Staniszewska, H. and Shishkoff, N. (1996) Distribution of mating types of *Sphaerotheca fuliginea* in the United States. *Plant Disease* 80, 1098-1102.

Mehta, Y.R. (1975) *Leptosphaeria nodorum* on wheat in Brazil and its importance. *Plant Disease Reporter* 59, 404-406.

Mittelstädt, A. and Fehrmann, H. (1987) Zum Auftreten der Hauptfruchtform von *Septoria nodorum* in der Bundesrepublik Deutschland. *Zeitschrift für Pflanzenkrankheiten und Pflanzenschutz* 94, 380-385.

Müller, E. (1952) Pilzliche Erreger der Getreideblattdürre. *Phytopathologische Zeitschrift* 19, 403-416.

Nelson, R.R. (1959a) *Cochliobolus carbonum*, the perfect stage of *Helminthosporium carbonum*. *Phytopathology* 49, 807-810.

Nelson, R.R. (1959b) Genetics of *Cochliobolus heterostrophus* III. Genetic factors inhibiting ascus formation. *Mycologia* 51, 132-137.

Nelson, R.R. (1960) The genetics of compatibility in *Cochliobolus carbonum*. *Phytopathology* 50, 158-160.

Nelson, R.R. (1964) Genetic inhibition of perithecial formation in *Cochliobolus carbonum*. *Phytopathology* 54, 876-877.

Nelson, R.R. (1970) Variation in mating capacities among isolates of *Cochliobolus carbonum*. *Canadian Journal of Botany* 48, 261-263.

Newton, A.C. and Caten, C.E. (1985) Heterokaryosis and heterokaryon incompatibility in *Septoria nodorum*. In: Scharen, A.L. (ed.) *Septoria of Cereals*. USA Department of Agriculture, ARS, pp. 13-15.

O'Reilly, P. and Bannon, E. (1988) *Leptosphaeria nodorum* on wheat in Ireland. *Plant Pathology* 37, 153-154.

Rapilly, F., Foucault, B. and Lacazedieux, J. (1973) Etudes sur l'inoculum de *Septoria nodorum* Berk. (*Leptosphaeria nodorum* Müller) agent de la septoriose du blé I. Les ascospores. *Annales de Phytopathologie* 5, 131-141.

Rapilly, F., Skajennikoff, M., Halama, P. and Touraud, G. (1992) La reproduction sexuée et l'agressivité de *Phaeosphaeria nodorum* Hedj (*Septoria nodorum* Berk). *Agronomie* 12, 639-649.

Sanderson, F.R. and Hampton, J.G. (1978) Role of the perfect states in the epidemiology of the common *Septoria* diseases of wheat. *New Zealand Journal of Agricultural Research* 21, 277-281.

Scharen, A.L. and Sanderson, F.R. (1982) *Leptosphaeria nodorum* and *Mycosphaerella graminicola* in North America. *Phytopathology* 72, 934.

Shipton, W.A., Boyd, W.R.J., Rosielle, A.A. and Shearer, B.I. (1971) The common *Septoria* diseases of wheat. *Botanical Review* 37, 231-262.

Wale, S.J. and Colhoun, J. (1979) Further studies on the aerial dispersal of *Leptosphaeria nodorum*. *Phytopathologische Zeitschrift* 94, 185-189.

Weber, G.F. (1922) *Septoria* diseases of cereals. *Phytopathology* 12, 449-470.

Westbrook, M.A. (1955) Observations on nuclear structure in the Florideae. Beih. *Botanisches Zentralblatt* 53, 564-585.

Whitehouse, H.L.K. (1949) Heterothallism and sex in the fungi. *Biological Review* 24, 411-447.

Chapter five:

Population Dynamics of *Septoria* in the Crop Ecosystem

M.W. Shaw
Department of Agricultural Botany, School of Plant Sciences, The University of Reading, 2 Earley Gate, Whiteknights, Reading RG6 6AU, UK

Introduction

The intention of this chapter is to give an overview of the population dynamics of *Mycosphaerella graminicola* and *Phaeosphaeria nodorum*, that is, the forces which determine the population size and the characteristics of the population that determine its response to those forces. This is primarily of interest because the population size is linked closely, although not absolutely, to crop damage. The forces operating include the environmental conditions; the host size, age, phenology and genetic composition; the other organisms present on and in the host; and adaptive change in the pathogen population. To stimulate discussion, I have been deliberately speculative in many places.

More is known, or can be inferred, about *M. graminicola* than about *P. nodorum*, so I shall focus first on this. I shall then discuss *P. nodorum* and possible interactions between the pathogens.

Mycosphaerella graminicola

Genetic evidence suggests that the world population of *M. graminicola* is a single linked entity (McDonald *et al.*, 1995). It is structured as a meta-population, made up of many sub-populations of distinct wheat-growing and climatic areas, each made up of other sub-populations, ultimately consisting of all the wheat fields in which the pathogen is present. The dynamics has two distinct phases: multiplication in a single host field; and transfer between sub-populations.

Before these two can be discussed, it is necessary to define what the population is a population **of**. One conceptually simple way to do this is to consider each independent infection to give rise to a mycelium inside the host; the population can then be regarded as composed of individuals which are the physiologically distinct mycelia within a leaf. This begs a number of questions about the interactions of mycelia within a leaf, because the dynamics of such a population will not be simple if mycelia regularly merge within a leaf, so that a population of a thousand on one day can subsequently become a population of one, much larger 'individual'. We have little evidence on this, but there are two or three lines of evidence which suggest the assumption is not unreasonable. First, distinct lesions on a leaf usually have pycnidia which are genetically uniform, but distinct from other lesions (McDonald and Martinez, 1990). However, this observation can only be made where lesions are distinct, so genetic exchange between mycelia would not be expected. Second, even on a fully infected leaf, 'generations' of lesions may be identifiable, with initial infections having distinct margins, suggesting that they have not merged with later infections (Shaw and Royle, 1993). Third, observations which suggest antagonistic interactions between certain distinct isolates have been made (Zelikovitch and Eyal, 1991). However, these observations do not show whether such interactions are general, nor do they show what happens when a leaf is multiply infected by spores of the **same** genotype. In the latter case, genetic exchange does not alter the composition of the population, so it may be equally valid to consider the population on the leaf to be one large individual, or many small identical units. Nonetheless, it would be valuable to understand more about the interactions of individual mycelia resulting from infection by single spores within leaves. This description should be adequate at a field or plant level, at least over a scale such that the population is well mixed; over larger scales it may be more appropriate to take a view of a field as an individual with a certain size (related to the population within it at the lower level); and over a larger scale still whole regions may be regarded as individuals. In each case, characteristic dynamics of the host and pathogen population exist.

The dynamics of such a population then consists of an accounting operation, allowing for infection processes which create new individuals, and the death of individuals (meaning that the individual has no further potential for spore production). Infection may be by both local and distant or immigrant spores. The dynamics are then controlled by changes in the host substrate available - growth of new tissue, death of old tissue and changes in susceptibility of tissue with age or infection load, or at a higher level harvesting and planting of fields; by environmental conditions; and by changes in the quality and genetic composition of the pathogen population.

The rest of this section will discuss the dynamics at each level, and the linkages between levels. First, I shall consider an individual plot or field. Before the crop appears, the population, if any, exists only as individuals on dead host tissue from previous harvests, and as wind-blown spores, which have a short lifetime unless they succeed in infecting fresh hosts. There is no evidence that any reproductively competent individuals survive on seed. The host then begins to grow and is infected, either by splash-borne conidia from crop residues or possibly weeds (Brokenshire, 1975; Djerbi, 1977; Shtienberg *et al.*, 1990), or by wind-blown ascospores. After the latent period these individuals produce pycnidia and begin to infect the available host population, which will have grown considerably in the intervening period. In the UK, for example, the leaf and tiller production interval of winter wheat is about 100°C-day (Porter, 1984), depending on planting time, and leaves survive for a total of three to five times this; initially the crop grows exponentially as new tillers and leaves appear (Goudriaan and Monteith, 1990). Typical latent periods are 40 days at 5°C, reducing with temperature to as little as 15 days at the optimum of around 15-20°C (Shaw, 1990). The initial lesions may occupy about 0.5 cm^2, each with 50-100 pycnidia, each capable of producing two to three cycles of 3000-5000 conidia: roughly 10^6 conidia per initial infection. These are dispersed by splash from a film of water on the leaf containing a high concentration of spores. Thus any infected leaf will subsequently be attacked by many spores. Surrounding leaves may receive spores to distances of 3-4 m by splash during wind (Brennan *et al.*, 1985a,b), and leaves within 50 cm or so of the source will receive many spores. Thus, any initial unevenness in the density of infection over the field on scales less than several metres will quickly be evened out, and the potential for multiplication, given suitable weather conditions, is very large. The infection efficiency of spores can be extremely density dependent, with as many as one in ten spores successfully forming infections which lead to pycnidia when applied at very low density, falling rapidly with the number of spores applied; a rule of thumb might be that reducing the number of spores applied by a factor of ten leads only to a reduction by a factor of two in the number of lesions resulting (Shaw, unpublished results). Once again, this will tend to lead to rapid evening out of any variation in the area density of lesions.

The weather conditions required for infection appear to be not very stringent and are regularly fulfilled during the early growth of winter crops (Shaw and Royle, 1993). In the absence of rain to provide splash, spores may move by rubbing, but the horizontal distance scale of this process must be less than splash and the efficiency seems likely to be poor. Thus, it is quite likely that a very high proportion of the available host tissue will be latently infected. However, the biomass of fungus in a latently infected and asymptomatic leaf appears to be small (Kema *et al.*, 1996d), so the effect on plant growth may be slight and given the rapid growth of the crop, the visibly infected crop remains a small proportion of the total. Dead tissue also decays quite rapidly under moist conditions, so the pathogen population on dead tissue does not build up rapidly. There are, therefore, four phases to the population cycle in a field (a) initial infection; (b) a rapid increase at perhaps 0.5 d^{-1} [using $r = \ln(R_0)/p$ with latent period $p \approx 30$ and basic reproductive rate $R_0 \approx 10^6$] until the population is limited by available host; (c) a phase of linear increase limited by the linear phase of host growth, with substantial population turn-over; and (d) an apparent rapid increase during the sexual phase of crop growth, when visible pathogen continues to appear, but no new vegetative tissue is growing (Shaw and Royle, 1993). Phase (d) is when the principal damage is done to yield except in crops unable to compensate for the effects during (b) and (c) because they are so thinly planted.

The spores which initiate new infections in the above scheme are assumed initially to be ascospores, and then predominantly conidia. This is based on observations that a typical crop in the UK will contain 10^5-10^8 conidia m^{-2} by early spring or late winter (Shaw and Royle, 1993, multiplying the figures there by 1000 tillers m^{-2}), whereas ascospore densities giving rise to at most 100 infections m^{-2} $week^{-1}$ in the newly sown crop do not subsequently increase much. Perithecia are present in standing crops for much of the year (Kema *et al.*, 1996c; Hunter *et al.*, Chapter 7 this volume), but pycnidia are undoubtedly much more common.

This basic pattern is modulated by the exact growth pattern of the crop; by the weather; and possibly by genetic changes to the population during the year. A late planted crop will be invaded more sparsely by ascospores, will have a more rapid leaf production interval and a shorter leaf life-time, and will encounter relatively more extreme environments (initially cold, subsequently hot) than an early planted crop. The latter factor tends to a lower equilibrium disease level on the green tissue, and both factors to a delayed approach to equilibrium. This picture suggests that a *M. graminicola* population is limited by four interacting factors: initial inoculum; crop growth rate and pattern; crop susceptibility; and weather. It should, therefore, be relatively simple to determine the relationship between final disease and weather conditions, at least for a crop of one cultivar planted at a given time of the year and with a known initial pathogen population. In fact, correlations of final disease with spore transport or infection factors during flag leaf production provide no more than a general guide as to important times to intervene in the population dynamics with fungicide.

The most effective use of triazole fungicides has consistently been found to result from application around or shortly after the emergence of the flag leaf (Cook, 1977; Thomas *et al.*, 1989; Hims and Cook, 1992; Jørgensen, 1994). This is partly a result of disproportionate damage by small amounts of disease (Shaw and Royle, 1989), but it also suggests that the population cannot subsequently compensate for greatly reduced or delayed infection at the start of the life of the uppermost leaves. This may be because conditions tend to become less and less suitable for the pathogen as it becomes hotter, and correspondingly drier, later in the summer.

Several statistical attempts to pick out weather factors conducive to severe disease during the grain fill period have been made, with diverse but always surprising results. Tyldesley and Thompson (1981), in England and Wales, showed a broad correlation of May rainfall with severity of *M. graminicola* and

P. nodorum combined (but at a time when the predominant disease was probably *P. nodorum*). They did not explore time periods before mid spring. A similar correlation with a similar variable was shown by Hansen *et al.* (1994) in Denmark, and in Israel by Shtienberg *et al.* (1990). Daamen and Stol (1992), in The Netherlands, showed a strong positive correlation with sunshine the previous August, suggesting an influence of the initial infection of the crop by ascospores from the previous crop. Coakley *et al.* (1985), in Indiana, showed a more proximate correlation with rainfall and temperature in April. Finally Royle *et al.* (1993), in the UK, showed a good correlation with the number of severe frosts the previous December. The correlations with spring rainfall are not unexpected and are in broad agreement with the mechanistic hypotheses and observations of Bahat *et al.* (1980), Shaw and Royle (1993), and Lovell *et al.* (1997). However, the longer-time correlations are more surprising, since they are often stronger than the correlations with rainfall, and it is hard to show correlations of spring inoculum with summer severity. I shall return to them later.

On a larger scale, fields can be regarded as hosts, with the population of the pathogen in them regarded as the 'size' of the individual pathogen. These hosts have a life-cycle of a year. A new population is 'born' each autumn, and infected by wind-blown spores, probably predominantly ascospores, at least in temperate regions. The origin of these spores is not certainly established, but seems very likely to be previous crop residues and volunteers. Although the pathogen is known from many graminaceous hosts, it appears to be strongly specialized between durum and bread wheats, and it is certainly hard to obtain cross-infection from *Poa annua* isolates to wheat (Shaw, unpublished results) and the population on *P. annua* appears to be genetically isolated from the population on wheat in North America (B.A. McDonald, Texas A & M University, 1994, personal communication). This suggests that the populations on different grass species are actually fairly strongly specialized to those hosts, so the likely sources of ascospores are residues and volunteers from previous crops. Although numbers will depend on weather conditions, tillage, cultivar and so forth, they must, on average, be proportional to the area density of such residues. In broad terms also, the average distance between sources and new fields must be proportional to the inverse of the square root of the number of fields per unit area. So if wheat growing becomes more extensive, the number of ascospore sources increases and the average distance between a source and a newly planted field decreases. This means that the initial number of infections in a newly planted field should be more than proportional to the density of wheat planted. Below some threshold level of initial infection, the population will never become limited by availability of host tissue, and the severity of disease in the summer will be related to the initial level of inoculum as well as to the summer weather. Above the threshold density, the phase during which the population is limited by host tissue availability will insulate the summer population from fluctuations in the initial inoculum level. Averaged over years, one would expect the disease to become more important as wheat was grown over a larger proportion of a region. This is certainly in line with recent experience worldwide.

Leath *et al.* (1994) discuss factors which appear worldwide to be correlated with severe attack by *M. graminicola.* The unexpected feature of this survey was the importance of precipitation during the season **without** host, which was the most important single factor differentiating mild from severe attacks of *M. graminicola.* This observation concurs with Daamen and Stol's (1992) report, referred to above, of a strong positive correlation between sunshine the previous August and the severity of *M. graminicola.* Both are explicable if the initial inoculum, or some feature of the initial inoculum, is strongly linked to the final severity of the disease, and if survival of host tissue containing perithecia is worse in wetter, cooler seasons. However, this implies that phase (c) of the within-field dynamics, referred to above, is relatively unimportant in most wheat-growing areas. If the initial area density of infections within a field is correlated with the final disease severity, then the inoculum available at the start of the sexual stage should be strongly correlated with final disease severity, yet this correlation seems to be weak (Shaw and Royle, 1993), although data are few.

The influence of crop genotype on disease is important. The factors which alter the population dynamics are those which alter the birth and death schedules of the pathogen: principally latent period, spore output, and infection establishment. These will vary between genotypes on a particular cultivar, and between cultivars. This will lead to adaptation of the pathogen population to the host on which it finds itself and to the environment (Eyal *et al.*, 1985; Ahmed *et al.*, 1995; Kema *et al.*, 1996a,b). Since there are several asexual generations per year within a crop and differences in intrinsic rate of increase between clones may be large (al-Hamar and Shaw, poster presented at the 15th Long Ashton International Symposium 'Understanding Pathosystems: A Focus on Septoria', Bristol, UK, 13-15 September 1997), there is potential for substantial adaptive change in the frequencies of different clones during the growing period of the crop. Because of the long latent period of the pathogen, the time required for this adaptive change, and the change in environment with season, the composition of the population is likely to be far from that giving the maximum rate of increase at any given time. This might explain some of the curious correlations mentioned above. For example, after a cold winter, isolates with relatively short latent periods and high infectivities at low temperatures will be abundant; this may (unless these same isolates are also well adapted at high temperatures) lead to much slower subsequent rates of population increase, given the same weather conditions, than if the winter had been warm. This could lead to a correlation of summer severity with winter conditions but not with spring severity.

Phaeosphaeria nodorum

At planting, the population of *P. nodorum* in a field may exist as mycelia on a proportion of the infected seed (Baker, 1971; Hewett, 1975; Cunfer, 1978; 1991; Cunfer and Johnson, 1981; Babadoost and Herbert, 1984), as ascospores from previous crops (Rapilly *et al.*, 1973; Mittelstadt and Fehrmann, 1987), or as conidia on debris from previous crops, volunteers, or weeds (Jenkyn and King, 1977). The genetic evidence suggests that the sexual phase must occur frequently (McDonald *et al.*, 1994). Characteristic pycnidia are only produced on dead tissue, and so study of the population buildup is hampered by much of the population existing on senescent tissue on which it is hard to identify. Immunological (Mittermeier *et al.*, 1990; Petersen *et al.*, 1990) and other work (Bannon, 1978; O'Reilly and Downes, 1986; Sieber *et al.*, 1988) suggests that an invisible mycelial population of *P. nodorum* may be much more common than usually recognized; our understanding of population changes will benefit greatly if it is possible to get better estimates of all phases of the life-cycle of this fungus. In wet weather, the latent period is shorter than that of *M. graminicola* at high temperatures (Shearer and Zadoks, 1972), but multiplication is slowed more rapidly at low temperatures. Thus, the initial buildup of the population may be slower during the winter than for *M. graminicola*. By the time of grain production, in areas where the disease occurs regularly, a substantial population of conidia may exist on lower leaves, and these can act as sources for infection of the uppermost leaves and the grain. However, where infected grain is rare, it appears that late ascospore immigration may substantially augment the indigenous population. The source of the ascospores is unknown. There is less evidence for pathogenic specialization in *P. nodorum* (Scharen *et al.*, 1985) than in *M. graminicola*, although forms specialized to wheat and barley have been shown (Osbourn *et al.*, 1987), so it remains possible that grass weeds could be a substantial source of ascospore or conidial inoculum. In support of this, survey and experimental work in several countries has shown an association between grass, as well as wheat, as a preceding crop and the severity of *P. nodorum* (Jenkyn and King, 1977; King, 1977; Polley and Thomas, 1991). It is, however, curious that Leath *et al.* (1994) showed a significant **negative** association of severe *P. nodorum* infection with minimum or reduced tillage. It is possible that this is another beneficial effect of long-term reduced tillage leading to rapid decay of inoculum-bearing host, like that shown by Jalaluddin and Jenkyn (1996) for eyespot caused by *Tapesia* spp. Severe infection of leaves or of ears may build up rapidly in warm, moist conditions, in which the intrinsic rate of increase of *P. nodorum* will be substantially greater than that of *M. graminicola*, because of the shorter latent period. This shorter latent period also means that the population development is less tied to host phenology than is the case with *M. graminicola*.

Royle *et al.* (1986) observed that it was common to have a predominance of either *P. nodorum* or *M. graminicola* in an individual crop, even in areas where both occurred. This might be because of sensitivity to minor differences in environmental or nutritional conditions, or indicate some antagonistic interaction between the species. Such interactions are possible, because both species show strong density dependence in the infectivity of individual spores, suggesting that small numbers of infections trigger plant defences making subsequent infections harder. The difference in niches of the two pathogens could accentuate the effect of this, since *M. graminicola* infects younger tissue more easily. Then if *M. graminicola* were common, leaves would tend to be infected and be particularly unsuitable for infection by *P. nodorum* until they were relatively senescent, by which time much of the available nutriment might have been pre-empted by the *M. graminicola* infection. To speculate on the basis of a single piece of datum, the threshold of infectivity shown by Jeger *et al.* (1985) might then account for exclusion in the other direction: once abundant, *P. nodorum* would be able to infect young tissue. With its longer latent period and slower intrinsic rate of increase, *M. graminicola* might then be excluded from the crop.

These speculations lead naturally to a consideration of why the relative importance of the two diseases seems to have recently changed in many countries in Western Europe and elsewhere. Many factors may be involved, and the simultaneous changes may, in fact, be a coincidence. In the UK, Bayles (1991) has shown that cultivar resistance to *M. graminicola* did decline at about the time that the disease increased in importance; at the same time, fungicide use during ear-filling became common, which would have reduced the frequency of *P. nodorum* infection in subsequent crops, and nitrogen application rates increased, which has been correlated with increased susceptibility to *M. graminicola* and reduced susceptibility to *P. nodorum*. I have not found similar information on cultivars in other Western European countries, but the other changes occurred in parallel, as did the increase in relative importance of *P. nodorum*. Worldwide, shorter cultivars of wheat have been grown over larger areas, with higher nitrogen levels. These factors would all tend to increase the relative importance of *M. graminicola.*

If, in some region, conditions are **on average** unsuitable for one of the diseases, then, in the long term, it will decline in an area. Persistence of the disease shows that, on average, buildup during the cropping season compensates for decline in the off-season. If *M. graminicola* and *P. nodorum* have different climatic requirements, this would lead to areas with both pathogens present, and areas with one or the other absent. However, in areas from which one pathogen was absent, conditions suitable for its multiplication might regularly occur - but not often enough to ensure inoculum being present when conditions were suitable; and modifications of the environment, such as commoner wheat growing, could turn an unsuitable area into a suitable area, without modification of the climate. In fact, this hypothesis can be elaborated still further by supposing that more favourable conditions are required for *P. nodorum* to dominate *M. graminicola* in the summer than for it to persist as a marginal pathogen of senescent tissue. In this case, we would expect four types of region, moving geographically from *M. graminicola* only, to *M. graminicola* the dominant pathogen, to *P. nodorum* the dominant pathogen to *P. nodorum* only being present. But in any given year, climatic variations would make it very difficult to pick out the variables determining which pathogen was present. That is, the same weather conditions in one particular year might produce all four outcomes, depending on the long-run history of the region in which the field is. In turn, this would make transportable forecasting criteria hard to find, which fits with experience.

If interactions between Septoria and Stagonospora leaf blotch diseases are possible, how important are other biotic interactions in the population dynamics of the two species? *Mycosphaerella graminicola* has a much lower intrinsic growth rate than rusts and mildews except during the winter. A yellow or brown rust-infected leaf could well be destroyed by two or even three generations of multiplication before simultaneous *M. graminicola* infections appeared. Thus, in the summer, good resistance to mildew and rust would be expected to favour *M. graminicola*. However, Brokenshire (1974) reported a very specific association of *M. graminicola* lesions with mildew pustules on otherwise resistant wheat. This physiological facilitation, however, would not necessarily counter a population level competition acting via accelerated host death. Adee *et al.* (1990) showed suppression of *P. nodorum* by *Pyrenophora tritici-repentis*, although the data are hard to relate to the field because fixed joint inoculum doses were used. Wheat has been shown to be more susceptible to infection by *P. nodorum* if already infected with *Gaeumannomyces graminis* (Jenkins and Jones, 1980), which could add to the association of the disease with wheat or grass as a preceding crop; it has also been shown to be more susceptible if already infected with mildew, *Erysiphe graminis* (Weber *et al.*, 1994), or with rust (van der Wal and Cowan, 1974). These associations are consistent with the disease being favoured on stressed plants and/or on senescent tissue; the population level consequences of competing populations of *P. nodorum* and other pathogens, however, remain unclear.

It is extraordinary how many open questions there are - and how difficult it is to phrase them in ways that can be answered by observation or experiment - in two of the best studied diseases in the world. This suggests that research in population biology, as much as in molecular biology, will be required to manage disease in future agro-ecosystems.

References

Adee, S.R., Pfender, W.F. and Hartnett, D.C. (1990) Competition between *Pyrenophora tritici-repentis* and *Septoria nodorum* in the wheat leaf as measured with de Wit replacement series. *Phytopathology* 80, 1177-1181.

Ahmed, H.U., Mundt, C.C. and Coakley, S.M. (1995) Host pathogen relationship of geographically diverse isolates of *Septoria tritici* and wheat cultivars. *Plant Pathology* 44, 838-847.

Babadoost, M. and Herbert, T.T. (1984) Incidence of *Septoria nodorum* in wheat seed and its effects on plant growth and grain yield. *Plant Disease* 68, 125-128.

Bahat, A., Gelernter, I., Brown, M.B. and Eyal, Z. (1980) Factors affecting the vertical progression of *Septoria* leaf blotch in short-statured wheats. *Phytopathology* 70, 179-184.

Baker, C.J. (1971) Reaction of wheat varieties in the field to ear and subsequent seedling infection by *Septoria (Leptosphaeria) nodorum*. *Journal of the National Institute of Agricultural Botany* 12, 279-285.

Bannon, E. (1978) A method of detecting *Septoria nodorum* on symptomless leaves of wheat. *Irish Journal of Agricultural Research* 17, 323-325.

Bayles, R.A. (1991) Research note; varietal resistance as a factor contributing to the increased importance of *Septoria tritici* Rob. and Desm. in the UK wheat crop. *Plant Varieties and Seeds* 4, 177-183.

Brennan, R.M., Fitt, B.D.L., Taylor, G.S. and Colhoun, J. (1985a) Dispersal of *Septoria nodorum* pycnidiospores by simulated rain and wind. *Journal of Phytopathology* 112, 291-297.

Brennan, R.M., Fitt, B.D.L., Taylor, G.S. and Colhoun, J. (1985b) Dispersal of *Septoria nodorum* pycnidiospores by simulated raindrops in still air. *Journal of Phytopathology* 112, 281-290.

Brokenshire, T. (1974) Predisposition of wheat to *Septoria* infection following attack by *Erysiphe*. *Transactions of the British Mycological Society* 63, 393-397.

Brokenshire, T. (1975) The role of graminaceous species in the epidemiology of *Septoria tritici* on wheat. *Plant Pathology* 24, 33-38.

Coakley, S.M., McDaniel, L.R. and Shaner, G. (1985) Model for predicting severity of *Septoria tritici* blotch on winter wheat. *Phytopathology* 75, 1245-1251.

Cook, R.J. (1977) Effect of timed fungicide sprays on yield of winter wheat in relation to *Septoria* infection periods. *Plant Pathology* 26, 30-34.

Cunfer, B.M. (1978) The incidence of *Septoria nodorum* in wheat seed. *Phytopathology* 68, 832-835.

Cunfer, B. (1991) Long term viability of *Septoria nodorum* in stored wheat seed. *Cereal Research Communications* 19, 247-349.

Cunfer, B.M. and Johnson, J.W. (1981) Relationship of glume blotch symptoms on wheat heads to seed infections by *Septoria nodorum*. *Transactions of the British Mycological Society* 76, 205-211.

Daamen, R.A. and Stol, W. (1992) Surveys of cereal diseases and pests in the Netherlands. 5. Occurrence of Septoria spp. in winter wheat. *Netherlands Journal of Plant Pathology* 98, 369-376.

Djerbi, M. (1977) Épidémiologie du *Septoria tritici* Rob. et Desm. Conservation et mode de formation de l'inoculum primaire. *Travaux Dédiés à Georges-Viennot Bourgin.* Société Française de Phytopathologie, Paris, pp. 91-101.

Eyal, Z., Scharen, A.L., Huffman, M.D. and Prescott, J.M. (1985) Global insights into virulence frequencies of *Mycosphaerella graminicola. Phytopathology* 75, 1456-1462.

Goudriaan, J. and Monteith, J.L. (1990) A mathematical function for crop growth based on light interception and leaf area expansion. *Annals of Botany* 66, 695-701.

Hansen, J.G., Secher, B.J.M., Jorgensen, L.N. and Welling, B. (1994) Thresholds for control of Septoria spp. in winter wheat based on precipitation and growth stage. *Plant Pathology* 43, 183-189.

Hewett, P.D. (1975) *Septoria nodorum* on seedling and stubble of winter wheat. *Transactions of the British Mycological Society* 65, 7-18.

Hims, M.J. and Cook, R.J. (1992) Disease epidemiology and fungicide activity in winter wheat. *Proceedings of the 1992 Brighton Crop Protection Conference - Pests and Diseases* 2, 615-620.

Jalaluddin, M. and Jenkyn, J.F. (1996) Effects of wheat crop debris on the sporulation and survival of *Pseudocercosporella herpotrichoides. Plant Pathology* 45, 1052-1063.

Jeger, M.J., Griffiths, E. and Jones, D.G. (1985) The effects of post-inoculation wet and dry periods, and inoculum concentration, on lesion numbers of *Septoria nodorum* in spring wheat seedlings. *Annals of Applied Biology* 106, 55-63.

Jenkins, P.D. and Jones, D.G. (1980) Predisposition to *Septoria nodorum* as a result of take-all (*Gaeumannomyces graminis*) infection of wheat. *Annals of Applied Biology* 95, 47-52.

Jenkyn, J.F. and King, J.E. (1977) Observations on the origins of *Septoria nodorum* infection of winter wheat. *Plant Pathology* 26, 153-160.

Jørgensen, L.J. (1994) Duration of effect of EBI fungicide when using reduced rates in cereals. *Proceedings of the 1994 Brighton Crop Protection Conference - Pests and Diseases* 2, 703-710.

Kema, G.H.J., Annone, J.G., Sayoud, R., van Silfhout, C.H., van Ginkel, M. and de Bree, J. (1996a) Genetic variation for virulence and resistance in the wheat-*Mycosphaerella graminicola* pathosystem. I. Interactions between pathogen isolates and host cultivars. *Phytopathology* 86, 200-212.

Kema, G.H.J., Sayoud, R., Annone, J.G. and van Silfhout, C.H. (1996b) Genetic variation for virulence and resistance in the wheat-*Mycosphaerella graminicola* pathosystem. II. Analysis of interactions between pathogen isolates and host cultivars. *Phytopathology* 86, 213-220.

Kema, G.H.J., Verstappen, E.C.P., Todorova, M. and Waalwijk, C. (1996c) Successful crosses and molecular tetrad and progeny analyses demonstrate heterothallism in *Mycosphaerella graminicola. Current Genetics* 30, 251-258.

Kema, G.H.J., Yu, D., Rijkenberg, F.H.J., Shaw, M.W. and Baayen, R.P. (1996d) Histology of the pathogenesis of *Septoria tritici* in wheat. *Phytopathology* 86, 777-786.

King, J.E. (1977) Surveys of diseases of winter wheat in England and Wales 1970-75. *Plant Pathology* 26, 8-20.

Leath, S., Scharen, A.L., Dietzholmes, M.E. and Lund, R.E. (1994) Factors associated with global occurrences of Septoria nodorum blotch and Septoria tritici blotch of wheat. *Plant Disease* 77, 1266-1270.

Lovell, D.J., Parker, S.R., Hunter, T., Royle, D.J. and Coker, R.R. (1997) Influence of crop growth and structure on the risk of epidemics by *Mycosphaerella graminicola* (*Septoria tritici*) in winter wheat. *Plant Pathology* 46, 126-138.

McDonald, B.A. and Martinez, J.P. (1990) DNA restriction fragment length polymorphisms among *Mycosphaerella graminicola* (anamorph *Septoria tritici*) isolates collected from a single wheat field. *Phytopathology* 80, 1368-1373.

McDonald, B.A., Miles, J., Nelson, L.R. and Pettway, R.E. (1994) Genetic variability in nuclear DNA in field populations of *Stagonospora nodorum*. *Phytopathology* 84, 250-255.

McDonald, B.A., Pettway, R.E., Chen, R.S., Boeger, J.M. and Martinez, J.P. (1995) The population genetics of *Septoria tritici* (teleomorph *Mycosphaerella graminicola*). *Canadian Journal of Botany* 73 (Supplement), S292-S301.

Mittelstadt, A. and Fehrmann, H. (1987) Zum Auftreten der Hauptfruchtform von *Septoria nodorum* in der Bundesrepublik Deutschland. *Zeitschrift für Pflanzenkrankheiten und Pflanzenschütz* 94, 380-385.

Mittermeier, L., Dercks, W., West, S. and Miller, S.A. (1990) Field results with a diagnostic system for the identification of *Septoria nodorum* and *Septoria tritici. Proceedings of the 1990 Brighton Crop Protection Conference - Pests and Diseases* 2, 757-762.

O'Reilly, P. and Downes, M.J. (1986) Form of survival of *Septoria nodorum* on symptomless winter wheat. *Transactions of the British Mycological Society* 82, 381-385.

Osbourn, A.E., Caten, C.E. and Scott, P.R. (1987) Adaptation of *Septoria nodorum* to wheat or barley on detached leaves. *Plant Pathology* 36, 565-576.

Petersen, F.P., Rittenburg, J.H., Miller, S.A. and Grothaus, G.D. (1990) Development of monoclonal antibody-based immunoassays for detection and differentiation of *Septoria nodorum* and *S. tritici* in wheat. *Proceedings of the 1990 Brighton Crop Protection Conference - Pests and Diseases* 2, 751-756.

Polley, R.W. and Thomas, M.R. (1991) Surveys of diseases of winter wheat in England and Wales, 1976-1988. *Annals of Applied Biology* 119, 1-20.

Porter, J.R. (1984) A model of canopy development in winter wheat. *Journal of Agricultural Science, Cambridge* 102, 383-392.

Rapilly, F., Foucault, B. and Lacazedieux, J. (1973) Études sur l'inoculum de *Septoria nodorum* Berk. (*Leptosphaeria nodorum* Müller) agent de la septoriose du blé. I. Les ascospores. *Annales de Phytopathologie* 5, 131-141.

Royle, D.J., Shaw, M.W. and Cook, R.J. (1986) Patterns of development of *Septoria nodorum* and *S. tritici* in some winter wheat crops in Western Europe, 1981-83. *Plant Pathology* 35, 466-476.

Royle, D.J., Lovell, D.J., Coakley, S.M. and Shaner, G. (1993) Predicting the effects of climate change on *Septoria tritici* in winter wheat. In: *Abstracts of the 6th International Congress of Plant Pathology*. Montreal, p. 97.

Scharen, A.L., Eyal, Z., Huffman, M.D. and Prescott, J.M. (1985) The distribution and frequency of virulence genes in geographically separated populations of *Leptosphaeria nodorum*. *Phytopathology* 75, 1463-1468.

Shaw, M.W. (1990) Effects of temperature, leaf wetness and cultivar on the latent period of *Mycosphaerella graminicola* on winter wheat. *Plant Pathology* 39, 255-268.

Shaw, M.W. and Royle, D.J. (1989) Estimation and validation of a function describing the rate at which *Mycosphaerella graminicola* causes yield loss in winter wheat. *Annals of Applied Biology* 115, 425-442.

Shaw, M.W. and Royle, D.J. (1993) Factors determining the severity of *Mycosphaerella graminicola* (*Septoria tritici*) on winter wheat in the UK. *Plant Pathology* 42, 882-899.

Shearer, B.L. and Zadoks, J.C. (1972) The latent period of *Septoria nodorum* in wheat. 1. The effect of temperature and moisture treatments under controlled conditions. *Netherlands Journal of Plant Pathology* 78, 231-241.

Shtienberg, D., Dinoor, A. and Marani, A. (1990) Wheat disease control advisory, a decision support system for management of foliar diseases of wheat in Israel. *Canadian Journal of Plant Pathology* 12, 195-203.

Sieber, T., Riesen, T., Muller, E. and Fried, P.M. (1988) Endophytic fungi in four winter wheat cultivars (*Triticum aestivum* L) differing in resistance against *Stagonospora nodorum* (Berk.) Cast. & Germ. = *Septoria nodorum* (Berk.) Berk. *Journal of Phytopathology* 122, 289-306.

Thomas, M.R., Cook, R.J. and King, J.E. (1989) Factors affecting development of *Septoria tritici* in winter wheat and its effect on yield. *Plant Pathology* 38, 246-257.

Tyldesley, J.B. and Thompson, N. (1981) Forecasting *Septoria nodorum* on winter wheat in England and Wales. *Plant Pathology* 29, 9-20.

van der Wal, A.F. and Cowan, M.C. (1974) An ecophysiological approach to crop losses exemplified in the system wheat leaf rust and glume blotch. II. Development, growth, and transpiration of uninfected plants and plants infected with *Puccinia recondita* f. sp. *triticina* and/or *Septoria nodorum* in a climate chamber experiment. *Netherlands Journal of Plant Pathology* 80, 192-214.

Weber, G.E., Gulec, S. and Kranz, J. (1994) Interactions between *Erysiphe graminis* and *Septoria nodorum* on wheat. *Plant Pathology* 43, 158-163.

Zelikovitch, N. and Eyal, Z. (1991) Reduction in pycnidial coverage after inoculation of wheat with mixtures of isolates of *Septoria tritici*. *Plant Disease* 75, 907-910.

Chapter six:

Analysing Epidemics of *Septoria tritici* for Improved Estimates of Disease Risk

S.R. Parker[1*], D.J. Lovell[1], D.J. Royle[1] and N.D. Paveley[2]
[1]*IACR-Long Ashton Research Station, Department of Agricultural Sciences, University of Bristol, Long Ashton, Bristol BS41 9AF, UK.* [2]*ADAS High Mowthorpe, Duggleby, Malton, North Yorkshire YO17 8BP, UK*

Introduction

Over the past two decades, *Septoria tritici* (anamorph of *Mycosphaerella graminicola*) has caused the most serious foliar disease of winter wheat in the UK. Considerable effort has, therefore, been devoted towards improving biological understanding of the dynamics of epidemic progress. Achievements towards this goal will enable improved quantification of disease risk and, hence, the development of reliable systems of disease management. Disease risk in this context means the likelihood of disease progression. The consequences of disease for yield are not considered here. This aspect is covered by Paveley, Chapter 16 this volume.

The classical disease triangle (host-pathogen-environment) remains a valuable conceptual framework for epidemiological studies. However, its elegant simplicity belies the complexity inherent to most pathosystems. Epidemics of *S. tritici* are favoured by wet weather; in particularly wet seasons, the gross effect of rainfall may conceal the outcome of interactions of the disease triangle on the progression of *S. tritici* epidemics. However, when rainfall is limited, the spread of disease can be more difficult to predict. Ironically, it is in these situations that greatest benefit from disease forecasting can be achieved, because opportunities for reduced fungicide application might arise. Despite this, most disease forecast models have been developed empirically from observations of rainfall and disease. Generally, the resultant systems provide reasonably accurate estimates of disease risk. However, the disregard of other elements of the disease triangle must inevitably reduce the precision of risk estimates.

---

[*] Present address: ADAS High Mowthorpe.

The aim of this chapter is to demonstrate some approaches being used to improve understanding of *S. tritici* epidemiology, so that robust estimates of disease risk can be provided. For convenience we consider the epidemiology over two distinct phases of crop growth. First, during the vegetative period, which we loosely describe as the winter epidemic. Second, the summer epidemic, defined as disease progress occurring after growth stage GSZ 30 (Zadoks *et al.*, 1974). In addition, the quantification of resistance for use in disease management systems is considered.

The Winter Epidemic

Surveys made annually since 1970 by ADAS and CSL provide a unique record of the severity of *S. tritici.* Each year (except 1983 and 1984) 300-400 crops have been sampled at GSZ 73-75 and the severity of *S. tritici* estimated on 50 fertile tillers with the aid of a standard area key (Anon., 1976). Such large data sets are valuable for determining fluctuations in disease, as, for example, attempted by Royle *et al.* (1995). They are also useful for investigating how seasonal weather is correlated with disease progress (Coakley *et al.*, 1988a).

WINDOW PANE is a software package enabling exhaustive searches of large data sets for weather variables correlated with disease (Coakley *et al.*, 1982). The software has been used to provide empirical disease prediction models of yellow rust (*Puccinia striiformis* [Coakley *et al.*, 1982; 1988b]), *S. tritici* (Coakley *et al.*, 1985) and rice blast (*Magnaporthe grisea* [Calvero *et al.*, 1996]). The principal advantage of the software is that searches are not constrained by calendar date, e.g. monthly mean temperatures. Instead, the iterative algorithm allows meteorological data to be examined in segments of variable length. Variables that correlate highly with disease are then used in regression analyses to provide predictive models of disease intensity.

The WINDOW PANE software used in this investigation was an amended version of that used by Calvero *et al.* (1996). The principal amendment was to facilitate rapid selection of parameters constraining the searches of weather variables. In this study, for example, searches were made for the number of days that the minimum temperature fell below a range of values from -5°C up to 10°C from 1 October to 31 March.

Twenty observations were selected from the ADAS/CSL survey database for estimates of *S. tritici* severity on leaf 2 of cv. Avalon at GSZ 73/75. Only sites with a synoptic weather station within 5 miles were selected. Estimates of disease severity were logit-transformed before analyses. Linear regression models were fitted for logit-transformed disease severity against factors identified to be highly correlated with disease intensity.

Septoria tritici is universally acknowledged to be a wet weather disease. Therefore, forecast models invariably quantify rainfall by some means, generally as the cumulative rainfall within some defined period (e.g. Hansen *et al.*, 1994). In the UK during winter and early spring, rainfall conditions are unlikely to be limiting to epidemic development. Trends for increased disease severity with increased rainfall were detected for several quantifications of rainfall. For example, Fig. 6.1 shows the relationship between disease severity and the number of days when precipitation was above 0.2 mm during the period 1-22 November. However, all winter rainfall trends were of low predictive value.

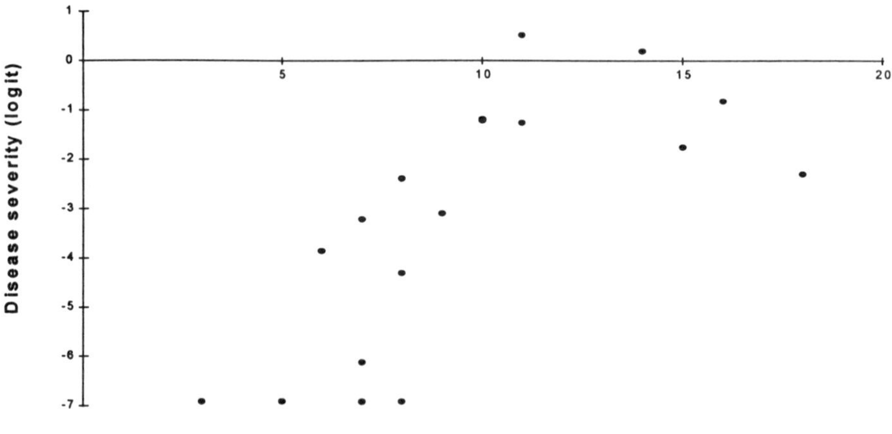

Fig. 6.1. Disease severity (logit) on cv. Avalon, leaf 2 at GSZ 75, in relation to the number of days that precipitation was greater than 0.2 mm between 1-22 November.

We found cold temperatures during winter were correlated strongly with reduced disease severity. Renfro and Young (1956) reported that infection by *S. tritici* failed to occur when temperature fell to 7°C or less for a 2 day period. Indeed, this threshold was used in a predictive model developed, using WINDOW software, from data collected at Lafayette, Indiana, USA (Coakley *et al.*, 1985). In the present study, 7°C was also negatively correlated with disease severity. However, the best model was provided by the frequency of days during the period 4 November-14 December when the minimum temperature fell below -2°C (Fig. 6.2). A large data set is presently being compiled from experiments designed specifically to investigate disease/weather relationships. This will comprise observations of *S. tritici* epidemics, with associated in-crop meteorological records, for some 30-site years and four cultivars variously susceptible to *S. tritici*. Preliminary analyses of these data suggest that the cold temperature correlation is valid, but less strong than for cv. Avalon.

In the UK, winter wheat crops are usually infected by *M. graminicola* soon after emergence (Shaw and Royle, 1989; Hunter *et al.*, Chapter 7 this volume). The inoculum pool for the summer epidemic is generally believed to be set during autumn and winter. Conditions that disrupt disease-cycling might, therefore, reduce disease intensity in the summer. Periods of cold temperature generally slow pathogen development. Thus, ascospore production and dispersal of *M. graminicola* might be slowed under the conditions identified. Similarly, latency of the asexual phase, *S. tritici,* can be prolonged substantially when temperatures fall below 5°C (Renfro and Young, 1956; T. Hunter, IACR-Long Ashton Research Station, 1997, personal communication). Alternatively, negative correlations, between cold winter temperatures and disease intensity in the summer, might arise because cold temperatures select isolates that are slower growing (Shaw, Chapter 5 this volume).

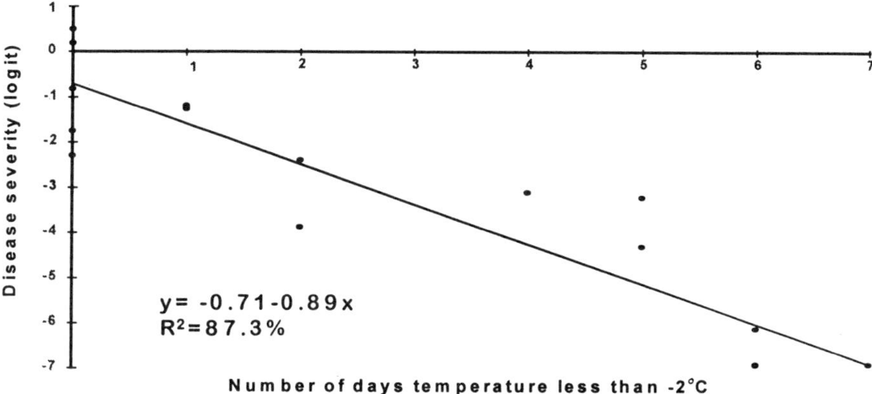

Fig. 6.2. Disease severity (logit) on cv. Avalon, leaf 2 at GSZ 75, in relation to the number of days that temperature was less than -2.0°C between 4 November-14 December.

The Summer Epidemic

During the mid 1980s, considerable progress was made towards mechanistic explanation of the progress of *S. tritici* epidemics. A key limit to epidemic development was shown to be dispersal of inoculum by rain splash from leaves at the base of the canopy to leaves 3-1 (flag leaf) (Shaw, 1987; Shaw and Royle, 1989; 1993). This knowledge provided the foundation for a forecast system to rationalize fungicide use, by recommending application only after critical rainfall events (Shaw and Royle, 1986; Shaw, 1987; Royle, 1990). Empirical correlations between rainfall during spring and disease intensity later in the season (Polley and Thomas, 1991; Hansen *et al.*, 1994) appeared to support the mechanistic description of disease progress.

Royle (1994) pointed out that, for many pathosystems, crop growth and structure might be an important influence on disease progress. Indeed, this possibility was acknowledged by Shaw and Royle (1989) for *S. tritici,* but they provided no corroborative data. Some quantitative evidence for the role of crop structure in *S. tritici* development is provided in the literature (e.g. Tavella, 1978; Bahat *et al.*, 1980), but this information has been relatively underdeveloped. In a recent report, however, Lovell *et al.* (1997a) demonstrated the substantial influence of crop growth and structure on disease progress.

Interpretation of disease intensity alone provides little clue to the role of crop structure in defining risk. Consider, for example, Fig. 6.3 which shows scale diagrams of the crop structure of three contrasting cultivars at GSZ 37. For all three, spore quantities decreased with increasing height in the canopy. The mean disease incidence on leaves 4 and 5 was almost identical on cvs Riband and Avalon. However, in comparison to cv. Avalon, the flag leaf of cv. Riband emerged much closer to these leaves. Therefore, whereas the flag leaf of cv. Avalon emerged into an area free of inoculum, the concentration of inoculum above the emerging flag leaf of cv. Riband was large (4.89×10^4 spores ml^{-1}). Moreover, due to the relative positions of leaf 5, there were sixfold more spores in the 10 cm of crop immediately below the emergence point of the flag leaf in cv. Riband compared with cv. Avalon. Thus, the amount of inoculum likely to reach the flag leaf of cv. Riband was substantially greater than for cv. Avalon. The spatial pattern of inoculum cannot, therefore, be explained by simply noting disease intensity on indicative leaf layers. Instead, some record must be made of the proximity of lesions to the emerging leaves. This adds weight to doubts concerning the usefulness of disease threshold values for *S. tritici* management (Parker *et al.*, 1997).

Studies by IACR-Long Ashton Research Station and ADAS have shown consistently that, due to crop growth and structure, cv. Avalon is able to escape disease. This may explain the strong correlations with winter temperature described earlier, because it suggests that disease progress will be more dependent on inoculum spread upwards from the basal leaves. Leaves can act as a source of inoculum only after they have been emerged for at least one latent period. Therefore, in cultivars with a prolonged latent period, there may be greater potential for emerging leaves to escape sources of inoculum. For example, Lovell *et al.* (1997a) demonstrated how, for cv. Apollo, the risk of inoculum transfer between leaf 4 and the emerging flag leaf was low in comparison with that for the susceptible cultivar Riband. They assumed a latent period of 21 days for cv. Riband and 27 days for cv. Apollo. However, recent studies suggest that these estimates of latency may be inappropriate. During the period of stem extension there appear to be only small differences in latency between cultivars (T. Hunter, IACR-Long Ashton Research Station, 1996, personal communication). A more significant effect of partial resistance, therefore, might be in the reduction of inoculum present close to emerging leaves (e.g. cv. Apollo, Fig. 6.3) resulting from the greater distance from sources of inoculum on leaves 4 and 5.

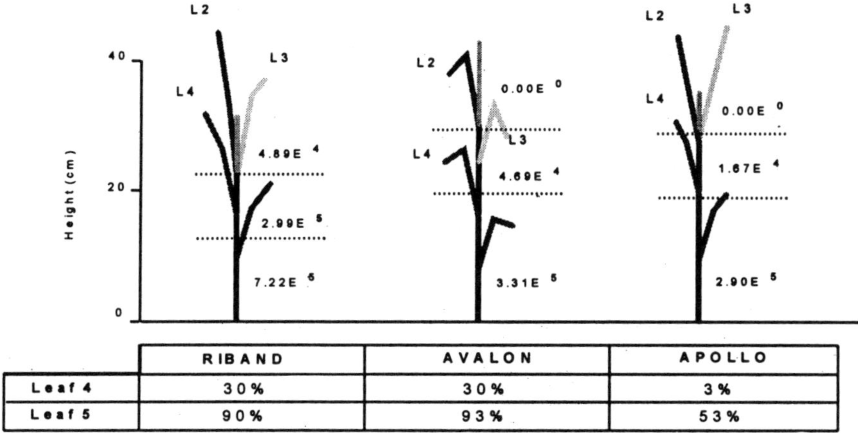

	RIBAND	AVALON	APOLLO
Leaf 4	30%	30%	3%
Leaf 5	90%	93%	53%

Fig. 6.3. Spore quantities in three horizontal sections of crop canopy, and percent incidence of disease on leaves 4 and 5, for three cultivars of winter wheat (from Lovell *et al.*, 1997a).

Quantification of inoculum location, as described above, requires intensive and time-consuming crop monitoring (Lovell *et al.*, 1997b). Such methods are practicable only in the context of fundamental research. Simplified and rapid methods are required, for example, in cultivar screening or in crop monitoring for disease management systems. The critical information is the position of inoculum in relation to the target leaves. Figure 6.4 shows the change in lesion proximity over time for cv. Chianti. The time axis represents the period for emergence of leaves 2 and 1 (flag). Negative distances indicate that the emerging leaf was above the nearest inoculum. Positive gradients occur when disease spreads up the canopy more rapidly than emerging leaves are moved upwards by stem extension. Negative gradients, therefore, indicate periods of disease escape. Risk was evidently greatest in 1996 when inoculum was much closer to the emerging leaves than in 1995, indicating that environment, as well as host genotype, might have a considerable influence on risk.

The number of spores moved vertically by rain splash declines exponentially with distance (Shaw, 1987; 1991). From the function defining this model, relative risk of inoculum transfer can be calculated for any distance of separation between inoculum and the target leaf (Fig. 6.5). Risk is unity when the distance between inoculum and target leaf equals the splash height.

Under identical conditions, the risk of inoculum transfer can differ substantially between crops. For example, Fig. 6.6 shows the variation in risk from a theoretical 20 cm splash distance for a range of cultivars in two seasons.

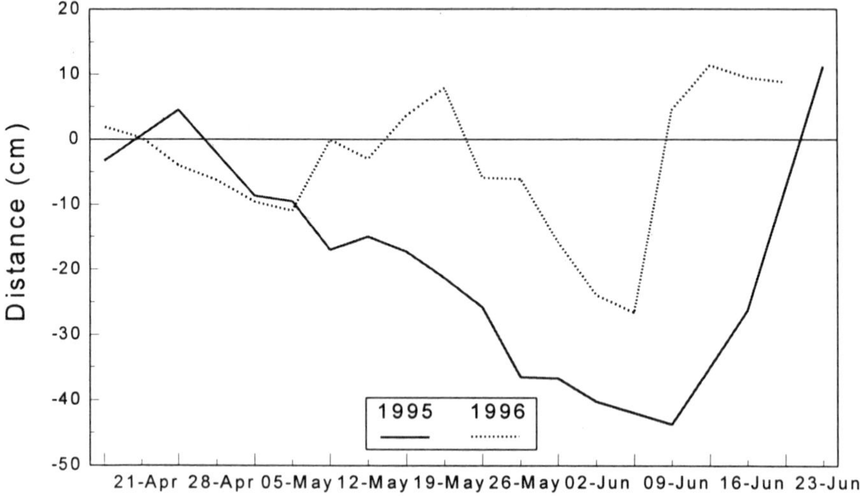

Fig. 6.4. Mean distance between emerging leaf and highest lesion in cv. Chianti, 1995 and 1996 (from Lovell *et al.*, 1997b).

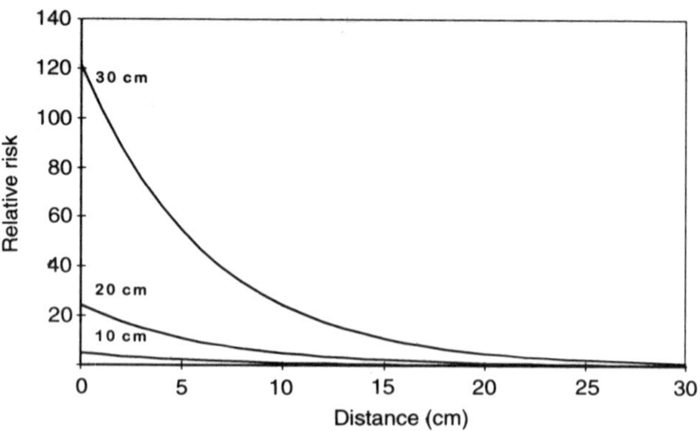

Fig. 6.5. Relative risk of inoculum transfer to the flag leaf for three splash heights (10, 20 and 30 cm). Risk is scaled around unity where splash height and the distance between inoculum and the target leaf is equal.

Shaw (1991) reported that the amount of rainfall (mm) correlates only poorly with splash distances. Although a simple monitoring device has been described for measuring splash thresholds (Shaw, 1987), this is unlikely to be adopted routinely for crop monitoring. A more sophisticated electronic device is being developed in Poland that will quantify the splash properties of rainfall more precisely. However, in the shorter term, this technology might best be applied to defining functions to estimate splash distances from less sophisticated rainfall measurements.

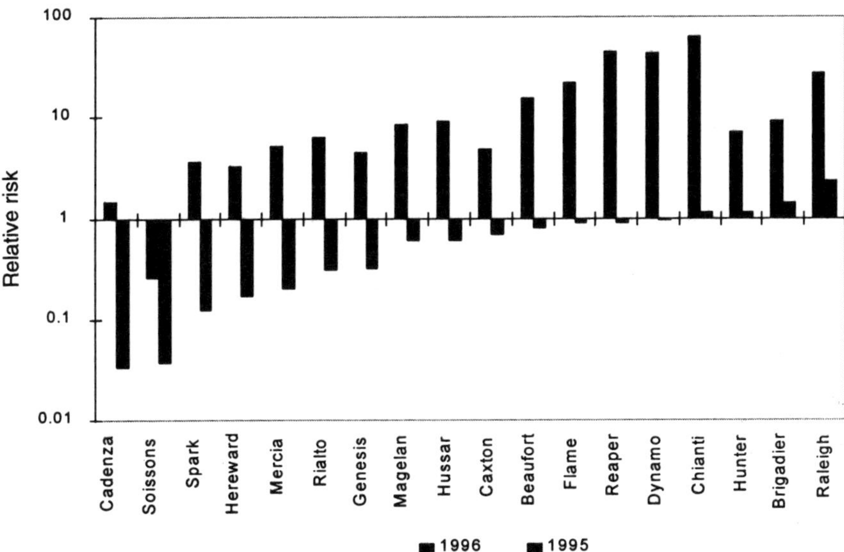

Fig. 6.6. Relative risk of inoculum transfer to the flag leaf for a 20 cm splash event at GSZ 37 (16-19 May) in 1995 and 1996. The cultivars are variously susceptible to *Septoria tritici.*

With improved quantification of rainfall, the estimates of risk described above could presently provide indications of when fungicide application was unnecessary, namely when conditions were unsuitable for inoculum movement to the target leaf layer. However, the estimate of risk accounts only for the likelihood of inoculum transfer between diseased and healthy portions of the crop. No account is taken of the quantity or quality of inoculum arriving at the target leaves. Genetic resistance will probably influence both of these factors. Wainshilbaum and Lipps (1991) demonstrated small, but significant, differences in pycnidial production (mm^{-2} of lesion) between two cultivars with different levels of *S. tritici* resistance. Potentially more important, however, was the threefold lower conidiospore production per pycnidium on the more resistant cultivar. This suggests that disease severity estimates, particularly given their suspect reliability (Parker *et al.*, 1995), will not quantify inoculum production effectively. Assuming better quantification of inoculum production is possible, more information on the responses of cultivars to inoculum density is required, so that reliable estimates of future disease progress can be given.

Accounting for Resistance

Disease resistance provides an obvious route by which crop losses to disease can be reduced. Indeed there is a strong link between the decrease in *S. tritici* resistance of the national winter wheat area and the upsurge of disease caused by this pathogen (Bayles, 1991).

Fungicide treatments are targeted towards protection of specific leaf layers. Accordingly, therefore, estimates of disease risk during the summer epidemic must be leaf-layer specific. The risk of inoculum transfer to individual leaf layers can now be estimated with greater accuracy than previously. However, the consequences of infection are dependent to a large extent on the effect of resistance. Decision support models will, therefore, require accurate estimates of the magnitude of the resistance effect.

In the UK, estimates of relative cultivar performance are produced by the National Institute of Agricultural Botany (NIAB) and updated annually as the UK Recommended List (Anon., 1997). Disease resistance ratings are calculated from estimates of disease in naturally-infected field experiments throughout the UK and in inoculated tests at NIAB headquarters, Cambridge. Cultivars are scored on a 1(susceptible)- 9(resistant) scale for each disease. In recent years, the cultivars most resistant to *S. tritici* have been rated 7 (e.g. cv. Spark). Ratings of 6-7 are considered moderately resistant, so that, whilst disease might develop under favourable conditions, yield loss is unlikely to be serious (Anon., 1997).

Studies to compare resistance effects over a large number of cultivars have indicated that resistance ratings may be poor predictors of disease intensity on individual leaf layers. For example, at a site in Devon in 1996, disease progress on leaf 3 of cv. Estica, rated moderately resistant (7), was similar to that on cv. Avalon which is rated susceptible (4) (Fig. 6.7). Similarly, disease progress on the susceptible (5) cultivar Rialto was within the range of disease progress observed on the moderately resistant cultivars (Fig. 6.7).

Current cultivar evaluation procedures do not distinguish between the contribution of disease escape mechanisms and true genetic resistance. Traditionally, resistance ratings have been used to provide general guidance about the likely field performance of cultivars. However, sophisticated decision support models may need to separate the effects of resistance and escape, not least because the expression of disease escape is strongly dependent on environmental conditions.

Cultivar screening is an expensive and time-consuming process. The information provided by present methodologies is appropriate for disease management approaches used currently. However, more precise timing of fungicides and adjustments of dose require accurate knowledge of the effects of resistance on pathogen latency and the severity of symptom expression. An immediate challenge, therefore, is to design inexpensive screening protocols that quantify and separate the effects of disease escape and genetic resistance over the range of UK conditions.

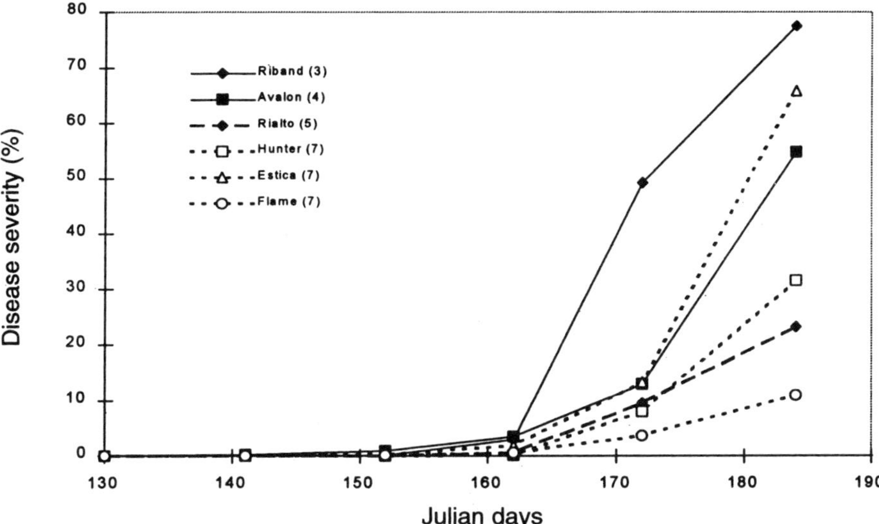

Fig. 6.7. Disease progress on leaf 3 for five cultivars variously susceptible to *Septoria tritici* (numbers in parentheses on the figure key are NIAB resistance ratings, 1=susceptible, 9=fully resistant).

Conclusions

Crude estimates of seasonal risk can be derived from knowledge of winter temperatures. These estimates may be of strategic value, in alerting growers to be vigilant in crop monitoring after a mild winter. They are inappropriate, however, for tactical planning of fungicide applications. Instead, accurate estimates of likely disease progress are needed during the period from the start of stem extension (GSZ 31) to ear emergence (GSZ 59). Estimates of the spread of inoculum around the canopy are now possible, and these have the potential to better inform treatment decisions. Additional improvements in precision could be possible if the consequences of resistance to disease progression, at the scale of individual leaf layers, were more accurately quantified.

Acknowledgements

The work was funded by the UK Ministry of Agriculture, Fisheries and Food. IACR receives grant-aided support from the Biotechnology and Biological Sciences Research Council of the UK.

References

Anon. (1976) *Manual of Plant Growth Stages and Disease Assessment Keys.* Central Science Laboratory, MAFF, Harpenden.

Anon. (1997) *NIAB Recommended List of Cereals.* National Institute of Agricultural Botany, Cambridge, UK.

Bahat, A., Gelernter, I., Brown, M.B. and Eyal, Z. (1980) Factors affecting the vertical progression of *Septoria* leaf blotch in short statured wheats. *Phytopathology* 70, 179-184.

Bayles, R.A. (1991) Varietal resistance as a factor contributing to the increased importance of *Septoria tritici* Rob. and Desm. in the UK wheat crop. *Plant Varieties and Seeds* 4, 177-183.

Calvero Jr, S.B., Coakley, S.M. and Teng, P.S. (1996) Development of empirical forecasting models for rice blast based on weather factors. *Plant Pathology* 45, 667-678.

Coakley, S.M., Boyd, W.S. and Line, R.F. (1982) Statistical models for predicting stripe rust on winter wheat in the Pacific Northwest. *Phytopathology* 72, 1539-1542.

Coakley, S.M., McDaniel, L.R. and Shaner, G. (1985) Model for predicting severity of *Septoria tritici* blotch on winter wheat. *Phytopathology* 75, 1245-1251.

Coakley, S.M., Line, R.F. and McDaniel, L.R. (1988a) Predicting stripe rust severity on winter wheat using an improved method for analyzing meteorological and rust data. *Phytopathology* 78, 543-550.

Coakley, S.M., McDaniel, L.R. and Line, R.F. (1988b) Quantifying how climatic factors affect variation in plant disease severity: a general method using a new way to analyze meteorological data. *Climatic Change* 12, 157-175.

Hansen, J.G., Secher, B.J.M., Jørgensen, L.N. and Welling, B. (1994) Thresholds for control of *Septoria* spp. in winter wheat based on precipitation and growth stage. *Plant Pathology* 43, 183-189.

Lovell, D.J., Parker, S.R., Hunter, T., Royle, D.J. and Coker, R.C. (1997a) The influence of crop growth and structure on the risk of epidemics by *Mycosphaerella graminicola* (*Septoria tritici*) in winter wheat. *Plant Pathology* 46, 126-138.

Lovell, D.J., Parker, S.R., Hunter, T., Royle, D.J. and Flind, A.C. (1997b) A novel method of monitoring risk from *Septoria tritici* leaf blotch in winter wheat. *Aspects of Applied Biology 48, Optimising Pesticide Applications,* 151-154.

Parker, S.R., Shaw, M.W. and Royle, D.J. (1995) The reliability of visual estimates of disease severity on cereal leaves. *Plant Pathology* 44, 856-864.

Parker, S.R., Lovell, D.J. and Royle, D.J. (1997) The importance of accurate risk prediction for reliable control of *Septoria tritici* leaf blotch in winter wheat. *Aspects of Applied Biology 48, Optimising Pesticide Applications,* 143-150.

Polley, R.W. and Thomas, M.R. (1991) Surveys of diseases of winter wheat in England and Wales 1976-1988. *Annals of Applied Biology* 119, 1-20.

Renfro, B.L. and Young, H.C. (1956) Techniques for studying varietal response to *Septoria* leaf blotch of wheat. *Phytopathology* 46, 23-24.

Royle, D.J. (1990) Advances towards integrated control of cereal diseases. *Pesticide Outlook* 1, 20-25.

Royle, D.J. (1994) Understanding and predicting epidemics: a commentary based on selected pathosystems. *Plant Pathology* 43, 777-789.

Royle, D.J., Parker, S.R., Lovell, D.J. and Hunter, T. (1995) Interpreting trends and risks for the control of *Septoria* in winter wheat. In: Hewitt, H.G., Tyson, D., Hollomon, D.W., Smith, J.M., Davis, W. and Dixon K.R. (eds) *A Vital Role for Fungicides in Cereal Production.* BIOS Scientific Publishers, Oxford, pp. 105-115.

Shaw, M.W. (1987) Assessment of upward movement of rainsplash using a fluorescent tracer method and its application to the epidemiology of cereal pathogens. *Plant Pathology* 36, 201-213.

Shaw, M.W. (1991) Variation in the height to which tracer is moved by splash during natural summer rain in the UK. *Agricultural and Forest Meteorology* 55, 1-14.

Shaw, M.W. and Royle, D.J. (1986) Saving *Septoria* fungicide sprays: the use of disease forecasts. *Proceedings of the 1986 British Crop Protection Conference - Pests and Diseases* 3, 1193-1200.

Shaw, M.W. and Royle, D.J. (1989) Airborne inoculum as a major source of *Septoria tritici* (*Mycosphaerella graminicola*) infections in winter wheat in the UK. *Plant Pathology* 38, 35-43.

Shaw, M.W. and Royle, D.J. (1993) Factors determining the severity of epidemics of *Mycosphaerella graminicola* (*Septoria tritici*) on winter wheat in the UK. *Plant Pathology* 42, 882-899.

Tavella, C.M. (1978) Date of heading and plant height of wheat varieties, as related to Septoria leaf blotch damage. *Euphytica* 27, 577-580.

Wainshilbaum, S.J. and Lipps, P.E. (1991) Effect of temperature and growth stage on development of leaf glume blotch caused by *Septoria tritici* and *S. nodorum*. *Plant Disease* 75, 993-998.

Zadoks, J.C., Chang, T.T. and Konzak, C.F. (1974) A decimal code for the growth stages of cereals. *Weed Research* 14, 415-421.

Chapter seven:

Studies on the Sexual Phase of Leaf Blotch in UK Winter Wheat

T. Hunter, R.R. Coker and D.J. Royle
IACR-Long Ashton Research Station, Department of Agricultural Sciences, University of Bristol, Long Ashton, Bristol BS41 9AF, UK

Introduction

Mycosphaerella graminicola (Fuckel) Schroeter (anamorph, *Septoria tritici* Rob. Ex Desm.), the pathogen of leaf blotch of wheat, is a disease of global importance. It often causes severe crop damage during summer which may lead to substantial yield losses (King *et al.*, 1983; Thomas *et al.*, 1989). These effects are attributable to repeated cycles of the asexual stage of the fungus, in which pycnidia give rise to splash-dispersed pycnidiospores, which eventually infect the upper leaves on whose photosynthetic activity crop yield is dependent. The sexual stage is also known to play a role in the disease cycle. It has been reported from various parts of the world, having first been described in New Zealand by Sanderson (1972) and in the UK by Scott *et al.* (1988). In the UK, ascospores of the sexual stage cause most of the routine, initial infection of winter wheat crops during the autumn (Shaw and Royle, 1989), and in the USA, ascospores are regarded as the source of primary inoculum (Schuh, 1990).

Following stem-extension, infection of the upper leaves of a crop has been thought to be due entirely to pycnidiospores, which are splash-dispersed from infected basal tissue by heavy rains. Indeed, disease forecasting methods have been developed on this premise and shown to be useful in timing fungicide applications (Shaw *et al.*, 1986; Shaw and Royle, 1993). More recent work, however, has shown that upward movement of inoculum can occur in the absence of splashy rainfall, being influenced by the position of developing leaves in relation to infected leaf layers (Lovell *et al.*, 1997). A third possible means of spread within a crop during summer is by airborne ascospores, which may play a more important role than previously recognized. This notion is based on: (i) the results of long-term ascospore trapping studies at IACR-Long Ashton Research Station, (ii) the demonstration in North America of large variations in the *M. graminicola* population and of the genetic structure being dominated by outcrossing (Chen and McDonald, 1996), and (iii) suggestions by Kema *et al.* (1996) that *S. tritici* may be able to produce several sexual cycles and, hence, ascospore generations, during a growing season.

The present work briefly describes methods and observations at IACR-Long Ashton Research Station on the occurrence of the sexual stage of *M. graminicola* in wheat crops, with the object of evaluating its role in the epidemiology of the disease during the main period of seasonal crop growth.

Materials and Methods

Ascospore Trapping

The dissemination patterns of *M. graminicola* from October 1991-April 1993 and October 1995-July 1997 were determined using Burkard volumetric spore traps set 0.3 m above ground level. Traps were operated at IACR-Long Ashton Research Station close to areas sown with winter wheat. Daily mean ascospore numbers were recorded. Ascospores were authenticated as *M. graminicola* by inverting the trap tape surface (from each 24 hour collection period) on to 1.2% water agar in Petri dishes and incubating at 17°C for 24-48 hours. Only ascospores (10-18 μm long and 3-4.5 μm wide) which had the distinctive germination pattern associated with *M. graminicola* (producing budding conidia) were included in spore counts.

Development Following Crop Harvest

The development of ascocarps of *M. graminicola* in leaf remains was monitored from July until the following March over two seasons, 1995-1996 and 1996-1997. Samples of plants from trials of unsprayed wheat cv. Riband, known to be previously infected, were collected just before harvest in July. They were suspended in bundles on metal frames, where they were exposed to ambient conditions. At that time, leaves 1 (flag)-5 were intact on all plant stems. Samples of these leaves were then taken at monthly intervals and recorded for the presence of mature ascocarps. For observations of their development, fruiting bodies were teased from leaf tissue using fine dissecting needles, placed on a microscope slide and stained with 0.25% Trypan Blue in lactic acid-glycerol-water (1:1:1). They were then examined with a light microscope (x400).

Production of Ascocarps in Developing Crop

Ascocarp development was monitored weekly during 1997 in trial plots of unsprayed winter wheat cv. Riband, from late April (GSZ 32) to early July (GSZ 85). Plants with leaves visibly infected with *S. tritici* were collected and returned to the laboratory for examination.

Temporal Development

Between October 1996 and July 1997, pot-grown seedlings (five per pot) of winter wheat cv. Longbow, which had been sown at weekly intervals and grown on in an unheated glasshouse, were inoculated at the three-leaf, fully-emerged stage with 5 ml suspension of *S. tritici* conidia (5 x 10^4). They were then transferred to an external standing ground and watered by drip-feed irrigation, and then examined at 2-3 day intervals for the development of symptoms of asexual and sexual fruiting bodies.

Results

Ascospore Catches

Between 1990 and 1993, the major periods of ascospore release occurred in autumn/winter (early October to December), with a secondary peak in early summer (June to July). More recently, in 1995-1997, there has been a trend for ascospores to be released throughout the year. Catches showed that ascospore numbers increased progressively during the period in summer when the wheat crop was actively growing. In the 1995-1996 season, there was an extended cold period during December to February, when there were ground frosts, and consequently, very few ascospores were detected. This contrasted with a relatively warmer 1996-1997, when ascospores were present throughout the winter months. In both of these seasons, ascospores were detected from the emergence of the upper leaves (GSZ 32) until harvest.

Natural Development of *Mycosphaerella graminicola* Following Harvest

In 1995-1996, no ascocarps were found until October in the remains of the upper five leaves of wheat plants which had been naturally infected by *S. tritici.* In October, mature ascocarps were detected on leaf 3 and immature ascocarps on both leaf 2 and the flag leaf. Mature ascocarps were first found on the flag leaf in November. Maximum numbers of ascocarps occurred in December, when there were least numbers of pycnidia. Subsequently, saprophytic colonization of the leaf tissues meant that fewer *M. graminicola* ascocarps were detected.

In 1996-1997, ascocarps were first observed on the earliest sampling date, in August, when mature fruiting bodies were detected on leaf 4; only immature ascocarps and pycnidia were found on leaves 3, 2 and the flag. By October, fully developed ascocarps were present on all leaf layers.

The estimated time from the first observed symptom of *S. tritici* in the upper leaf layers of the crop to first detection of ascocarps in the sampled plants was approximately 140 days in 1995-1996, but only 94 days in 1996-1997.

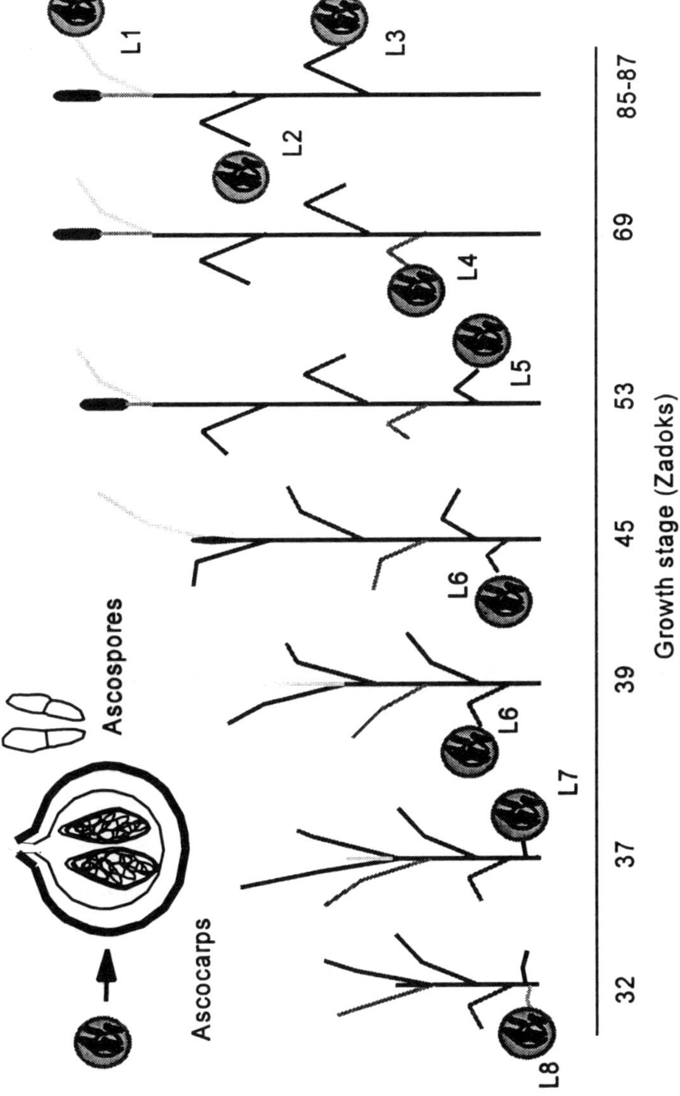

Fig. 7.1. Development of *Mycosphaerella graminicola* ascocarps in an unsprayed winter wheat cv. Riband crop in 1997. Uppermost leaf layer (L) bearing ascospores.

Ascocarps in the Developing Crop

Mature ascocarps were first found in the crop, on leaf 8, on 24 April 1997. As the season progressed, they were found progressively on emergent leaves until 15 July, when they appeared on the flag leaf (Fig. 7.1).

Ascocarp Production on Inoculated Plants

Inoculations of wheat leaves resulted in formation of typical *S. tritici* lesions with pycnidia after 44 days in winter and 14 days in summer. Pycnidial production was followed by the appearance of ascocarps, their maximum development time occurring, in winter, 132 days after inoculation and, in summer, after 84 days. The longest interval between pycnidium and ascocarp production was following inoculation in early January, when pycnidia appeared in early February and ascocarps in mid May (95 days). The shortest time (62 days) was after inoculation in early February.

Discussion

In studies of airborne populations of *M. graminicola* ascospores from wheat stubble and straw in Victoria, Australia, Brown *et al.* (1978) demonstrated a seasonal liberation, with greatest numbers during early growth of autumn-sown crops. They showed that, under favourable conditions, ascospores retain sufficient viability to survive aerial dispersal. The occurrence of the sexual stage of this pathogen in Texas (Garcia and Marshall, 1992) was also considered to be an important factor in the epidemiology of the disease in that region.

We have shown that ascospores can be released throughout the year in the UK. It was felt appropriate to confirm the authenticity of suspected *M. graminicola* ascospores on trap tapes, because of the reported colonization of wheat leaves and debris by a range of other fungi producing perithecia and pycnidia (Scott *et al.*, 1988). Furthermore, we have confirmed the observation of Kema *et al.* (1996) that ascospores do not originate exclusively from overwintering stubble or volunteer plants, as previously thought, but also from ascocarps produced on infected leaves of the current wheat crop. Data (unpublished) from 1993-1994 setaside stubble showed that ascocarps produced on old wheat leaves and stems were exhausted by mid winter and, therefore, could no longer contribute to infection of crops in the following spring and summer. To account for ascospore release over extended periods, as shown from trap collections, requires an additional source probably provided by ascocarps produced on infected volunteer wheat plants or perhaps, more importantly, on leaves of the developing crop. According to Kema *et al.* (1996), *M. graminicola* is able to complete a sexual cycle within 5 weeks under conducive conditions and, thus, could provide such a source. However, our results obtained from south-west England, do not support such rapid sexual development. Nevertheless, they do indicate that the sexual phase is possibly sufficiently active as the crop develops to complement the secondary infections caused by asexual pycnidiospores. Being dispersed in air currents rather than by splash, this could have important implications for risk-based decisions on disease control.

The precise conditions for ascospore infection have not yet been investigated, but Garcia and Marshall (1992) showed that infection by ascospores requires a longer incubation period and produces fewer pycnidia than infections resulting from pycnidiospores. Disease development and final severity may, therefore, be less from ascospores than from pycnidiospores. Using weekly batches of seedling wheat plants placed at different locations remote from obvious sources of inoculum, Shaw and Royle (1989) showed that *S. tritici* was 'trapped' throughout the autumn, and occasionally, during winter and spring. They attributed this infection to ascospores. Hunter (unpublished results) showed that plant trap seedlings, exposed similarly, could be infected throughout the year, although no correlation could be found between levels of infection and weekly numbers of ascospores caught in volumetric traps. However, different conditions are likely to be required for ascospore dispersal and infection. The development of ascocarps on previously infected leaves exposed outdoors varied considerably between the growing seasons 1995 and 1996, possibly due to weather conditions and different levels of disease (more severe in the unsprayed crop in 1996). In 1997, the sexual phase was already present on the flag leaf on unsprayed crops in July.

Our results show that the development of ascocarps throughout the season on leaf layers, which were becoming senescent at different rates, would serve as a threat to newly sown autumn crops over an extended range of sowing dates. Neither the relative importance of the infections on various leaf layers to overall ascospore production, nor the exact role of the sexual phase in disease development is understood. It is likely that the development of ascocarps during the growing season would contribute substantially to the genetic variation in the population structure of *M. graminicola* in the field. The continuous development of the sexual phase sustains genetic exchange in *M. graminicola* throughout the growing season and would increase opportunities for the pathogen to respond quickly to the selection pressures exerted either by the introduction of new wheat cultivars or from the application of fungicides.

Acknowledgements

IACR-Long Ashton Research Station receives grant-aided support from the Biotechnology and Biological Sciences Research Council of the United Kingdom. The authors thank the farm staff at IACR-Long Ashton Research Station for maintaining field trials and Mr D.J. Lovell for help in collating the ascospore trapping data.

References

Brown, J.S., Kellock, A.W. and Paddick, R.G. (1978) Distribution and dissemination of *M. graminicola* (Fuckel) Schroeter in relation to the epidemiology of speckled leaf blotch of wheat. *Australian Journal of Agricultural Research* 29, 1139-1145.

Chen, R.S. and McDonald, B.A. (1996) Sexual reproduction plays a major role in the genetic structure of populations of the fungus *Mycosphaerella graminicola. Genetics* 142, 1119-1127.

Garcia, C. and Marshall, D. (1992) Observations on the ascogenous stage of *Septoria tritici* in Texas. *Mycological Research* 96, 65-70.

Kema, G.H.J., Verstappen, E.C.P., Todorova, M. and Waalwijk, C. (1996) Successful crosses and molecular tetrad and progeny analyses demonstrate heterothallism in *Mycosphaerella graminicola. Current Genetics* 30, 251-258.

King, J.E., Cook, R.J. and Melville, S.C. (1983) A review of *Septoria* diseases of wheat and barley. *Annals of Applied Biology* 103, 345-373.

Lovell, D.J., Parker, S.R., Hunter, T., Royle, D.J. and Coker, R.R. (1997) Influence of crop growth on the risk of epidemics by *Mycosphaerella graminicola* (*Septoria tritici*) in winter wheat. *Plant Pathology* 46, 126-138.

Sanderson, F.R. (1972) A *Mycosphaerella* species as the ascogenous state of *Septoria tritici* Rob. and Desm. *New Zealand Journal of Agricultural Research* 21, 277-281.

Schuh, W. (1990) Influence of tillage systems on disease and spatial pattern of *Septoria* Leaf Blotch. *Phytopathology* 80, 1337-1340.

Scott, P.R., Sanderson, F.R. and Benedikz, P.W. (1988) Occurrence of *Mycosphaerella graminicola*, teleomorph of *Septoria tritici*, on wheat debris in the UK. *Plant Pathology* 37, 285-290.

Shaw, M.W. and Royle, D.J. (1989) Airborne inoculum as a major source of *Septoria tritici* (*Mycosphaerella graminicola*) infections in winter wheat crops in the UK. *Plant Pathology* 38, 35-43.

Shaw, M.W. and Royle, D.J. (1993) Factors determining the severity of epidemics of *Mycosphaerella graminicola* (*Septoria tritici*) on winter wheat in the UK. *Plant Pathology* 42, 882-889.

Shaw, M.W., Royle, D.J. and Cook, R.J. (1986) Patterns of development of *Septoria nodorum* and *S. tritici* in some winter wheat crops in Western Europe, 1981-83. *Plant Pathology* 35, 466-476.

Thomas, M.R., Cook, R.J. and King, J.E. (1989) Factors affecting the development of *Septoria tritici* in winter wheat and its effect on yield. *Plant Pathology* 38, 246-257.

Chapter eight:

Population Genetics and Host Resistance

C.C. Mundt, M.E. Hoffer, H.U. Ahmed*, S.M. Coakley, J.A. DiLeone and C. Cowger

Department of Botany and Plant Pathology, 2082 Cordley Hall, Oregon State University, Corvallis, OR 97331-2902, USA

Introduction

Terminology

We will use Van der Plank's (1968; 1978) classification of pathogenicity (the amount of disease caused by a genotype or population of a pathogen) divided into the subcomponents of virulence and aggressiveness. Virulence will be considered as genetic variation among pathogen isolates or populations that interact differentially with host genotypes, while aggressiveness will be considered as genetic variation among pathogen isolates or populations that do not interact differentially with host genotypes. Such a classification makes no assumptions concerning the degree of expression or the number of genes controlling either virulence or aggressiveness (Van der Plank, 1978).

Analysis of Plant Pathogenic Variation

The analysis of plant pathogenic variation is itself an evolving process. Stakman and Levine (1922) were the first to describe pathogenic variation in a plant pathogen, and developed a concept of physiological races that was based on reactions of a series of wheat (*Triticum aestivum*) genotypes to isolates of *Puccinia graminis* f. sp. *tritici*. Later, Flor (1956) put plant host-parasite interactions on a genetic basis with his development of the gene-for-gene hypothesis in the flax (*Linum usitatissimum*)/rust (*Melampsora lini*) system. The subsequent application of the gene-for-gene concept to other diseases led to calls for the abandonment of the physiological race concept in favour of analyses of avirulence/virulence allele frequencies at the population level (Day,

*Present address: Plant Pathology Division, Bangladesh Rice Research Institute, Joydepur, Gazipur 1701, Bangladesh.

1974; Wolfe and Schwarzbach, 1975; Van der Plank, 1982). Further demonstrations of intraracial, quantitative variation for pathogenicity have led to suggestions that physiological race is not an adequate category of classification (e.g. Michelmore *et al.*, 1984).

Recently, molecular techniques have proved highly useful in studying pathogenicity, as they allow for an analysis of DNA that is not under the direct influence of host selection. For example, the Philippines population of *Xanthomonas oryzae* pv. *oryzae*, causal agent of bacterial blight of rice, was divided into four to five distinct clonal lineages based on RFLP analyses. Associations between lineage and virulence led to useful inferences concerning the evolutionary pathways of physiological races (Leach *et al.*, 1992; Leung *et al.*, 1993; Nelson *et al.*, 1994). As with physiological races, however, one cannot assume lack of variation within a lineage. Some lineages contain more than one race, and a given race can arise convergently from different lineages (Levy *et al.*, 1991; 1993; Leach *et al.*, 1992; Xia *et al.*, 1993). Further, evolutionary lineages are often based on a degree of similarity of electrophoretic bands of less than 100%. Thus, there may be considerable potential within a lineage for relevant pathogenic variation, either qualitative or quantitative. It is probably safe, therefore, to assume that molecular analyses are critically important for determining the evolutionary structure of a pathogen population, but that phenotypic analyses are also required to determine the amount of pathogenic variation within that structure (Mundt *et al.*, 1994).

Analysis of Quantitative Variation for Virulence

Several studies have been done to determine if pathogens can become quantitatively adapted to the genetic background of their host, i.e. if there is selection for increased pathogen virulence that is quantitatively expressed. Results have been contradictory, and may depend strongly on the starting population utilized. In some cases (Jeffrey *et al.*, 1962; Jinks and Grindle, 1963; Caten, 1974; Clifford and Clothier, 1974), isolates have been found to be more virulent to the cultivar they were isolated from in the field, but subsequent cycling of single isolates failed to show increases in virulence. Such results should be expected, as large field populations may harbour significant variation, while individual isolates are likely to be invariant, or at least highly uniform, for pathogenicity. Conversely, when Leonard (1969) cycled a bulk, heterogeneous population of *P. graminis* f. sp. *avenae* on different oat (*Avena sativa*) genotypes, adaptation to host genotype was found. Adaptation of individual isolates of *Erysiphe graminis* was detected after 30 cycles of asexual selection on two of three cultivars studied (Newton and McGurk, 1991) and on all three cultivars after 76 cycles (Newton, 1992). Thus, it may be possible to generate selectable pathogenic variation within isolates, perhaps through mutation, in such asexual cycling experiments. However, adaptation is more likely to be seen if genetically diverse starting populations representative of natural variation are utilized. Such host-specific adaptation may be important to the stability of quantitative host plant resistance to disease (Clifford and Clothier, 1974; Latin *et al.*, 1981; Newton, 1989).

In contrast to pure stands, such quantitative variation in pathogen populations could contribute to stability of host resistance when host diversification schemes, such as multilines and cultivar mixtures (Wolfe, 1985; Mundt, 1994), are used. For example, Chin and Wolfe (1984) found that a race of powdery mildew (*E. graminis*) virulent on two barley (*Hordeum vulgare*) cultivars grown in mixture had reduced infection efficiency as compared to populations of the same race obtained from pure stands of the same two cultivars. This result suggests that pathogen adaptation to its host resulted in a form of disruptive selection that prevented the complex race from obtaining high fitness on both host components in the mixture.

Genetic Control and Selection for Pathogen Aggressiveness

Less emphasis has been given to the genetic control and selection of aggressiveness in pathogen populations. Controlled crosses with several plant pathogens have shown that aggressiveness is heritable (Brasier, 1977; Hill and Nelson, 1982; Caten *et al.*, 1984; Kolmer and Leonard, 1986), and may be polygenically controlled (Brasier, 1977; Caten *et al.*, 1984). Mew *et al.* (1992) presented data indicating that populations of *X. oryzae* pv. *oryzae* in the Philippines may have increased substantially in aggressiveness from the 1970s to the 1980s. Anecdotal evidence suggests that the same may be true in Oregon for *Mycosphaerella graminicola* on wheat (C.C. Mundt, unpublished results). Kolmer and Leonard (1986) found that both virulence and aggressiveness of *Cochliobolus heterostrophus* to maize (*Zea mays*) could be increased through artificial selection over three sexual generations of the pathogen.

Mycosphaerella graminicola as a Model System for Studying Quantitative Variation in Pathogenicity

Biology of *Mycosphaerella graminicola* in the Willamette Valley of Oregon, USA

Mycosphaerella graminicola in the Willamette Valley of Oregon provides an ideal system for studying the selective influence of host genotype on pathogen populations. Conditions are optimum for Septoria tritici leaf blotch development, with frequent rains beginning in November and continuing through June. In recent years, naturally occurring epidemics have varied from moderate to severe, and commercial fields of susceptible cultivars are sprayed every year. In Oregon, *M. graminicola* predominates over *Phaeosphaeria nodorum*, causal agent of Septoria nodorum leaf blotch, probably because ascospore showers of *M. graminicola* occur much earlier than those of *P.*

nodorum (DiLeone *et al.*, 1997), resulting in competitive exclusion. Differential temperature optima for the two species may also play a role. Finally, epidemics of Septoria tritici leaf blotch occur over a period of more than 6 months, thus allowing a large number of pathogen generations for host selection to act upon.

As in some other parts of the world (Shaw and Royle, 1989; Kema *et al.*, 1996), Septoria tritici leaf blotch epidemics in Oregon begin from a very dominant sexual stage. Ascocarps are abundant on wheat straw in summer and autumn. Infections on trap plants, used to monitor ascospore infections in Oregon, often show a peak of ascospore-derived infections in November, after autumn rains have begun, but continue at some level throughout the crop season. Even in isolated fields never previously planted to wheat, every tiller will be infected by February, and infection patterns are remarkably uniform. Further, molecular studies indicate patterns of genetic variation expected from a sexual process (Boeger *et al.*, 1993; Chen *et al.*, 1994). A result of this predominance of sexual reproduction is that essentially each isolate collected from a plot is a different haplotype (Boeger *et al.*, 1993; Chen *et al.*, 1994), unless obtained in very close proximity.

There is considerable variation in susceptibility to *M. graminicola* among Pacific Northwest wheat cultivars. 'Complete' resistance against *M. graminicola* has been available in only one Pacific Northwest cultivar, but this resistance has eroded to susceptibility over the last several years (see below). Other cultivars, however, differ quantitatively in their resistance to *M. graminicola*. Plant height is often associated with resistance to *M. graminicola*, but physiological resistance also occurs separately from any physical effects of plant height (King *et al.*, 1983). In fact, resistance ratings of Pacific Northwest wheat cultivars rank the same in the field as in monocyclic greenhouse inoculations, where plant height can play no direct role in resistance.

Pathogen Adaptation to Host - Field Observations

The susceptible cultivar Stephens dominated the wheat area of the Willamette Valley during the 1980s. The popularity of cvs Gene and Madsen increased in the early and mid 1990s, almost solely due to their lesser susceptibility to Septoria tritici leaf blotch. Figure 8.1 represents an attempt to quantify changes in Septoria tritici leaf blotch on these two cultivars during that time, based on disease readings taken in experimental field plots. Data are expressed as disease severity relative to cv. Stephens, in an attempt to eliminate variation in favourableness for epidemics among years. Methodologies used to quantify disease severity varied somewhat among years, and data were sometimes collected from different sites in the Corvallis area over years. Further, in 1990, 1991, 1992 and 1997, no attempt was made to distinguish between disease caused by *P. nodorum* and by *M. graminicola*, though the latter was the dominant species present. In 1993-1996, the two pathogen species were

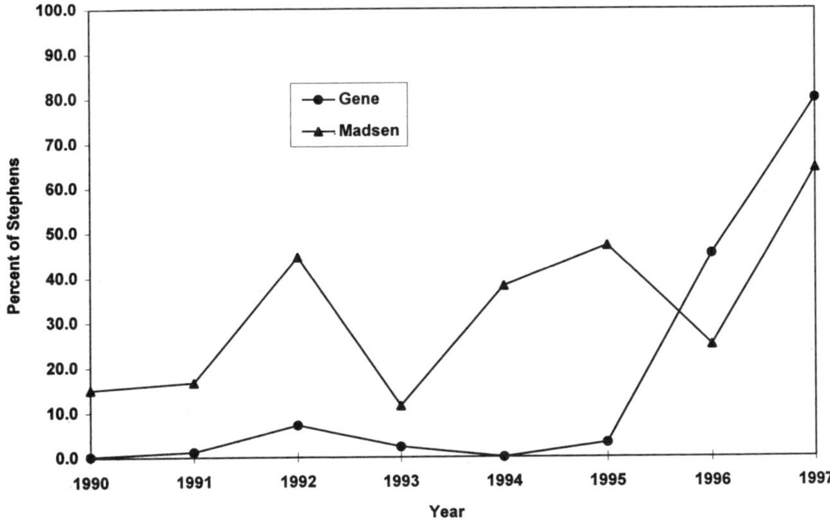

Fig. 8.1. Estimates of Septoria tritici leaf blotch severity for a highly resistant (Gene) and a moderately resistant (Madsen) wheat cultivar in the Willamette Valley of Oregon from 1990-1997. Data are expressed relative to the susceptible cultivar Stephens.

separated when making visual assessments. Thus, Fig. 8.1 provides only a very general picture of virulence dynamics during this time.

Figure 8.1 demonstrates clearly the 'breakdown' of resistance in cv. Gene that has been very obvious in both experimental plots and in commercial fields. The curve for cv. Gene looks no different from that observed many times for cultivars initially resistant to biotrophic pathogens such as rusts and mildews. For biotrophs, the breakdown of resistance is usually caused by an increased frequency of pathogen genotypes with a high degree of virulence on the formerly resistant host. For *M. graminicola* on cv. Gene, however, field observations have suggested a very different process of resistance breakdown that involved a gradual change in reaction type over years. In 1990, lesions found on cv. Gene were very infrequent and small, and pycnidia were rarely seen. In the subsequent few years, lesions increased both in size and in frequency and pycnidia appeared, although their density was less than on susceptible cultivars. This change of reaction type has gradually continued until the present, at which time there is no obvious difference in reaction type of *M. graminicola* on cv. Gene compared to susceptible cultivars. We are currently in the process of conducting studies to compare isolates collected from cv. Gene in different years for their disease severity and reaction type in a common environment in the greenhouse. In California, resistance to *M. graminicola* in spring wheat cv. Express appears to be breaking down in the same manner as described for cv. Gene in Oregon, though cv. Express has not yet become fully susceptible (L.F. Jackson, University of California at Davis, 1997, personal communication). It may be possible in the near future to determine the inheritance of such quantitative adaptation to host genotype in *M. graminicola*, as the first successful sexual crosses under experimental conditions have recently been reported for the pathogen (Kema *et al.*, 1996).

A second impact of the commercial use of cv. Gene was an increased incidence of *P. nodorum*. The initially high level of resistance of cv. Gene to *M. graminicola* provided an available niche for infection by *P. nodorum*, which disperses its ascospores later in the season (DiLeone *et al.*, 1997) and is usually competitively excluded by *M. graminicola*. Now that more virulent strains of *M. graminicola* have appeared, however, it seems that *P. nodorum* is again being replaced by *M. graminicola* on cv. Gene.

The curve for the moderately resistant cv. Madsen in Fig. 8.1 shows considerable variability among years, but with some suggestion of erosion of resistance over time. Given the diversity of methodology utilized and the variability in the data, however, it is not possible to draw any conclusions at this time regarding the stability of moderate resistance to *M. graminicola*. Clearly, this question deserves more attention in the future.

Pathogen Adaptation to Host - Evaluations of Populations Within a Region

Field and greenhouse studies with *M. graminicola* indicate that selection for both virulence and aggressiveness to wheat occurs within a single growing season in the field (Ahmed *et al.*, 1996). Bulk populations of isolates of *M. graminicola* collected from wheat cultivars in a replicated experiment at the end of the season, after selection had been operative, were inoculated on the same cultivars in the greenhouse. There were statistically significant increases in virulence of the populations on the cultivars from which they were obtained in the field (Table 8.1). More surprising, however, was the effect of the host on pathogen aggressiveness (mean level of disease averaged over all test cultivars). Populations collected from susceptible cultivars were significantly more aggressive (caused more disease averaged over testers) than populations obtained from resistant or moderately resistant cultivars (Table 8.1). Very similar results were obtained with populations collected from a second, smaller experiment that was conducted 2 years later and 110 km distant from the first (Table 8.2). Although questions could reasonably be raised regarding the relevance of greenhouse studies to the field, it is significant that the relative susceptibilities of cultivars that we measured in the greenhouse (Ahmed *et al.*,

Table 8.1. Area under disease progress curve (percent days) caused by populations of *Mycosphaerella graminicola* (20 isolates per population) collected from four winter wheat cultivars in the field and tested on the same wheat cultivars in the greenhouse. Data from Ahmed *et al.* (1996).

Tester cultivar	Pathogen population derived from cultivar				
	Gene	Madsen	Malcolm	Stephens	Mean[a]
Gene	2.6 **2.0**[b]	2.7 **1.5**	1.3 **3.0**	2.7 **2.7**	2.3d
Madsen	55.9 **52.7**	43.5 **39.5**	86.8 **77.3**	52.5 **69.2**	59.7c
Malcolm	100.5 **99.0**	75.6 **74.2**	149.2 **145.2**	123.0 **130.0**	112.1a
Stephens	79.0 **84.2**	56.5 **63.1**	111.5 **123.4**	134.0 **110.4**	95.3b
Mean[a]	59.5B	44.6C	87.2A	78.1A	

Mean of homologous[c] combinations = 82.3
Mean of heterologous[d] combinations = 62.3[e]

[a]Means followed by the same letter are not significantly different at P=0.05 according to Fisher's protected LSD.
[b]Bold figures are expected values under the assumption of no pathogen population x cultivar interaction, and were calculated as: expected disease = (pathogen population mean over all cultivars x cultivar mean over all pathogen/populations) grand mean.
[c]Homologous indicates disease reaction of *M. graminicola* populations on cultivars from which the isolates were originally obtained in the field.
[d]Heterologous indicates disease reactions of *M. graminicola* populations on the cultivars from which the isolates were not obtained in the field.
[e]Homologous vs. heterologous means are significantly different at P=0.0015 (linear contrast).

1996) are in close accord with those we have observed in the field (e.g. Mundt *et al.*, 1995).

Interestingly, molecular analyses of isolates collected from the same treatments over a 3-year period showed no evidence of selection either within or between seasons (Chen *et al.*, 1994). This is likely because the large minimum population size of *M. graminicola* and the strong role of sexual recombination resulted in pathogenicity alleles being randomly distributed over a large number of genotypes, thus precluding the possibility of detecting changes of frequency for any combination of selectively neutral RFLP markers (McDonald *et al.*, 1996).

A similar effect of host susceptibility on aggressiveness was found for *X. oryzae* pv. *oryzae* on rice (C.C. Mundt, M.R. Finckh, H.U. Ahmed and R.F.

Alfonso, unpublished results). In this case, 80 individual isolates of the pathogen collected from each of two cultivars in replicated field experiments at each of two locations were tested individually in the greenhouse. For both locations, the mean lesion length caused by isolates from a susceptible genotype was significantly longer than that produced by isolates collected from the cultivar with moderate resistance. Thus, similar results have been obtained both for a fungal pathogen that undergoes sexual recombination annually (Boeger *et al.*, 1993; Chen *et al.*, 1994; DiLeone *et al.*, 1997) and for a bacterial pathogen that is primarily clonal (Leach *et al.*, 1992; Nelson *et al.*, 1994).

Table 8.2. Percentage of leaf area covered by lesions of *Mycosphaerella graminicola* for pathogen populations (five isolates per population) collected from two winter wheat cultivars in the field and tested on the same wheat cultivars in the greenhouse. Data are from Ahmed *et al.* (1996).

Tester cultivar	Pathogen population derived from cultivar		
	Madsen	Stephens	Mean[a]
Madsen	12.6 **9.4**[b]	12.4 **15.6**	12.5[b]
Stephens	11.4 **14.6**	27.4 **24.2**	19.4[a]
Mean[a]	12.0A	19.9B	

Mean of homologous[c] combinations = 20.0
Mean of heterologous[d] combinations = 11.9[e]

Superscripts a-d are as for Table 8.1.
[e]Homologous vs. heterologous means are significantly different at P=0.0034 (linear contrast).

One explanation for the greater aggressiveness of pathogens collected from susceptible cultivars as compared to less susceptible ones is that the variance of reproduction rate (lesions produced per lesion per day) among pathogen genotypes, in either a sexually or an asexually reproducing population, differs when the pathogen reproduces on different host genotypes. As variances are commonly observed to be positively correlated with their corresponding means, and gain due to selection is directly proportional to genetic variance, one might expect greater gains due to selection for increased pathogen reproductive ability on susceptible cultivars than on resistant or moderately resistant ones. In fact, Falconer (1960, p. 296) postulated that the positive correlation between mean and variance may explain why there is often a greater rate of gain due to selection for quantitative traits in the upward than in the downward direction in reciprocal selection experiments.

The variance of pathogen reproductive rate (estimated by percent diseased leaf area) among four isolates of *M. graminicola* increased with increasing susceptibility of the tester host (Table 8.3). Percent diseased leaf area accounts for the fitness components of infection efficiency and lesion expansion, and is also highly correlated with pycnidial coverage in our experiments. Similarly, variance for lesion length among isolates of *X. oryzae* pv. *oryzae* was greater when tested on a susceptible genotype than when tested on a moderately resistant cultivar (C.C. Mundt, unpublished results). Carson (1987) reported a very strong and positive correlation between level of host resistance and variance for aggressiveness among isolates of *Pyrenophora tritici-repentis* when inoculated on wheat.

Table 8.3. Percentage severity of disease caused by four isolates of *Mycosphaerella graminicola* when inoculated separately on four wheat cultivars in the greenhouse (Hoffer and Mundt, unpublished results).

Cultivar	Field rating[a]	Greenhouse severity	
		Mean[b]	Standard deviation
Stephens	S	9.0	7.4
Malcolm	S	8.7	7.0
Madsen	MR	7.6	4.7
Gene	R	0.4	0.08

[a]S = susceptible; MR = moderately resistant; R = resistant.
[b]Mean of four pots with ten seedlings each.

It has previously been suggested that new, virulent genotypes of pathogens arise more frequently under conditions of rapid pathogen increase caused either by favourable weather (Groth, 1984) or by a susceptible cultivar (MacKenzie, 1979; Kiyosawa, 1989). However, the relationship between pathogen reproduction rate and response to selection for increased pathogenicity has not, to our knowledge, been addressed experimentally for either qualitative or quantitative pathogenicity traits.

An alternative explanation for our results with *M. graminicola* is that shorter pathogen generation times allowed for more generations of selection for aggressiveness on susceptible than on moderately resistant or resistant cultivars. However, analysis of our greenhouse data (C.C. Mundt, unpublished results) suggests little or no variation among cultivars for the latent period. Similarly,

Shaw (1990) found small differences in latent period among 14 cultivars when using controlled inoculations of plants in the open.

Pathogen Adaptation to Host - Evaluation of Populations Between Regions

A selective influence of the host also may result in differentiation of a pathogen among regions in which different cultivars are grown. Greenhouse evaluation of isolates from California and Oregon, and tested on Oregon cultivars, indicated that these two populations are very different for pathogenicity (Ahmed *et al.*, 1995). This is true despite the fact that analysis of isolates from the two regions showed that the populations were remarkably similar for selectively neutral RFLP markers (Boeger *et al.*, 1993). Populations of *M. graminicola* in California and Oregon possibly share a common origin, accounting for similarity based on selectively neutral markers. However, lack of significant overlap in cultivars grown in the two regions may have resulted in very different selective influences of the host for pathogenicity alleles.

Additional evidence for a host influence on pathogenicity in California vs. Oregon resulted from an evaluation of six isolates each of *M. graminicola* from California and Oregon when tested on two cultivars from both states (Ahmed *et al.*, 1995). This study showed that, on average, the isolates caused more disease on cultivars from their own state than on those from the other state. Although use of only two cultivars from each state is an obvious weakness of this study, the results provide some support of host genotype as an influence on genetic differences of the pathogen in the two states.

Host Mixtures

Relatively little work has been done with cultivar mixtures and multiline cultivars for the control of non-obligate pathogens. A 1:1 mixture of susceptible cv. Stephens with highly resistant cv. Gene resulted in a 35% reduction of Septoria leaf blotch at grain filling, averaged over three seasons (Mundt *et al.*, 1995). In contrast, a 1:1 mixture of cv. Stephens with moderately resistant cv. Madsen showed no disease reduction (Mundt *et al.*, 1995), and the same result has been found in replicated on-farm trials (C.C. Mundt, M.E. Hoffer and L.S. Brophy, unpublished results). These studies all involved data collected at only one time during the season. In contrast, a more recent study followed disease progression throughout the season on cvs Stephens and Madsen, and on their 1:1 mixture (Fig. 8.2). One set of plots received naturally occurring inoculum and a second set was artificially inoculated with ten isolates in November, so as to competitively exclude naturally occurring inoculum. Thus, the artificially inoculated plots had an extremely low level of diversity as compared to the naturally inoculated ones, where there are likely to be hundreds of genotypes per square metre (C.C. Mundt, unpublished results).

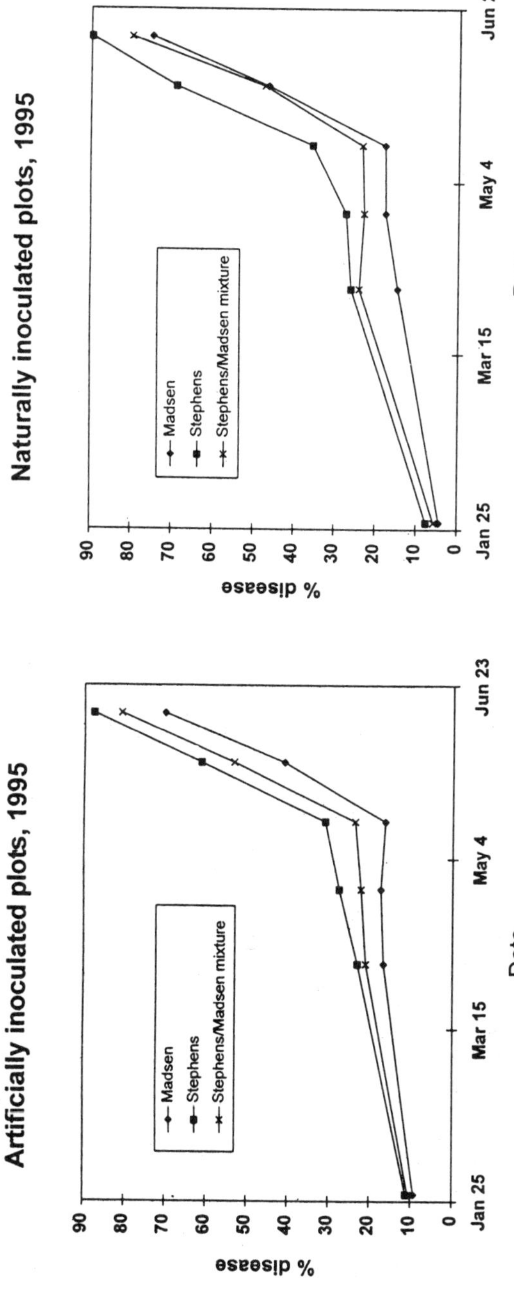

Fig. 8.2. Disease progress curves for Septoria tritici leaf blotch epidemics in plots of a susceptible (Stephens) and a moderately resistant (Madsen) wheat cultivar, and for a 1:1 mixture of Madsen:Stephens in 1995. Curves in the left-hand panel are for plots artificially inoculated with asexual spores of ten isolates of *Mycosphaerella graminicola*. Curves in the right-hand panel are for plots in which naturally occurring ascospores initiated epidemics.

In early season, no reduction in disease severity was evident for the mixture as compared to the pure stands for either inoculation treatment (Fig. 8.2). However, as the season progressed the epidemic in the mixture receiving natural inoculum became suppressed to the level of the most resistant host component. If *M. graminicola* adapts to a cultivar during the season, as suggested in Table 8.2, then disruptive selection may have reduced pathogen fitness in mixtures. Lack of genetic variability may have prevented such selection from occurring in the artificially inoculated plots. We have investigated this effect in detail for only a single location and season, and stronger conclusions await further experimentation. However, the study suggests that host/pathogen interactions occurring in host mixtures may be complex.

Summary and Conclusions

Quantitative variation for pathogenicity in *M. graminicola* is high and can be influenced strongly by host selection. Comparison of our field and greenhouse data with published molecular studies indicates that molecular and phenotypic analyses provide different and complementary information regarding the population genetics of the pathogen. Clearly, both approaches are critical to gaining a full understanding of the population biology of *M. graminicola.*

Mycosphaerella graminicola is able to adapt to qualitative resistance at the field level, and such resistance appears unstable when used in monoculture. Greenhouse analysis of field-collected isolates suggests that quantitative resistance may also erode over time. However, field observations suggest that such erosion, if it occurs, will be slower than with qualitative resistance. Host/pathogen interactions in cultivar mixtures are likely to be complex, and more research is needed to evaluate the potential for mixtures to control *M. graminicola.*

The culture of susceptible host genotypes may invite accelerated selection for pathogen aggressiveness over time. Such selection may have played a role in the Willamette Valley of Oregon, where Septoria tritici leaf blotch epidemics seem to have increased in severity over the years during which the wheat area was dominated by susceptible cv. Stephens. However, evidence for increased severity of the disease over time is still anecdotal, and it is not yet possible to dissociate impacts of host selection from other variables that may have influenced Septoria leaf blotch severity, such as build up of inoculum, increased use of fertilizer nitrogen, reduced height of modern cultivars, changes in crop rotation, etc.

References

Ahmed, H.U., Mundt, C.C. and Coakley, S.M. (1995) Host-pathogen relationship of geographically diverse isolates of *Septoria tritici* and wheat cultivars. *Plant Pathology* 44, 838-847.

Ahmed, H.U., Mundt, C.C., Hoffer, M.E. and Coakley, S.M. (1996) Selective influence of wheat cultivars on pathogenicity of *Mycosphaerella graminicola* (anamorph *Septoria tritici*). *Phytopathology* 86, 454-458.

Boeger, J.M., Chen, R.-S. and McDonald, B.A. (1993) Gene flow between geographic populations of *Mycosphaerella graminicola* (anamorph *Septoria tritici*) detected with restriction fragment length polymorphism markers. *Phytopathology* 83, 1148-1154.

Brasier, C.M. (1977) Inheritance of pathogenicity characters in *Ceratocystis ulmi*; hybridization of protoperithecial and non-aggressive strains. *Transactions of the British Mycological Society* 68, 45-52.

Carson, M.L. (1987) Assessment of six models of host-pathogen interaction in horizontal pathosystems. *Phytopathology* 77, 241-246.

Caten, C.E. (1974) Intra-racial variation in *Phytophthora infestans* and adaptation to field resistance for potato late blight. *Annals of Applied Biology* 77, 259-270.

Caten, C.E., Person, C., Groth, J.V. and Dhahi, S.J. (1984) The genetics of pathogenic aggressiveness in three dikaryons of *Ustilago hordei*. *Canadian Journal of Botany* 62, 1209-1219.

Chen, R.-S., Boeger, J.M. and McDonald, B.A. (1994) Genetic stability in a population of a plant pathogenic fungus over time. *Molecular Ecology* 3, 209-218.

Chin, K.M. and Wolfe, M.S. (1984) Selection on *Erysiphe graminis* in pure and mixed stands of barley. *Plant Pathology* 33, 535-546.

Clifford, B.C. and Clothier, R.B. (1974) Physiologic specialization of *Puccinia hordei* on barley hosts with nonhypersensitive resistance. *Transactions of the British Mycological Society* 63, 421-430.

Day, P.R. (1974) *Genetics of Host-Parasite Interactions.* W.H. Freeman & Co., San Francisco, 238pp.

DiLeone, J.A., Karow, R.S., Coakley, S.M. and Mundt, C.C. (1997) Biology and control of Septoria diseases of winter wheat in Western Oregon. *Oregon State University Extension Service Crop Science Bulletin* 109, 17pp.

Falconer, D.S. (1960) *Introduction To Quantitative Genetics.* Ronald Press, New York, 365pp.

Flor, H.H. (1956) The complementary genic systems in flax and flax rust. *Advances in Genetics* 8, 29-54.

Groth, J.V. (1984) Virulence frequency dynamics of cereal rust fungi. In: Bushnell, W.R. and Roelfs, A.P. (eds) *The Cereal Rusts*, Volume 1, Academic Press, Orlando, pp. 231-252.

Hill, J.P. and Nelson, R.R. (1982) The heritability of three parasitic fitness attributes of *Helminthosporium maydis* race T. *Phytopathology* 72, 525-528.

Jeffrey, S.I.B., Jinks, J.L. and Grindle, M. (1962) Intraracial variation in *Phytophthora infestans* and field resistance to potato late blight. *Genetica* 32, 323-328.

Jinks, J.L. and Grindle, M. (1963) Changes induced by training in *Phytophthora infestans*. *Heredity* 18, 245-264.

Kema, G.H.J., Vertappen, E.C.P., Todorova, M. and Waalwijk, C. (1996) Successful crosses and molecular tetrad and progeny analyses demonstrate heterothallism in *Mycosphaerella graminicola*. *Current Genetics* 30, 251-258.

King, J.E., Cook, R.J. and Melville, S.C. (1983) A review of Septoria diseases of wheat and barley. *Annals of Applied Biology* 103, 345-373.

Kiyosawa, S. (1989) Breakdown of blast resistance in relation to general strategies of resistance gene deployment to prolong effectiveness of resistance in plants. In: Leonard, K.J. and Fry, W.E. (eds) *Plant Disease Epidemiology*, Volume 2. McGraw-Hill, New York, pp. 251-283.

Kolmer, J.A. and Leonard, K.J. (1986) Genetic selection and adaptation of *Cochliobolus heterostrophus* to corn hosts with partial resistance. *Phytopathology* 76, 774-777.

Latin, R.X., MacKenzie, D.R. and Cole Jr, H. (1981) The influence of host and pathogen genotypes on the apparent infection rates of potato late blight epidemics. *Phytopathology* 71, 82-85.

Leach, J.E., Rhoads, M.L., Vera Cruz, C.M., White, F.F., Mew, T.W. and Leung, H. (1992) Assessment of genetic diversity and population structure of *Xanthomonas oryzae* pv. *oryzae* using a repetitive DNA element. *Applied and Environmental Microbiology* 58, 2188-2195.

Leonard, K.J. (1969) Selection in heterogeneous populations of *Puccinia graminis* f. sp. *avenae*. *Phytopathology* 59, 1851-1857.

Leung, H., Nelson, R.J. and Leach, J.E. (1993) Population structure of plant pathogenic fungi and bacteria. *Advances in Plant Pathology* 10, 157-205.

Levy, M., Romao, J., Marchetti, M.A. and Hamer, J.E. (1991) DNA fingerprinting with a dispersed repeated sequence resolves pathogenic diversity in the rice blast fungus. *Plant Cell* 3, 95-102.

Levy, M., Correa-Victoria, F.J., Zeigler, R.S., Xu, S. and Hamer, J.E. (1993) Genetic diversity of the rice blast fungus in a disease nursery in Colombia. *Phytopathology* 83, 1427-1433.

MacKenzie, D.R. (1979) The multiline approach to the control of some cereal diseases. In: *Proceedings of the Rice Blast Workshop*. International Rice Research Institute, Los Banos, Philippines, pp. 199-216.

McDonald, B.A., Mundt, C.C. and Chen, R.-S. (1996) The role of selection on the genetic structure of pathogen populations: Evidence from field experiments with *Mycosphaerella graminicola*. *Euphytica* 92, 73-80.

Mew, T.W., Vera Cruz, C.M. and Medalla, E.S. (1992) Changes in race frequencies of *Xanthomonas oryzae* pv. *oryzae* in response to the planting of rice cultivars. *Plant Disease* 76, 1029-1032.

Michelmore, R.W., Norwood, J.M., Ingram, D.S., Crute, I.C. and Nicholson, P. (1984) The inheritance of virulence in *Bremia lactucae* to match resistance factors 3, 4, 5, 6, 8, 9, 10, and 11 in lettuce (*Lactuca sativa*). *Plant Pathology* 33, 301-315.

Mundt, C.C. (1994) Use of host genetic diversity to control cereal diseases: Implications for rice blast. In: Leong, S.A., Zeigler, R.S. and Teng, P.S. (eds) *Rice Blast Disease*. CAB International, Cambridge, and the International Rice Research Institute, Manila, pp. 293-307.

Mundt, C.C., Ahmed, H.U., Finckh, M.R., Leach, J.E., Nelson, R.J. and McDonald, B.A. (1994) Integrating molecular and phenotypic analyses of plant pathogen populations. *American Journal of Botany* 81, 39 (abstract 109).

Mundt, C.C., Brophy, L.S. and Schmitt, M.E. (1995) Choosing crop cultivars and mixtures under high versus low disease pressure: A case study with wheat. *Crop Protection* 14, 509-515.

Nelson, R.J., Baraoidan, M.R., Vera Cruz, C.M., Yap, I.V., Leach, J.E., Mew, T.W. and Leung, H. (1994) Relationship between phylogeny and pathotype for the bacterial blight pathogen of rice. *Applied Environmental Microbiology* 60, 3275-3283.

Newton, A.C. (1989) Genetic adaptation of *Erysiphe graminis* f. sp. *hordei* to barley with partial resistance. *Journal of Phytopathology* 126, 133-148.

Newton, A.C. (1992) Selection for aggressiveness in *Erysiphe graminis* f. sp. *hordei* towards partial resistance in barley. *Journal of Phytopathology* 136, 165-169.

Newton, A.C. and McGurk, L. (1991) Recurrent selection for adaptation of *Erysiphe graminis* f. sp. *hordei* to partial resistance and the effect of environment on expression of partial resistance of barley. *Journal of Phytopathology* 132, 328-338.

Shaw, M.W. (1990) Effects of temperature, leaf wetness and cultivar on the latent period of *Mycosphaerella graminicola* on winter wheat. *Plant Pathology* 39, 255-268.

Shaw, M.W. and Royle, D.J. (1989) Airborne inoculum as a major source of *Septoria tritici* (*Mycosphaerella graminicola*) infections in winter wheat crops in the UK. *Plant Pathology* 38, 35-43.

Stakman, E.C. and Levine, M.N. (1922) The determination of biologic forms of *Puccinia graminis* on *Triticum* spp. *University of Minnesota Agricultural Experiment Station Technical Bulletin* 8, 3-10.

Van der Plank, J.E. (1968) *Disease Resistance in Plants.* Academic Press, New York, 206pp.

Van der Plank, J.E. (1978) *Genetic and Molecular Basis of Plant Pathogenesis.* Springer-Verlag, Berlin, 167pp.

Van der Plank, J.E. (1982) *Host-Pathogen Interactions in Plant Disease.* Academic Press, New York, 207pp.

Wolfe, M.S. (1985) The current status and prospects of multiline cultivars and variety mixtures for disease resistance. *Annual Review of Phytopathology* 23, 251-273.

Wolfe, M.S. and Schwarzbach, E. (1975) The use of virulence analysis in cereal mildews. *Phytopathologische Zeitschrift* 82, 297-307.

Xia, J.Q., Correll, J.C., Lee, F.N., Marchetti, M.A. and Rhoads, D.D. (1993) DNA fingerprinting to examine microgeographic variation in the *Magnaporthe grisea (Pyricularia grisea)* population in two rice fields in Arkansas. *Phytopathology* 83, 1029-1035.

Chapter nine:

Host-Pathogen Interactions in the *Septoria*-Disease Complex

H.J.L Jørgensen and V. Smedegaard-Petersen
Plant Pathology Section, Department of Plant Biology, The Royal Veterinary and Agricultural University, Thorvaldsensvej 40, DK-1871 Frederiksberg C, Copenhagen, Denmark

Introduction

The two major pathogens in the *Septoria*-disease complex are *Stagonospora nodorum* (Berk.) Castellani and E.G. Germano [syn. *Septoria nodorum* (Berk.) Berk. in Berk. and Br.], teleomorph *Phaeosphaeria nodorum* (E. Müller) Hedjaroude (syn. *Leptosphaeria nodorum* Müller), causing glume blotch or Septoria nodorum blotch, and *Septoria tritici* Rob. ex Desm., teleomorph *Mycosphaerella graminicola* (Fuckel) Schröter in Cohn., causing speckled leaf blotch or Septoria tritici blotch.

Both diseases are of great economic importance in wheat-growing areas throughout the world (Shipton *et al.*, 1971; King *et al.*, 1983; Eyal *et al.*, 1987). They often occur simultaneously in a crop where it may be difficult to visually distinguish between them, especially at the onset of infection (Beck and Ligon, 1995). In Scandinavia, *S. tritici* has historically been present in virtually all wheat crops during the early spring with only little or no progress of the pathogen later in the growth season, and consequently resulting in little or no impact on yield (Welling *et al.*, 1984). However, from the beginning of the 1980s, *S. tritici* has frequently been found on the upper leaves, including the flag leaves (Welling *et al.*, 1984), and in Denmark, *S. tritici* is now considered as more damaging in wheat than *S. nodorum*.

Attempts to control the diseases by breeding for resistance have, so far, not been very successful. There may be several reasons for this, among which are the great variability in the two pathogens and the fact that the nature of resistance is not well understood. Many important aspects of resistance expression rely on a thorough understanding of how the pathogen infects and develops in a host. Only direct observations can determine at which steps in the infection process pathogen growth is inhibited or arrested, to give valuable information on resistance mechanisms at the cellular level. The objective of this chapter is to describe the interactions between *S. nodorum* and *S. tritici* and their hosts, with particular emphasis on pathogenic specialization, infection processes and host defence.

Host Specialization

Both *S. nodorum* and *S. tritici* vary considerably in their ability to infect different hosts. Two levels of this physiological specialization may be recognized, namely specialization between different genera and species of grasses and specialization on cultivars or other types of a particular grass species. Isolates of the two pathogens from different parts of the world have been studied in both respects but whether different mechanisms of action are responsible for the two levels of specialization have not.

Specialization on Different Species of Grasses

Stagonospora nodorum and *S. tritici* are primarily known as pathogens of wheat but the former is also known as a pathogen of barley (Holmes and Colhoun, 1970; Smedegård-Petersen, 1974; Newton and Caten, 1991; Cunfer *et al.*, 1992; Ueng *et al.*, 1995). However, both *S. nodorum* (Sprague, 1950a,b; Mäkelä, 1977; Rufty *et al.*, 1981; Krupinsky, 1982; 1986; 1997b; Cunfer and Youmans, 1983; Khokhar and Pacumbaba, 1987; Newton and Caten, 1991; Ueng *et al.*, 1995) and, to some extent, *S. tritici* (Sprague, 1950b) have been reported as pathogens on several other grasses. Whether specific isolates of the two pathogens from wheat, barley or other economically important hosts are able to infect other grasses and *vice versa* has been studied in order to examine whether such grasses could act as inoculum sources for cultivated crops.

Stagonospora nodorum

Cross-inoculation studies have shown that, under controlled conditions, isolates of *S. nodorum* from one particular host may infect several other grass species in different genera. Weber (1922) observed that *S. nodorum* from wheat could infect *Triticum* spp., *Secale cereale* and *Poa pratensis;* Shearer and Zadoks (1972a) observed that an isolate from wheat could infect barley, wheat, *Elytrigia repens*, *Lolium perenne* and *Poa annua*; and Harrower (1977) observed that an isolate from wheat could infect *Lolium multiflorum*, *Phleum pratense* and *Bromus sterilis*. Ao and Griffiths (1976) also showed that *S. nodorum* from wheat could infect and produce pycnidia on several grasses in the genera *Agropyron*, *Agrostis*, *Avena*, *Bromus*, *Festuca*, *Holcus*, *Hordeum*, *Lolium*, *Phleum*, *Poa* and *Triticum*. In all cases, *S. nodorum* spores could be reisolated from pycnidia and the reisolates infect wheat again. Similarly, Rufty *et al.* (1981) found that a wheat and a barley isolate of *S. nodorum* could infect species of the genera *Agropyron*, *Elymus*, *Fectuca*, *Hordeum*, *Hystrix*, *Lolium*, *Poa* and *Triticum* although not always to a high degree. However, the fungus could be reisolated from most of the species and the reisolates infect wheat and barley again.

Cross-inoculation studies have also included fungal isolates obtained from other host species. Krupinsky (1982) found that isolates of *S. nodorum* from *Agropyron* spp. infected wheat but induced less damage than isolates from wheat, and Krupinsky (1986) showed that generally *S. nodorum* isolated from *Bromus inermis* more readily attacked that species than wheat, and that wheat-derived isolates damaged wheat more than *B. inermis*. However, isolates have also been found which react differently from the general pattern. Khokhar and Pacumbaba (1987) tested *S. nodorum* isolates from wheat, triticale, *L. perenne*, *Aegilops cylindrica*, *B. inermis*, *E. repens* and *Hordeum pusillum* and observed that they readily infected wheat whereas an isolate from *Cynodon dactylon* caused only minor infections on wheat. When the isolates were tested on the seven species of grasses they were originally isolated from, only limited infections resulted. Also, Krupinsky (1997b) observed that varying levels of infection were obtained in wheat when inoculated with isolates of *S. nodorum* from 13 different grass species in the genera *Agropyron*, *Bromus*, *Elymus* and *Hordeum*.

There has been some controversy about whether or not *S. nodorum* from wheat could attack barley and *vice versa*. Luthra *et al.* (1938) showed that isolates from wheat could not attack barley. On the other hand, Holmes and Colhoun (1970) reported that *S. nodorum* from wheat infected barley to a minor degree whereas *S. nodorum* from barley did not infect wheat. Martin and Cooke (1979) reported that *S. nodorum* from barley readily infected wheat and, to a lesser degree, *vice versa*. Smedegård-Petersen (1974) demonstrated that isolates from barley and wheat would only infect members of the genera *Hordeum* and *Triticum*, respectively and, therefore, suggested that *S. nodorum* was divided in two *formae speciales*, f.sp. *nodorum* (wheat pathogen) and f.sp. *hordei* (barley pathogen). The observation that wheat isolates of *S. nodorum* do not readily infect barley is supported by histopathological observations. Keon and Hargreaves (1984) found that fungal growth normally stopped shortly after penetration of the barley leaf cuticle although sometimes a few hyphae developed subcuticularly or grew into the pectin layer between the anticlinal cell walls. However, no further tissue colonization was observed. Other studies have shown that *S. nodorum* from wheat, when inoculated on to barley, was generally stopped by papilla formation without any further growth (H.J.L. Jørgensen, P.S. Lübeck, H. Thordal-Christensen, E. de Neergaard and V. Smedegård-Petersen, unpublished results).

The specialization of *S. nodorum* to either wheat or barley was questioned by results from other investigations which suggested that even though *S. nodorum* from wheat was strongly specialized to wheat and *S. nodorum* from barley to barley, this specialization could be changed through repeated inoculations and reisolations of one type of isolate on the opposite host, i.e. the pathogenicity towards the other host could increase while decreasing towards the original host (Rufty *et al.*, 1981; Fitzgerald and Cooke, 1982; Sharma *et al.*, 1982; Sharma and Brown, 1983b; Cunfer, 1984). Changes in specialization of *S. nodorum* isolates have also been observed for other grass species. Harrower (1977) reported that a wheat isolate of *S. nodorum* increased its virulence on *B. sterilis*, *L. multiflorum* and *P. pratense,* while its virulence on wheat decreased. Krupinsky (1997c), however, did not observe increased specialization to wheat for 11 *S. nodorum* isolates from six grasses after five passages through wheat.

From experiments conducted under stringent conditions of isolation, Osbourn *et al.* (1986; 1987) also found apparently similar changes in host specialization of *S. nodorum* to wheat or barley. However, when they performed detailed analyses of the fungal isolates after host passages, including studies of genetic markers, heterokaryon compatibility, colony morphology and fluorescence, they concluded that specialization to either wheat or barley is a stable trait and that the apparent changes in host specialization were due to cross-contaminations of the isolates used, followed by selection of adapted contaminants in their, as well as in previous, investigations.

Other comprehensive studies on the specialization of *S. nodorum* to wheat or barley (Cunfer and Youmans, 1983; Newton and Caten, 1991; Ueng *et al.*, 1995) have confirmed generally that isolates from wheat only infect wheat and that isolates from barley only infect barley. In addition to host specialization, other characteristics of the isolates have been recorded such as growth on artificial media, fluorescence, rate of production of various isozymes, and restriction fragment length polymorphisms (RFLPs). Results have further substantiated the distinction of two separate groups within *S. nodorum*, now termed wheat (W) biotype and barley (B) biotype. Isolates of *S. nodorum* from wheat, triticale and rye are usually of the W biotype whereas isolates from barley are usually of the B biotype (Cunfer and Youmans, 1983; Newton and Caten, 1991; Ueng *et al.*, 1995). However, isolates deviating from the pattern of traits for the W and B biotypes are occasionally found and may not be classified to either biotype (Newton and Caten, 1991; Ueng *et al.*, 1995; cf. Sharma and Brown, 1983b). In addition, isolates with all the characteristics of the W biotype may be found on barley and be considered as the W biotype growing on a non-adapted host; however, B biotype isolates have not been observed on either wheat or triticale (Cunfer and Youmans, 1983; Newton and Caten, 1991; Ueng *et al.*, 1995).

Although there now seems to be general agreement over the existence of two separate and fairly well-characterized groups within *S. nodorum*, there are particular strains of the fungus with all the characteristics of one of the biotypes which, in fact, are found on the other host. Even although the W and B biotypes are now stable groups, the existence of such deviant isolates indicates that a change of specialization must indeed have occurred at some time.

Septoria tritici

Sprague (1950b) reported that three specialized forms, besides the one on wheat, could be distinguished: f. *avenae* (on *Avena* spp.); f. *holci* (on *Holcus lanatus*); and var. *lolicola* (on *Lolium* spp.). However, this distinction between different forms of the pathogen appears now to have been mostly abandoned (cf. Shipton *et al.*, 1971).

In cross-inoculation studies, Weber (1922) only obtained infection by *S. tritici* in *Triticum* spp., *S. cereale* and *P. pratensis* but not in several other grass species tested. Similarly, Hilu and Bever (1957) were able only to infect *Triticum* species when they inoculated species in the genera *Aegilops*, *Agropyron*, *Agrostis*, *Avena*, *Brachypodium*, *Dactylis*, *Poa*, *Secale* and *Triticum*. On the other hand, Brokenshire (1975) was able to demonstrate infection and pycnidium formation by a mixture of six *S. tritici* isolates in 11 grass species belonging to the genera *Agropyron*, *Arrhenatherum*, *Bromus*, *Festuca*, *Hordeum*, *Phleum*, *Poa* and *Vulpia*. Pycnidia were produced on all 11 species and isolated pycnidiospores were able to reinfect wheat. Likewise, Ao and Griffiths (1976) found that a wheat isolate of *S. tritici* infected 11 species of grasses in the genera *Agrostis*, *Bromus*, *Festuca*, *Hordeum*, *Lolium*, *Phleum*, *Poa* and *Triticum*. Also in their study, the fungus was reisolated from the grasses and was able to reinfect wheat.

In a manner similar to the specialization of *S. nodorum* isolates to either wheat or barley, *S. tritici* isolates are more or less specialized to either bread or durum wheat (van Ginkel and Scharen, 1988b; Kema *et al.*, 1996a,b), although some isolates may be specialized to the opposite host or infect both (Kema *et al.*, 1996a). *Septoria tritici* isolates from durum wheat producing little disease on bread wheat and *vice versa*, have been reported by Ballantyne and Thomson (1995). So far, it appears that this specialization has not been characterized further than to examine whether or not fungal isolates will produce necrosis and pycnidia on the hosts.

Cross-inoculation experiments have demonstrated that isolates of both pathogens from one host may infect other grass species. This is particularly well documented for *S. nodorum* but less so for *S. tritici*. However, care should be taken when considering which grass species may be alternative hosts and their importance as inoculum sources for wheat (cf. Krupinsky, 1997b,c). Only species found to be naturally infected can be considered as natural hosts. Cross-inoculation studies may demonstrate that infection of other species is possible under conditions highly conducive for fungal infection and development. Hence, barley has been found to be infected by *S. tritici* (Brokenshire, 1975; Ao and Griffiths, 1976) although this species is not normally considered a host of the pathogen (Sprague, 1950b; cf. Mäkelä, 1977). Whether the pattern of grass species infected by *S. nodorum* under controlled conditions reflects that of species infected by strains in nature remains to be examined before it is possible to say whether or not such species belong to the natural host spectrum.

Although certain specializations to particular hosts have been observed, only limited studies have been performed to assign isolates of *S. nodorum* from hosts other than cereals to the W and B biotypes (cf. Krupinsky, 1982; 1986). Ueng *et al.* (1995) found that *S. nodorum* isolates from *H. pusillum* and *Lolium* spp. belonged to the W biotype. However, it has not been determined whether or not isolates of *S. tritici* from hosts other than wheat are specialized to bread or durum wheat.

Additional studies are needed to characterize non-cereal isolates and to determine which biotypes they belong to and if more than the W and B biotypes can be distinguished for *S. nodorum* and the bread and durum wheat types for *S. tritici*. The reason why isolates specialized to one host may infect some species but not others also needs to be studied.

Specialization Within One Host Species

Stagonospora nodorum

There are several reports of marked variation in pathogenicity of *S. nodorum* isolates tested on a varying number of wheat cultivars (Krupinsky *et al.*, 1973; Rufty *et al.*, 1981; Scharen and Eyal, 1983; Scharen *et al.*, 1985; Krupinsky, 1986; 1997a,b). Cultivars may differ in the degree of infection by specific isolates and isolates may differ in their ability to infect specific cultivars. Since differences in the degree of infection are quantitative rather than qualitative, results of such tests are analysed by analysis of variance (Rufty *et al.*, 1981) and the presence of significant statistical interactions between cultivars and isolates indicates specific resistance (Van der Plank, 1968). Significant interactions, and thus indications of specific resistance, have been reported (Rufty *et al.*, 1981; Scharen and Eyal, 1983; Krupinsky, 1986). The degree of specificity is, however, low because all cultivars are infected to a varying degree (Rufty *et al.*, 1981; Scharen and Eyal, 1983; Krupinsky, 1986). Low levels of specificity in the interaction between host and pathogen, considerable variation in pathogenicity, and cultural characteristics as well as high environmental influence on the host response has made the classification of isolates into races dubious (Scharen and Krupinsky, 1970; Rufty *et al.*, 1981; Scharen and Eyal, 1983).

Lack of, or low levels of, specificity in the interaction between *S. nodorum* and wheat has also been reported (Scharen and Eyal, 1983; Scharen *et al.*, 1985; Krupinsky, 1997a,b,c). Variation in the ability of isolates to infect particular wheat cultivars has been ascribed to differences in aggressiveness although some significant interactions have been regularly observed between cultivars and isolates (Krupinsky, 1997a,b,c).

Septoria tritici

Eyal *et al.* (1973) reported a clear differential interaction and, therefore, clear physiological specialization of five isolates of *S. tritici* on 14 cultivars of wheat in Israel. However, as for *S. nodorum*, most investigations report quantitative differences in pathogenicity of isolates of *S. tritici* and, hence, indications of a certain degree of physiological specialization of these isolates to particular wheat cultivars (Yechilevich-Auster *et al.*, 1983; Eyal *et al.*, 1985; Eyal and Levy, 1987;

Saadaoui, 1987; Danon and Eyal, 1990; Perelló *et al.*, 1990; Ballantyne and Thomson, 1995; Ahmed *et al.*, 1996; Kema *et al.*, 1996a). Significant interactions between cultivars and isolates also indicate the presence of specific virulence and resistance (Yechilevich-Auster *et al.*, 1983; Eyal *et al.*, 1985; Eyal and Levy, 1987; Saadaoui, 1987; Danon and Eyal, 1990; Ballantyne and Thomson, 1995; Ahmed *et al.*, 1996; Kema *et al.*, 1996a,b; Kema and van Silfhout, 1997), and a gene-for-gene relationship for resistance in the host and virulence in the pathogen has been suggested in bread and durum wheat (Yechilevich-Auster *et al.*, 1983; Eyal *et al.*, 1985; Eyal and Levy, 1987; Danon and Eyal, 1990; Kema *et al.*, 1996a,b). Additional evidence that specific resistance against *S. tritici* may be present in certain cultivars was provided by Kema *et al.* (1996b), who refer to the Dutch wheat cultivar Obelisk. When released in 1985 this cultivar was resistant but about ten years later was susceptible, suggesting a shift in the pathogen population leading to a 'break-down' of resistance. Similar observations have been made in Australia (Ballantyne and Thomson, 1995) and in Israel (Eyal *et al.*, 1973).

Assuming a gene-for-gene relationship, the presence of several resistance genes in the host and corresponding virulence genes in the pathogen has been estimated by computer analyses in studies involving varying numbers of fungal isolates and host cultivars (Yechilevich-Auster *et al.*, 1983; Eyal *et al.*, 1985; Eyal and Levy, 1987). However, the validity of results from such analyses is questionable since the results depend on the numbers of cultivars and isolates tested (cf. Kema *et al.*, 1996a) and because the reaction of the cultivars to infection is divided into only two classes, resistant and susceptible, the largely quantitative nature of the disease response is ignored.

In addition to the qualitative aspects, the resistance may also be quantitative and is rarely complete (Jlibene *et al.*, 1994; Kema *et al.*, 1996a,b; Kema and van Silfhout, 1997). Thus, there are reports on the absence of gene-for-gene relationships, based on constant ranking of cultivars to several fungal isolates, therefore suggesting that the isolates vary in aggressiveness and not in virulence (van Ginkel and Scharen, 1988a,b). Based on analyses of variance, Ahmed *et al.* (1996) found indications that both race-specific and race-non-specific resistance were present in the interactions they studied. Also, Eyal *et al.* (1985) apparently observed that the cultivars they tested maintained the same ranking of responses for all isolates, normally indicating race-non-specific resistance. Likewise, van Ginkel and Scharen (1986) did not find evidence for classical gene-for-gene relationships although no further details were given.

Due to the quantitative aspects of the interaction between *S. tritici* and the hosts, race designation has been avoided in most investigations; one exception being Saadaoui (1987) who designated three races in Morocco.

There may be various reasons for conflicting results regarding the presence of specificity in the interactions between wheat and *S. nodorum* and wheat and *S. tritici*. For *S. tritici*, Kema and van Silfhout (1997) suggested that although the 'break-down' of resistance was not readily observed, the general incompleteness of resistance along with fungal diversity of the pathogen might mask the expression of race-specific resistance. Similar circumstances may apply for *S. nodorum*. Also, different experimental conditions, including environmental (cf. Kema *et al.*, 1996a; Krupinsky, 1997a,b), may be partly responsible for discrepancies between results from different investigations. Thus, varying numbers of different host cultivars and fungal isolates, differing in resistance and virulence respectively, have been tested under varying environmental conditions. In particular, if the number of host differentials is too small and do not represent different sources of resistance with a range of disease reactions, any specificity will be very difficult to disclose. The fungal isolates will also have to represent a wide spectrum of pathogenicity (cf. Krupinsky, 1997a,b). Finally, scoring of the level of resistance, i.e. of necrosis or pycnidium formation, may influence the results (Kema *et al.*, 1996a, cf. section on Expression of Resistance) as may the method of data analysis (Kema *et al.*, 1996a).

Infection Biology in Compatible Interactions

All studies on infection biology and resistance expression are strongly influenced by the environment and are very sensitive to alterations in the environmental conditions. Therefore, in order to understand fully how plant and pathogen interact, it is necessary to know the environmental conditions under which the experiments were conducted.

One approach to study the influence of environmental factors on disease development is to correlate weather conditions to severe disease epidemics. This has been done both for *S. nodorum* (Scharen, 1964; Dickopp and Hindorf, 1994) and for *S. tritici* (Shaner and Finney, 1976; Welling *et al.*, 1984; Hess and Shaner, 1987b; Murray *et al.*, 1990; Dickopp and Hindorf, 1994), and the importance of appropriate temperature and prolonged periods of humidity demonstrated. Several investigations have studied the influence of environmental factors on pathogen development under more or less controlled conditions and found that light, physiological tissue age, inoculum concentration, temperature and prolonged humid periods etc. are important for *S. nodorum* (Scharen, 1964; Baker, 1970; Holmes and Colhoun, 1971; 1974; Shearer and Zadoks, 1972b; Eyal *et al.*, 1977; Aust and Hau, 1981; 1983; Babadoost and Herbert, 1984; Wainshilbaum and Lipps, 1991) and for *S. tritici* (Fellows, 1962; Benedict, 1971; Holmes and Colhoun, 1974; Hess and Shaner, 1987a; Wainshilbaum and Lipps, 1991; Magboul *et al.*, 1992).

A comprehensive review of the influence of environmental factors on *S. nodorum* and *S. tritici* is beyond the scope of this chapter. However, of particular interest when studying aspects of infection, fungal development and expression of resistance is the observation that environmental factors may compensate for each other. Therefore, if one environmental parameter is not optimal for fungal development, one of the other factors may compensate for the suboptimal level of the former, resulting in, for example, a final latent period which is not changed.

This has been demonstrated for *S. nodorum* (Aust and Hau, 1981; 1983) and to some extent for *S. tritici* (Hess and Shaner, 1987a). For example, suboptimal temperatures may be compensated for by longer wetting periods and/or by higher inoculum concentrations (Aust and Hau, 1981). Under natural conditions, the latent period may be very long, such as during winter when conditions are unfavourable for the pathogen. In Ireland, O'Reilly and Downes (1986) observed conidia of *S. nodorum* on wheat plants at the end of November while the first pycnidia were observed in mid February. Under controlled conditions, the latent period may also vary considerably. Shearer and Zadoks (1972b) found latent periods for *S. nodorum* to be as short as 6 days under optimal conditions but as long as 50 days under suboptimal conditions. The duration of the incubation period, and hence the latent period, may, however, not only be determined by the environment but may also be partly under genetic control (Rapilly *et al.*, 1981).

Whereas infection and development of *S. nodorum* and *S. tritici* in wheat leaves have been studied, infection of other plant parts, such as glumes, has not. Also, although the initial stages of the interaction between wheat and *S. nodorum* is well studied, sporulation and its initiation have not been described in detail, at least from a histopathological point of view. On the other hand, all processes from infection to sporulation have been described for *S. tritici*, although in fewer investigations than for *S. nodorum*.

Stagonospora nodorum

Under controlled conditions, the first signs of infection may appear only a few days after inoculation (dai). Bird and Ride (1981) saw the first symptoms three dai in a susceptible interaction under glasshouse conditions. Subsequently, necroses and chloroses rapidly developed and spread beyond the inoculation sites nine dai. Eleven dai, the lesions coalesced, and 15 dai, the leaves were totally diseased.

Conidia may start to germinate within 2 hours of inoculation (Bird and Ride, 1981; Zinkernagel *et al.*, 1988), the germ tubes growing in close contact with the surface (Karjalainen and Lounatmaa, 1986; Zinkernagel and Riess, 1987; Zinkernagel *et al.*, 1988). The contact is mediated either by amorphous material or by a mucilaginous sheet (Karjalainen and Lounatmaa, 1986; Zinkernagel and Riess, 1987). Often polar germination of conidia is observed, resulting in two germ tubes which occasionally branch (Tiedemann and Firsching, 1993). Superficial hyphal growth may occur, as observed by O'Reilly and Downes (1986) for naturally infected plants during winter. These hyphae were found to grow in very close contact with the leaf surface, apparently being embedded in the surface wax.

Penetration of the host may commence about 10 hours after inoculation (Bird and Ride, 1981; Zinkernagel *et al.*, 1988). One to several appressoria form on the germ tubes (Bird and Ride, 1981; Karjalainen and Lounatmaa, 1986; Tiedemann and Firsching, 1993), and penetration occurs, from a penetration peg, directly through the surface of the epidermal cells (Bird and Ride, 1981; Karjalainen and

Lounatmaa, 1986). Weber (1922) also observed direct cuticular penetration and reported that the penetrating hypha was very small, enlarging somewhat after penetration. Appressoria are chiefly formed over the anticlinal cell walls (Baker and Smith, 1978; Tiedemann and Firsching, 1993). Bird and Ride (1981) observed that most successful penetrations took place in the thin-walled cells over the sclerenchyma fibres. Zinkernagel and Riess (1987) and Zinkernagel *et al.* (1988), however, observed that most often penetration took place from conidia or germ tubes without discernible appressoria and that only a few germ tubes actually formed appressoria. If penetration was not successful, the germ tubes could resume growth on the leaf surface.

The cuticle appears to be dissolved during penetration, probably by enzymes (Baker and Smith, 1978; Karjalainen and Lounatmaa, 1986; Zinkernagel and Riess, 1987; Zinkernagel *et al.*, 1988). The germ tubes grow into a subcuticular position and form a mycelial mat during the next few days (Zinkernagel and Riess, 1987; Zinkernagel *et al.*, 1988). Subcuticular growth has only been observed to a limited extent by Bird and Ride (1981) and by Karjalainen and Lounatmaa (1986). The latter concluded that penetration most often took place directly into the lumen of epidermal cells. However, they did not study the very early stages of infection only starting to take samples three dai. Ingress of the fungus through stomata is generally rarely observed (Zinkernagel and Riess, 1987; Zinkernagel *et al.*, 1988; Tiedemann and Firsching, 1993) although O'Reilly and Downes (1986) reported this mode of entry to be quite common.

When substantial amounts of mycelia have formed subcuticularly (about four dai), *S. nodorum* starts to penetrate into epidermal cells (Zinkernagel and Riess, 1987; Zinkernagel *et al.*, 1988) and then grows intercellularly into the mesophyll (Bird and Ride, 1981; Karjalainen and Lounatmaa, 1986; Zinkernagel and Riess, 1987; Zinkernagel *et al.*, 1988) by lysing the middle lamellae and forcing the cell walls apart (Zinkernagel and Riess, 1987; Zinkernagel *et al.*, 1988). Growth in the mesophyll is sparse (Bird and Ride, 1981; Karjalainen and Lounatmaa, 1986; Zinkernagel and Riess, 1987). However, O'Reilly and Downes (1986), studying naturally infected leaves, and Weber (1922) observed abundant intercellular hyphal growth and Baker and Smith (1978) observed abundant intracellular and intercellular hyphal growth in the mesophyll when studying the infection of *S. nodorum* in detached wheat leaves. Hyphae may grow all the way to the vascular bundles although they are not colonized (Bird and Ride, 1981; Zinkernagel and Riess, 1987; Zinkernagel *et al.*, 1988).

Mesophyll cells in contact with, or in advance of, progressing hyphae degenerate and collapse (Bird and Ride, 1981; Zinkernagel and Riess, 1987; Zinkernagel *et al.*, 1988), this process being accompanied by lignification of cell walls and cytoplasm together with development of extracellular lignin deposits (Bird and Ride, 1981). No effects on host cells were observed by Weber (1922) until almost all intercellular spaces were colonized by the pathogen. After limited intercellular growth in the mesophyll, intracellular penetration starts into the epidermal and mesophyll cells (Zinkernagel and Riess, 1987; Zinkernagel *et al.*,

1988). Only dead cells are penetrated (Zinkernagel and Riess, 1987), penetration taking place by localized lysis of the cell walls (Zinkernagel and Riess, 1987; Zinkernagel *et al.*, 1988). However, in the mesophyll, extensive surface lysis of the cell walls may also occur where a hypha is in contact with a dead cell (Zinkernagel and Riess, 1987; Zinkernagel *et al.*, 1988). Cell wall degrading enzymes have been isolated and studied *in vitro* and *in vivo* (Magro, 1984; Sieber, 1989; Lehtinen, 1993). The cell wall is gradually digested until the hypha reaches the cell lumen (Zinkernagel and Riess, 1987). After penetration into a cell, a kind of vesicle develops from which an intracellular hypha is formed (Zinkernagel and Riess, 1987; Zinkernagel *et al.*, 1988). This occurs especially in the epidermis (Zinkernagel *et al.*, 1988).

Cytological changes in the host during the progress of *S. nodorum* infection, studied by transmission electron microscopy (TEM), have been published (Bird and Ride, 1981; Karjalainen and Lounatmaa, 1986; Zinkernagel and Riess, 1987; Zinkernagel *et al.*, 1988).

Pycnidia are formed from hyphae aggregating in the substomatal cavities (Weber, 1922; Baker and Smith, 1978). The time required from the initiation of their formation to the release of spores may be only a few days (Scharen, 1966). The ostioles of the pycnidia are formed directly under stomata, and host cells adjacent to the pycnidia may be covered with fungal mycelium and eventually collapse and die (Weber, 1922).

Collapse and death of host cells was thought to be associated with the production of toxins by *S. nodorum*. A toxic compound, proposed to be an unstable, low molecular weight cationic acid, was isolated by Kent and Strobel (1976). This phytotoxin was unspecific since it caused similar symptoms on several species in the genus *Triticum* as well as on barley, rye and oats. The symptoms resembled the typical necrotic lesions found in the later stages of disease development in the field. Bousquet and Skajennikoff (1974) isolated and partly characterized a phytotoxic substance from culture filtrates of *S. nodorum*, later found to be ochracine (=melleine), dihydro-3,4-hydroxy-8-methyl-3-isocoumarine (Devys *et al.*, 1974). Bousquet and Skajennikoff (1974) found that ochracine could completely inhibit root growth of several plant species at concentrations of 200 ppm. At 50 ppm, root and coleoptile growth of wheat was strongly inhibited but at 5 ppm root growth was stimulated. They concluded that ochracine was not a primary determinant of disease because it did not produce the necrotic symptoms characteristic of disease initiated by *S. nodorum* and because it was not specific in its action. Later, Bousquet *et al.* (1977) demonstrated that ochracine caused a reduction in net CO_2 assimilation, probably because it affected CO_2 transfer through stomata. This alone, however, could not explain the decrease in photosynthesis and Bousquet *et al.* (1977) suggested that intracellular diffusion of CO_2 was inhibited. Ochracine has no direct effect on stomata (Bethenod *et al.*, 1982; Laffray *et al.*, 1982) and Bethenod *et al.* (1982) concluded that the decrease in CO_2 assimilation, due to closure of stomata, was a consequence of reduced photosynthesis caused by the toxin. The direct inhibiting effect on photosynthesis was suggested to be on the Calvin cycle or on reactions linked to it (Bethenod *et al.*, 1982). Ochracine also strongly influences mitosis (Essad *et al.*, 1981).

Another toxic substance, septorine (Devys *et al.*, 1978), was also found to reduce the growth of wheat plants and in mitochondria isolated from wheat plants the toxin affected respiration by reducing oxidative phosphorylation (Bousquet *et al.*, 1980).

Physiologically, plants are affected in several ways by infection by *S. nodorum*. Photosynthesis (net CO_2 assimilation) is drastically reduced in infected plant parts (Scharen and Taylor, 1968; Scharen and Krupinsky, 1969; Krupinsky *et al.*, 1973), whereas respiration is not significantly altered compared to healthy plants (Scharen and Taylor, 1968). Reduced photosynthesis might partly be explained by a decrease in the chlorophyll content of the plant which is coupled to a reduction in the production and content of carbohydrates, such as glucose, fructose, saccharose and starch (Verreet *et al.*, 1987). Also, the distribution of nitrogen in the plant, but not total nitrogen uptake (Verreet and Hoffmann, 1986; 1990), and amino acid metabolism (Verreet and Hoffmann, 1990) is changed. Thus, nitrogen is retained in infected plant parts and is diverted from grain production (Verreet and Hoffmann, 1986; 1990). Reduced photosynthesis (net CO_2 assimilation) and reduced nitrogen uptake in the ears result in reduced yield, grain weight and numbers of grains per ear (Scharen and Taylor, 1968; Verreet and Hoffmann, 1986). Seedling fresh weight and height may also be reduced (Holmes and Colhoun, 1971), but the effect on these parameters was not investigated later in the growth season.

Septoria tritici

Pycnidiospores usually germinate either from the ends or from the septa (Hilu and Bever, 1957; Cohen and Eyal, 1993; Kema *et al.*, 1996c) and this may start as early as 2 hours after inoculation (Cohen and Eyal, 1993). The germ tubes often branch (Cohen and Eyal, 1993; Kema *et al.*, 1996c) and form hyphal aggregates over stomatal apertures (Kema *et al.*, 1996c) but may also grow across the leaf surface and stomata (Hilu and Bever, 1957; Cohen and Eyal, 1993; Kema *et al.*, 1996c). For penetration into the host, two principally different routes have been described; either through stomata or directly through the intact leaf cuticle. Thus, penetration was reported to be strictly (Benedict, 1971; Kema *et al.*, 1996c) or primarily (Hilu and Bever, 1957; Cohen and Eyal, 1993) through open or closed stomatal apertures (Hilu and Bever, 1957). Direct cuticular penetration has previously been reported to occur only occasionally (Hilu and Bever, 1957; Cohen and Eyal, 1993), contrasting with results for *S. nodorum*. Stomatal penetration by *S. tritici* was observed to take place by single germ tubes or hyphal aggregates resulting in multiple penetrations (Kema *et al.*, 1996c). Sometimes, appressorium-like structures were formed just above stomatal apertures (Cohen and Eyal, 1993; Kema *et al.*, 1996c) and occasionally over anticlinal cell walls or at the base of trichomes (Cohen and Eyal, 1993). Additional studies are needed to clarify the relative importance of these two different modes of penetration and their possible dependence on environmental conditions.

After invasion of the host, hyphae grow intercellularly (Weber, 1922; Hilu and Bever, 1957; Benedict, 1971; Cohen and Eyal, 1993; Kema *et al.*, 1996c), with no subcuticular phase as with *S. nodorum*. Hilu and Bever (1957) observed intercellular growth after stomatal penetration which took place 24 hours post-inoculation (hpi) and Kema *et al.* (1996c) observed hyphae growing intercellularly in the substomatal cavities 12 hpi, the intercellular proliferation continuing and increasing for the next 8 days. The hyphae grow in close contact with the cells (Hilu and Bever, 1957; Kema *et al.*, 1996c). Hilu and Bever (1957) observed that germ tube diameters were about 1 μm, increasing slightly following invasion of the host, but increasing to about 2.5 μm after extensive tissue colonization. During proliferation in the tissue, no macroscopic signs of infection are seen (Hilu and Bever, 1957; Kema *et al.*, 1996c).

Hyphal growth remains mostly intercellular with only very few branches penetrating into dead or living cells (Hilu and Bever, 1957). Hyphae ramify through all leaf tissue except the vascular bundles which, however, may be colonized or killed in cases of extreme hyphal proliferation (Weber, 1922; Hilu and Bever, 1957). Intercellular hyphae may spread considerable distances from areas showing macroscopic symptoms (Hilu and Bever, 1957).

Chloroplasts may become swollen, change shape and their affinity for histochemical stains (Kema *et al.*, 1996c). Chloroplasts may even completely disappear in the later stages of disease development (Hilu and Bever, 1957). Intercellular hyphal growth also affects the mesophyll cells, an increasing proportion of which collapse (Kema *et al.*, 1996c). Hilu and Bever (1957) and Weber (1922) observed very few collapsed cells during fungal tissue colonization while more extensive cell death was seen after initiation of pycnidial formation. Cells die, particularly those in contact with hyphae (Kema *et al.*, 1996c), and may degenerate in the late stages of infection (Hilu and Bever, 1957). Cytological details involving TEM have been published by Kema *et al.* (1996c).

Pycnidial initials form in substomatal cavities (Weber, 1922; Hilu and Bever, 1957; Cohen and Eyal, 1993; Kema *et al.*, 1996c) with usually, only one pycnidium developing per cavity (Kema *et al.*, 1996c). The formation may start 8-9 days post-inoculation (Hilu and Bever, 1957; Kema *et al.*, 1996c) with one or more hyphae first bordering and later growing into the spaces (Hilu and Bever, 1957). The hyphae branch and fuse to form pycnidial initials, which develop into a globular primordium with a pointed beak growing towards a stomatal aperture (Hilu and Bever, 1957). As the pycnidium matures, it fills out the substomatal chamber (Weber, 1922; Hilu and Bever, 1957) and the beak develops into an ostiole that assumes the same shape as the stomatal aperture by which it is limited (Weber, 1922; Hilu and Bever, 1957; Kema *et al.*, 1996c). If pycnidial density increases, pycnidia become smaller (Eyal and Brown, 1976). Kema *et al.* (1996c) reported that there are no internal or external appendages on pycnidia whereas Weber (1922) observed that the outer walls were covered with mycelial strands connected to nearby hyphae. The walls of mature pycnidia consist of three to five layers of pseudo-parenchymatic cells (Hilu and Bever, 1957). The time from initiation of a pycnidium's formation to maturity may be only 2-3 days (Kema *et al.*, 1996c) and the appearance of macroscopic symptoms may coincide with its formation (Hilu and Bever, 1957), although individual pycnidia may develop over a long period of time (Hilu and Bever, 1957).

Pycnidiospores are borne as aseptate structures from conidiophores on the inside of pycnidia walls (Hilu and Bever, 1957; Kema *et al.*, 1996c). A spore starts as a bud, which, as it matures, become septate (Hilu and Bever, 1957; Kema *et al.*, 1996c) and filiform (Hilu and Bever, 1957). Spores are extruded through the ostiole in a cirrhus (Hilu and Bever, 1957; Kema *et al.*, 1996c).

The cell collapse observed suggests the action of toxic substances (Kema *et al.*, 1996c). Hilu and Bever (1957) found that a culture filtrate of *S. tritici* caused necrotic lesions in wheat leaves in a manner similar to a spore suspension of the fungus. No further characterization of the toxic metabolite was performed. A phytotoxin produced by *S. tritici* was also mentioned by Kent and Strobel (1976) but no further details were given.

Infection of wheat by *S. tritici* has been reported to reduce photosynthesis (net CO_2 assimilation) (Shtienberg, 1992; Zuckerman *et al.*, 1997) and transpiration (Cornish *et al.*, 1990; Shtienberg, 1992). Transpiration was more or less reduced as expected by the reduction in the percentage of healthy leaf area (Cornish *et al.*, 1990; Shtienberg, 1992) whereas the reduction in photosynthesis was larger than expected (Shtienberg, 1992). Generally, Cornish *et al.* (1990) did not observe a reduction in transpiration per unit of leaf area or in leaf water potential, indicating that infection did not affect water uptake or stomatal control. Water use and plant growth were reduced only if the leaf area index in a well-watered crop was reduced below the value required for full interception of solar radiation. Reduced photosynthetic activity results from a reduced green leaf area and reduced chlorophyll content (Zuckerman *et al.*, 1997). However, the CO_2 assimilation per unit of chlorophyll may increase and, thus, compensate to a certain degree for the reductions in the green leaf area and chlorophyll content (Zuckerman *et al.*, 1997). Leaf infection may also result in reduced photosynthesis and respiration of the awns and glumes (Zuckerman *et al.*, 1997).

Expression of Resistance

There are many reports of resistance against *S. nodorum* in wheat, rye and triticale (Scharen and Krupinsky, 1970; Krupinsky *et al.*, 1977; Scharen and Eyal, 1980; Arseniuk *et al.*, 1995; Woś and Makowiak, 1995) and in barley (Cunfer *et al.*, 1992), and against *S. tritici* in wheat and triticale (Rosielle, 1972; Krupinsky *et al.*, 1977; Eyal *et al.*, 1983; Mielke, 1995). Resistance against *S. nodorum* (Krupinsky *et al.*, 1977; Scharen and Eyal, 1980; Trottet *et al.*, 1983; Rubiales *et al.*, 1996) and against *S. tritici* (Krupinsky *et al.*, 1977; Gough and Tuleen, 1979; Yechilevich-Auster *et al.*, 1983; Brown-Guedira *et al.*, 1996) have also been found, and in some cases even transferred to wheat from other wild grass or wheat species.

The wheat plant has evolved different ways of avoiding or reducing the risk of infection and pathogen development by *S. nodorum* and *S. tritici* and certain correlations between various plant characters and resistance have been observed. In addition, tolerance (the ability of the host to endure severe epidemics of the pathogen while only suffering minor yield losses) to *S. nodorum* (Scott, 1973; Brönnimann, 1975) and to *S. tritici* (Zuckerman *et al.*, 1997) has been reported as a way of circumventing some of the detrimental effects of infection on yield.

Stagonospora nodorum

Resistance against *S. nodorum* has mainly been characterized as polygenically inherited (Brönnimann, 1975; Mullaney *et al.*, 1982; Scott *et al.*, 1982), although monogenic or oligogenic resistance have also been reported (Kleijer *et al.*, 1977). Mainly additive gene effects, but also some non-additive effects, have been found (Brönnimann, 1975; Nelson, 1980; Mullaney *et al.*, 1982; Scott *et al.*, 1982). Epistasis (Nelson, 1980; Mullaney *et al.*, 1982), as well as dominant or partial dominant gene action, may be present (Kleijer *et al.*, 1977; Scott *et al.*, 1982) or absent (Nelson, 1980).

Several factors, including reduced areas of infected leaf (Baker and Smith, 1979; Jeger *et al.*, 1983), reduced lesion numbers (Baker and Smith, 1979) reduced infection frequency (Lancashire and Jones, 1985; Loughman *et al.*, 1996), reduced lesion size and rate of growth (Baker and Smith, 1979; Lancashire and Jones, 1985; Griffiths and Jones, 1987; Cunfer *et al.*, 1988), reduced sporulation (Jeger *et al.*, 1983; Lancashire and Jones, 1985; Griffiths and Jones, 1987; Cunfer *et al.*, 1988; Loughman *et al.*, 1996), and increased incubation and latent periods (Lancashire and Jones, 1985; Griffiths and Jones, 1987; Cunfer *et al.*, 1988; Loughman *et al.*, 1996), have been implicated and studied as expressions of resistance against *S. nodorum* and found to be of varying significance in different wheat cultivars. The application of different statistical procedures demonstrated significant interrelationships between resistance and each parameter as well as positive, negative or lack of correlation between the components of resistance (Jeger *et al.*, 1983; Lancashire and Jones, 1985; Griffiths and Jones, 1987; Cunfer *et al.*, 1988; Loughman *et al.*, 1996). Furthermore, expression of resistance was found to be influenced by, or correlated to, differences in environmental conditions (Baker and Smith, 1979; Cunfer *et al.*, 1988).

In some investigations, the word 'partial' has been used to describe resistance. Jeger *et al.* (1983) used it to describe resistance which was characterized by a reduced rate of epidemic development in the host population whereas Lancashire and Jones (1985) referred to resistance where cultivars were neither highly susceptible nor completely resistant (cf. Cunfer *et al.*, 1992). However, both Jeger *et al.* (1983) and Lancashire and Jones (1985) assumed that there was no physiological specialization in the pathogen. Conversely, Griffiths and Jones (1987) assumed that resistance was inherited polygenically and was analogous to horizontal or race-non-specific resistance (Van der Plank, 1968). Cunfer *et al.* (1988) referred to cultivars with a long lasting resistance, assuming polygenic inheritance. Cultivar ranking for disease severity was similar for six isolates, indicating race-non-specific resistance (cf. Van der Plank, 1968). Finally, Loughman *et al.* (1996) referred to resistance based on small differences in components of resistance accumulated over several pathogen generations. Whether the cultivars examined possessed race-non-specific resistance *sensu* Van der Plank (1968) is, however, not known, possibly excepting those studied by Cunfer *et al.* (1988).

Based on microscopic evidence, attempts have been made to characterize the resistance further and to elucidate the mechanisms responsible for inhibition of the pathogen.

Pre-penetration growth of *S. nodorum* is not, or is only slightly, influenced by resistance in wheat (Baker and Smith, 1978; Bird and Ride, 1981). The latter authors found that the percentage of germinating conidia, the number of germ tubes per conidium, and the number of appressoria per germ tube were not affected, although germ tube length was shorter on the more resistant of the four cultivars studied. They further suggested that reduced germ tube length could be due either to pre-formed inhibitory mechanisms or to an early release of induced inhibitory compounds since their data indicated that *de novo* protein synthesis took place.

Papillae are formed where *S. nodorum* attempts to penetrate the host directly (Bird and Ride, 1981; Keon and Hargreaves, 1984; Hargreaves and Keon, 1986; Karjalainen and Lounatmaa, 1986; Zinkernagel and Riess, 1987; Zinkernagel *et al.*, 1988; Tiedemann and Firsching, 1993) and are associated with the failure of these penetration attempts (Bird and Ride, 1981; Keon and Hargreaves, 1984; Karjalainen and Lounatmaa, 1986). Papillae with haloes and containing lignin appear identical to those produced after penetration attempts by other organisms (Bird and Ride, 1981).

Although *S. nodorum* grows into a subcuticular position after penetration and is arrested there, papillae are formed on the inside of the cell wall and papilla formation appears to be elicited by physical damage to the cuticle and/or by a chemical stimulus from the pathogen (Keon and Hargreaves, 1984). Therefore, it appears that host resistance is not caused by papillae acting as a physical barrier. Similar observations have been reported for barley infected by *Rhynchosporium secalis*, which also grows subcuticularly after penetration (Jørgensen *et al.*, 1993). The histochemical composition of papillae formed after penetration attempts by *S. nodorum* from wheat has been studied in barley, wheat and oats (Keon and Hargreaves, 1984; Hargreaves and Keon, 1986). After penetration, hyphal growth is reduced. Baker and Smith (1978) observed fewer hyphae in a resistant than in a susceptible interaction between *S. nodorum* and wheat, although no quantitative data were presented.

In some studies comparing compatible and incompatible interactions between wheat and *S. nodorum*, certain spectacular histological and/or histochemical differences have been observed. It has been suggested that these differences may be part of the resistance response of the host, but often without evidence to explain how resistance might be influenced by these putative defence reactions. Baker and Smith (1978) further observed a necrotic and browning reaction both in a resistant and in a susceptible cultivar, although the reaction appeared earlier, was more pronounced and was confined to the intercellular spaces in the former. In the susceptible cultivar, the reaction affected whole cells. Bird and Ride (1981) found partial or complete lignification of epidermal and mesophyll cells following infection by *S. nodorum* and suggested this to be a general resistance response of wheat against the pathogen, fungal growth being slowed down, but not completely stopped by, for example, the blocking of intercellular spaces. Baker and Smith (1978) did not observe any induced lignin formation maybe because they studied the infection of *S. nodorum* in detached resistant and susceptible wheat leaves. By using detached leaves, only a small piece of host tissue is available, with little metabolic activity and without any possibility of drawing resources from the whole plant to use for defence against pathogens (cf. Verreet and Hoffmann, 1986; 1990). The presence of antifungal compounds in wheat plants was investigated by Baker and Smith (1977) who extracted the benzoxazinone DIMBOA (2,4-dihydroxy-7-methoxy-2H-1,4-benzoxazin-3/4H/-one) from seedlings up to 3 weeks old. However, the role of DIMBOA was not studied further. Additional, slightly antifungal compounds were found in older seedlings and in field-grown plants but Baker and Smith (1977) suggested that the role of these compounds in resistance could be debated.

In addition to resistance expressed after infection of the host, the wheat plant has various ways to avoid infection. There are several reports that wheat cultivars heading or maturing early in the growth season are more heavily infected or damaged by *S. nodorum* than cultivars maturing late (Brönnimann *et al.*, 1973; Scott, 1973; Scott *et al.*, 1982; Cunfer *et al.*, 1988; Woś and Maćkowiak, 1995). Likewise, it has been observed that cultivars with long straw are often less prone to serious disease attacks than cultivars with short straw (Brönnimann *et al.*, 1973; Scott, 1973; Scott *et al.*, 1982). However, in other studies, poor correlation between resistance and long straw (Cunfer *et al.*, 1988; Arseniuk *et al.*, 1995) or late heading (Arseniuk *et al.*, 1995) was observed. Scott *et al.* (1982) conducted an extensive genetic study of the relationship between resistance against *S. nodorum* and straw length as well as heading date. They found that resistance and long straw segregated together and, to a lesser degree, resistance and late heading date. This suggested pleiotropy in the genetic control of these traits whereas linkage was possibly not involved. However, in addition to part of the resistance segregating together with straw length and heading date, another considerable part was found to be independent of these traits.

Tvaruzek (1994) observed, from studies of 25 wheat cultivars and breeding lines, that genotypes with long and heavy ears were generally the most susceptible, perhaps because they provided the best nutritional base for fungal development.

Septoria tritici

Resistance against *S. tritici* in certain wheat cultivars appears to be controlled by relatively few genes (Gough and Tuleen, 1979; Rosielle and Brown, 1979; Danon *et al.*, 1982; van Ginkel and Scharen, 1986; Danon and Eyal, 1990; Somasco *et al.*, 1996) although there are also reports of quantitative resistance, but this was not further specified (Jlibene *et al.*, 1994). Dominant, partially dominant, recessive, additive and non-additive gene effects, as well as epistasis, have been reported (Danon *et al.*, 1982; van Ginkel and Scharen, 1986; 1988a; Danon and Eyal, 1990; Jlibene *et al.*, 1994; Somasco *et al.*, 1996).

The expression of resistance against *S. tritici* has been studied in different cultivars of bread and durum wheat inoculated with different isolates of the fungus, for example by using a bread wheat cultivar as a susceptible and a durum wheat cultivar as a resistant host (Hilu and Bever, 1957).

Amongst the most conspicuous macroscopic differences in fungal development between resistant and susceptible hosts are reduced chlorotic and necrotic leaf area (Shaner *et al.*, 1975; Eyal *et al.*, 1985; Ballantyne and Thomson, 1995), reduced lesion numbers (Hilu and Bever, 1957), reduced infection frequency (Loughman *et al.*, 1996), reduced numbers of leaves showing infection (Hilu and Bever, 1957), reduced lesion size (Hilu and Bever, 1957) and reduced formation of pycnidia (Shaner *et al.*, 1975; Danon *et al.*, 1982; Yechilevich-Auster *et al.*, 1983; Eyal and Levy, 1987; Hess and Shaner, 1987a; Cohen and Eyal, 1993; Ballantyne and Thomson, 1995; Loughman *et al.*, 1996). The incubation and latent period may also be longer in resistant than in susceptible cultivars (Hilu and Bever, 1957; Loughman *et al.*, 1996).

The formation of necrosis and pycnidia appears to be under the control of different genes (Cohen and Eyal, 1993; Kema *et al.*, 1996a,b). Due to the high environmental influence on necrosis formation, formation of pycnidia has been suggested as the most reliable measure of resistance (Kema *et al.*, 1996a; Somasco *et al.*, 1996). However, Ballantyne and Thomson (1995) concluded that pycnidia formation did not allow a distinction between fungal isolates when infecting hosts showed only a small variation in resistance. Saadaoui (1987) recommended that formation of both necrosis and pycnidia be included in disease assessments.

Resistance against *S. tritici* appears not to be manifested until after penetration into the host. Thus, pre-penetration growth, including conidial germination, germ tube length, and branching and formation of appressorium-like structures, has been reported to be similar in resistant and in susceptible interactions (Hilu and Bever, 1957; Cohen and Eyal, 1993). The latter also reported that the extent of stomatal penetration was unaffected.

After penetration, hyphal growth was slow in resistant interactions but occurred in a similar manner to a susceptible interaction, except for limited extension and branching (Hilu and Bever, 1957). However, when symptoms began to appear, Hilu and Bever (1957) observed similar amounts of hyphal growth per unit area of killed tissue in both susceptible and resistant interactions. Almost no tissue colonization was observed by Kema *et al.* (1996c) in resistant interactions, where only a few hyphae were seen intercellularly in the mesophyll near substomatal cavities.

Resistance may also prevent pycnidia maturation. Cohen and Eyal (1993) observed that in a resistant interaction there was no macroscopic pycnidia formation. However, microscopic examination revealed that although pycnidia formation had commenced, their development stopped at various stages. Conversely, Hilu and Bever (1957) found a tendency for resistance to reduce pycnidia size although their numbers were unaffected.

Very limited attempts have been made to explain and characterize the mechanisms behind the observed differences in resistance of wheat cultivars to *S. tritici*. Cohen and Eyal (1993) observed that autofluorescent materials accumulated in resistant interactions, most often associated with stomata, epidermal cell walls and mesophyll cells around substomatal cavities. Callose accumulation was also observed, most often in epidermal cell walls, but there was no indication of lignin formation. Kema *et al.* (1996c) found no evidence either that resistance was associated with accumulation of polyphenolic compounds and no indication of papilla formation, compartmentalization or hypersensitive responses. They, therefore, suggested that resistance was a result of the production of compounds inhibiting fungal colonization and pycnidia formation.

The causes of tolerance to *S. tritici* in wheat were studied by Zuckerman *et al.* (1997). They found that in the wheat cultivar Miriam, tolerance resulted from a more effective fixation of CO_2 per unit of infected leaf, per tiller and per unit of chlorophyll than by remobilization of carbohydrates stored previously.

With *S. nodorum*, a correlation has been observed between resistance against *S. tritici* and long straw of wheat (Tavella, 1978; Rosielle and Brown, 1979; Danon *et al.*, 1982; Eyal *et al.*, 1983; Jlibene *et al.*, 1994; Somasco *et al.*, 1996) and late maturity (Shaner *et al.*, 1975; Tavella, 1978; Rosielle and Brown, 1979; Danon *et al.*, 1982), although this association was not always found (Danon and Eyal, 1990). Certain genetic correlations have also been found between resistance and long straw and late heading (Rosielle and Brown, 1979). However, based on extensive crossings of wheat cultivars with long or short straw and early or late heading, Danon *et al.* (1982) concluded that resistance, heading date and straw length were controlled by different genes and that there was no evidence for genetic linkage or pleiotropy between long straw and resistance although there were indications that linkage existed between late heading date and resistance. Other studies involving crosses between cultivars varying in resistance, straw length and heading date have also revealed that the connection between these traits may not be genetically determined (Jlibene *et al.*, 1994; Somasco *et al.*, 1996).

With *S. nodorum*, the observed correlation between resistance and long straw thus seems to be mainly the result of disease escape and environmental influence. Hence, inoculum may face more difficulties in moving upwards in a tall plant than in a short plant. Inoculation is most easily achieved either by horizontal transfer or by rain splash (cf. Somasco *et al.*, 1996; Lovell *et al.*, 1997). Furthermore, the microclimate in a tall crop is drier due to the more open structure of the plants, thus hampering infection (cf. Brönnimann *et al.*, 1973; Somasco *et al.*, 1996). The correlation between late maturity and resistance may be explained because early heading often coincides with weather more favourable for disease development than later heading.

Characterization of the mechanisms responsible for resistance against *S. nodorum* and *S. tritici* has not yet progressed very far. In many of the wheat cultivars examined, it has been claimed that resistance to *S. nodorum* was 'partial', probably because there is no consensus about the presence of specific resistance in this host-pathogen interaction. The macroscopic expression of resistance of some wheat cultivars has been studied to determine at which developmental stages infection ceases, but failed to reveal the resistance mechanisms responsible. Hence, the consequences rather than the causes of resistance have been investigated.

However, since only limited studies have been made on resistance expression against *S. nodorum* and *S. tritici* and since results from these investigations are not detailed, it is not possible to state that any particular component of resistance is a result of race-non-specific or of race-specific resistance. The fact that a cultivar possesses resistance at all must be reflected in an influence on parameters examined (e.g. reduced lesion numbers or reduced sporulation). The best documented resistance mechanism appears to be papilla resistance against infection attempts by *S. nodorum* in the outer epidermal cell wall. Papilla formation has not been reported for infection attempts by *S. tritici,* mainly because penetration by this pathogen is considered to occur largely through stomata. Apart from papilla formation, only observations, or suggestions, of accumulations of different compounds with a putative role in defence have been made. The precise role of these accumulations has, however, not been studied further.

Investigations on expression and mechanisms of resistance are complicated by the fact that the genetics of resistance are not well understood. Examination of differences in fungal development between two cultivars may, therefore, reflect a difference between resistant interactions (cf. expression of resistance against *S. nodorum*) and not between resistant and susceptible interactions, thereby perhaps overlooking some of the manifestations of resistance.

Furthermore, cultivars may also vary in resistance during their lifetime as certain differences in the reaction of seedlings and of adult plants have been reported both for *S. nodorum* (Aust and Hau, 1981; 1983; Sharma and Brown, 1983a; Verreet *et al.*, 1987; Cunfer *et al.*, 1992; Arseniuk *et al.*, 1995) and for *S. tritici* (Kema and van Silfhout, 1997), although some of the differences may be due to environmental influence. It has been demonstrated that young wheat plants are often more resistant against *S. nodorum* than when they become older (Aust and Hau, 1981; 1983; Sharma and Brown, 1983a; Verreet *et al.*, 1987), probably because the fungus prefers to colonize physiologically mature tissue (Sharma and Brown,1983a; Wainshilbaum and Lipps, 1991). Scharen (1963) did not observe any difference in susceptibility of leaf tissue of a spring wheat cultivar aged between 20 and 106 days, possibly because he inoculated the plants by injecting the conidia which meant that any resistance to penetration was ineffective.

Clearly, much work is still needed to investigate the mechanisms of resistance in the interactions between wheat (and other hosts) and *S. nodorum* and *S. tritici*. This should be initiated with histopathological investigations which form the basis for further biochemical and molecular characterization of resistance. Characterization of resistance against *S. nodorum* by molecular techniques has apparently only recently started (Stevens *et al.*, 1996; H.J.L. Jørgensen, P.S. Lübeck, H. Thordal-Christensen, E. de Neergaard and V. Smedegaard-Petersen, unpublished results).

References

Ahmed, H.U., Mundt, C.C., Hoffer, M.E. and Coakley, S.M. (1996) Selective influence of wheat cultivars on pathogenicity of *Mycosphaerella graminicola* (anamorph *Septoria tritici*). *Phytopathology* 86, 454-458.

Ao, H.C. and Griffiths, E. (1976) Change in virulence of *Septoria nodorum* and *S. tritici* after passage through alternative hosts. *Transactions of the British Mycological Society* 66, 337-340.

Arseniuk, E., Czembor, H.J., Sowa, W., Krysiak, H. and Zimny, J. (1995) Genotypowa reakcja pszenzyta, pszenicy i zyta na inokulacje *Stagonospora* (= *Septoria*) *nodorum* w warunkach polowych oraz *S. nodorum* i *Septoria tritici* w warunkach kontrolowanych. *Biuletyn Instytutu Hodowli i Aklimatyzacji Roslin* No. 195/196, 209-246.

Aust, H.-J. and Hau, B. (1981) Einfluss kompensatorischer Wirkungen auf die Dauer der Latenzzeit von *Septoria nodorum*. *Zeitschrift für Pflanzenkrankheiten und Pflanzenschutz* 88, 655-664.

Aust, H.-J. and Hau, B. (1983) Die Latenzzeit von *Septoria nodorum* in Abhängigkeit von der ontogenetisch bedingten Anfälligkeit des Sommerweizens. *Zeitschrift für Pflanzenkrankheiten und Pflanzenschutz* 90, 55-62.

Babadoost, M. and Herbert, T.T. (1984) Factors affecting infection of wheat seedlings by *Septoria nodorum*. *Phytopathology* 74, 592-595.

Baker, C.J. (1970) Influence of environmental factors on development of symptoms on wheat seedlings grown from seed infected with *Leptosphaeria nodorum*. *Transactions of the British Mycological Society* 55, 443-447.

Baker, E.A. and Smith, I.M. (1977) Antifungal compounds in winter wheat resistant and susceptible to *Septoria nodorum*. *Annals of Applied Biology* 87, 67-73.

Baker, E.A. and Smith, I.M. (1978) Development of resistant and susceptible reactions in wheat on inoculation with *Septoria nodorum*. *Transactions of the British Mycological Society* 71, 475-482.

Baker, E.A. and Smith, I.M. (1979) Components of *Septoria nodorum* infection in winter wheat: lesion number and lesion size. *Transactions of the British Mycological Society* 73, 57-63.

Ballantyne, B. and Thomson, F. (1995) Pathogenic variation in Australian isolates of *Mycosphaerella graminicola*. *Australian Journal of Agricultural Research* 46, 921-934.

Beck, J.J. and Ligon, J.M. (1995) Polymerase chain reaction assays for the detection of *Stagonospora nodorum* and *Septoria tritici* in wheat. *Phytopathology* 85, 319-324.

Benedict, W.G. (1971) Differential effect of light intensity on the infection of wheat by *Septoria tritici* Desm. under controlled environmental conditions. *Physiological Plant Pathology* 1, 55-66.

Bethenod, O., Bousquet, J.-F., Laffray, D. and Louguet, P. (1982) Réexamen des modalités d'action de l'ochracine sur la conductance stomatique des feuilles de plantules de blé, *Triticum aestivum* L., cv. 'Etoile de Choisy'. *Agronomie* 2, 99-101.

Bird, P.M. and Ride, J.P. (1981) The resistance of wheat to *Septoria nodorum*: fungal development in relation to host lignification. *Physiological Plant Pathology* 19, 289-299.

Bousquet, J.-F. and Skajennikoff, M. (1974) Isolement et mode d'action d'une phytotoxine produite en culture par *Septoria nodorum* Berk. *Phytopathologische Zeitschrift* 80, 355-360.

Bousquet, J.-F., Skajennikoff, M., Bethenod, O. and Chartier, P. (1977) Action dépressive de l'ochracine, phytotoxine synthétisée par le *Septoria nodorum* (Berk.) Berk., sur l'assimilation du CO_2 par des plantules de blé. *Annales de Phytopathologie* 9, 503-510.

Bousquet, J.-F., Belhomme de Franqueville, H., Kollmann, A. and Fritz, R. (1980) Action de la septorine, phytotoxine synthétisée par *Septoria nodorum*, sur la phosphorylation oxydative dans les mitochondries isolées de coléoptiles de blé. *Canadian Journal of Botany* 58, 2575-2580.

Brokenshire, T. (1975) The rôle of graminaceous species in the epidemiology of *Septoria tritici* on wheat. *Plant Pathology* 24, 33-38.

Brönnimann, A. (1975) Beitrag zur Genetik der Toleranz auf *Septoria nodorum* Berk. bei Weizen (*Triticum aestivum*). *Zeitschrift für Pflanzenzüchtung* 75, 138-160.

Brönnimann, A., Fossati, A. and Häni, F. (1973) Ausbreitung von *Septoria nodorum* Berk. und Schädigung bei künstlich induzierten Halmlänge-Mutanten der Winterweizensorte 'Zenith' (*Triticum aestivum*). *Zeitschrift für Pflanzenzüchtung* 70, 230-245.

Brown-Guedira, G.L., Gill, B.S., Bockus, W.W., Cox, T.S., Hatchett, J.H., Leath, S., Peterson, C.J., Thomas, J.B. and Zwer, P.K. (1996) Evaluation of a collection of wild timopheevi wheat for resistance to disease and arthropod pests. *Plant Disease* 80, 928-933.

Cohen, L. and Eyal, Z. (1993) The histology of processes associated with the infection of resistant and susceptible wheat cultivars with *Septoria tritici*. *Plant Pathology* 42, 737-743.

Cornish, P.S., Baker, G.R. and Murray, G.M. (1990) Physiological responses of wheat (*Triticum aestivum*) to infection with *Mycosphaerella graminicola* causing septoria tritici blotch. *Australian Journal of Agricultural Research* 41, 317-327.

Cunfer, B.M. (1984) Change of virulence of *Septoria nodorum* during passage through barley and wheat. *Annals of Applied Biology* 104, 61-68.

Cunfer, B.M. and Youmans, J. (1983) *Septoria nodorum* on barley and relationships among isolates from several hosts. *Phytopathology* 73, 911-914.

Cunfer, B.M., Stooksbury, D.E. and Johnson, J.W. (1988) Components of partial resistance to *Leptosphaeria nodorum* among seven soft red winter wheats. *Euphytica* 37, 129-140.

Cunfer, B.M., Johnson, J.W. and Brown, A.R. (1992) Resistance in semihardy winter barley to leaf and glume blotch caused by *Stagonospora nodorum*. *Plant Disease* 76, 1227-1230.

Danon, T. and Eyal, Z. (1990) Inheritance of resistance to two *Septoria tritici* isolates in spring and winter bread wheat cultivars. *Euphytica* 47, 203-214.

Danon, T., Sacks, J.M. and Eyal, Z. (1982) The relationships among plant stature, maturity class, and susceptibility to Septoria leaf blotch of wheat. *Phytopathology* 72, 1037-1042.

Devys, M., Bousquet, J.-F., Skajennikoff, M. and Barbier, M. (1974) L'ochracine (melléine), phytotoxine isolée du milieu de culture de *Septoria nodorum* Berk. *Phytopathologische Zeitschrift* 81, 92-94.

Devys, M., Bousquet, J.-F., Kollmann, A. and Barbier, M. (1978) La septorine, nouvelle pyrazine substituée, isolée du milieu de culture de *Septoria nodorum* Berk. champignon phytopathogéne. *Comptes Rendu Hebdomadaires des Séances de L'Académie des Sciences, Série C* 286, 457-458.

Dickopp, S. and Hindorf, H. (1994) Befallsverlauf und Pyknidienentwicklung von *Septoria* spp. auf Winterweizen in Abhängigkeit von der Witterung. *Mededelingen-Faculteit Landbouwkundige en Toegepaste Biologische Wetenschappen, Universiteit Gent* 59, 859-867.

Essad, S., Bousquet, J.-F. and Maunoury, C. (1981) Action de l'ochracine, phytotoxine de *Septoria nodorum* Berk., sur le cycle mitotique de *Triticum aestivum* L. *Agronomie* 1, 689-694.

Eyal, Z. and Brown, M.B. (1976) A quantitative method for estimating density of *Septoria tritici* pycnidia on wheat leaves. *Phytopathology* 66, 11-14.

Eyal, Z. and Levy, E. (1987) Variations in pathogenicity patterns of *Mycosphaerella graminicola* within *Triticum* spp. in Israel. *Euphytica* 36, 237-250.

Eyal, Z., Amiri, Z. and Wahl, I. (1973) Physiologic specialization of *Septoria tritici*. *Phytopathology* 63, 1087-1091.

Eyal, Z., Brown, J.F., Krupinsky, J.M. and Scharen, A.L. (1977) The effect of postinoculation periods of leaf wetness on the response of wheat cultivars to infection by *Septoria nodorum*. *Phytopathology* 67, 874-878.

Eyal, Z., Wahl, I. and Prescott, J.M. (1983) Evaluation of germplasm response to Septoria leaf blotch of wheat. *Euphytica* 32, 439-446.

Eyal, Z., Scharen, A.L., Huffman, M.D. and Prescott, J.M. (1985) Global insights into virulence frequencies of *Mycosphaerella graminicola*. *Phytopathology* 75, 1456-1462.

Eyal, Z., Scharen, A.L., Prescott, J.M. and van Ginkel, M. (1987) *The Septoria Diseases of Wheat: Concepts and Methods of Disease Management*. Centro Internacional de Mejoramiento de Maíz y Trigo (CIMMYT), México, D.F., 46pp.

Fellows, H. (1962) Effects of light, temperature, and fertilizer on infection of wheat leaves by *Septoria tritici*. *Plant Disease Reporter* 46, 846-848.

Fitzgerald, W. and Cooke, B.M. (1982) Response of wheat and barley isolates of *Septoria nodorum* to passage through barley and wheat cultivars. *Plant Pathology* 31, 315-324.

Gough, F.J. and Tuleen, N. (1979) Septoria leaf blotch resistance among *Agropyron elongatum* chromosomes in *Triticum aestivum* 'Chinese Spring'. *Cereal Research Communications* 7, 275-280.

Griffiths, H.M. and Jones, D.G. (1987) Components of partial resistance as criteria for identifying resistance. *Annals of Applied Biology* 110, 603-610.

Hargreaves, J.A. and Keon, J.P.R. (1986) A comparison of the reaction sites associated with the resistance of coleoptiles of barley, oats and wheat to infection by *Septoria nodorum*. *Journal of Phytopathology* 115, 72-82.

Harrower, K.M. (1977) Specialization of *Leptosphaeria nodorum* to alternative graminaceous hosts. *Transactions of the British Mycological Society* 68, 101-103.

Hess, D.E. and Shaner, G. (1987a) Effect of moisture and temperature on development of Septoria tritici blotch in wheat. *Phytopathology* 77, 215-219.

Hess, D.E. and Shaner, G. (1987b) Effect of moisture on Septoria tritici blotch development on wheat in the field. *Phytopathology* 77, 220-226.

Hilu, H.M. and Bever, W.M. (1957) Inoculation, oversummering, and suscept-pathogen relationship of *Septoria tritici* on *Triticum* species. *Phytopathology* 47, 474-480.

Holmes, S.J.I. and Colhoun, J. (1970) *Septoria nodorum* as a pathogen of barley. *Transactions of the British Mycological Society* 55, 321-325.

Holmes, S.J.I. and Colhoun, J. (1971) Infection of wheat seedlings by *Septoria nodorum* in relation to environmental factors. *Transactions of the British Mycological Society* 57, 493-500.

Holmes, S.J.I. and Colhoun, J. (1974) Infection of wheat by *Septoria nodorum* and *S. tritici* in relation to plant age, air temperature and relative humidity. *Transactions of the British Mycological Society* 63, 329-338.

Jeger, M.J., Jones, D.G. and Griffiths, E. (1983) Components of partial resistance of wheat seedlings to *Septoria nodorum*. *Euphytica* 32, 575-584.

Jlibene, M., Gustafson, J.P. and Rajaram, S. (1994) Inheritance of resistance to *Mycosphaerella graminicola* in hexaploid wheat. *Plant Breeding* 112, 301-310.

Jørgensen, H.J.L., de Neergaard, E. and Smedegaard-Petersen, V. (1993) Histological examination of the interaction between *Rhynchosporium secalis* and susceptible and resistant cultivars of barley. *Physiological and Molecular Plant Pathology* 42, 345-358.

Karjalainen, R. and Lounatmaa, K. (1986) Ultrastructure of penetration and colonization of wheat leaves by *Septoria nodorum*. *Physiological and Molecular Plant Pathology* 29, 263-270.

Kema, G.H.J. and van Silfhout, C.H. (1997) Genetic variation for virulence and resistance in the wheat-*Mycosphaerella graminicola* pathosystem. III. Comparative seedling and adult plant experiments. *Phytopathology* 87, 266-272.

Kema, G.H.J., Annone, J.G., Sayoud, R., van Silfhout, C.H., van Ginkel, M. and de Bree, J. (1996a) Genetic variation for virulence and resistance in the wheat-*Mycosphaerella graminicola* pathosystem. I. Interactions between pathogen isolates and host cultivars. *Phytopathology* 86, 200-212.

Kema, G.H.J., Sayoud, R., Annone, J.G. and van Silfhout, C.H. (1996b) Genetic variation for virulence and resistance in the wheat-*Mycosphaerella graminicola* pathosystem. II. Analysis of interactions between pathogen isolates and host cultivars. *Phytopathology* 86, 213-220.

Kema, G.H.J., Yu, D., Rijkenberg, F.H.J., Shaw, M.W. and Baayen, R.P. (1996c) Histology of the pathogenesis of *Mycosphaerella graminicola* in wheat. *Phytopathology* 86, 777-786.

Kent, S.S. and Strobel, G.A. (1976) Phytotoxin from *Septoria nodorum*. *Transactions of the British Mycological Society* 67, 354-358.

Keon, J.P.R. and Hargreaves, J.A. (1984) The response of barley leaf epidermal cells to infection by *Septoria nodorum*. *The New Phytologist* 98, 387-398.

Khokhar, L.K. and Pacumbaba, R.P. (1987) Alternative gramineous hosts of *Leptosphaeria nodorum* including two new records for the U.S.A. *Journal of Phytopathology* 120, 75-80.

King, J.E., Cook, R.J. and Melville, S.C. (1983) A review of *Septoria* diseases of wheat and barley. *Annals of Applied Biology* 103, 345-373.

Kleijer, G., Brönnimann, A. and Fossati, A. (1977) Chromosomal location of a dominant gene for resistance at the seedling stage to *Septoria nodorum* Berk. in the wheat variety 'Atlas 66'. *Zeitschrift für Pflanzenzüchtung* 78, 170-173.

Krupinsky, J.M. (1982) Comparative pathogenicity of *Septoria nodorum* isolated from *Triticum aestivum* and *Agropyron* species. *Phytopathology* 72, 660-661.

Krupinsky, J.M. (1986) Virulence on wheat of *Leptosphaeria nodorum* isolates from *Bromus inermis*. *Canadian Journal of Plant Pathology* 8, 201-207.

Krupinsky, J.M. (1997a) Aggressiveness of *Stagonospora nodorum* isolates obtained from wheat in the northern Great Plains. *Plant Disease* 81, 1027-1031.

Krupinsky, J.M. (1997b) Aggressiveness of *Stagonospora nodorum* isolates from perennial grasses on wheat. *Plant Disease* 81, 1032-1036.

Krupinsky, J.M. (1997c) Stability of *Stagonospora nodorum* isolates from perennial grass hosts after passage through wheat. *Plant Disease* 81, 1037-1041.

Krupinsky, J.M., Scharen, A.L. and Schillinger, J.A. (1973) Pathogenic variation in *Septoria nodorum* (Berk.) Berk. in relation to organ specificity, apparent photosynthetic rate and yield of wheat. *Physiological Plant Pathology* 3, 187-194.

Krupinsky, J.M., Craddock, J.C. and Scharen, A.L. (1977) *Septoria* resistance in wheat. *Plant Disease Reporter* 61, 632-636.

Laffray, D., Bousquet, J-F., Bethenod, O. and Louguet, P. (1982) Mécanismes de l'action dépressive de l'ochracine, phytotoxine synthétisée par *Septoria nodorum* Berk., sur le degré d'ouverture des stomates des feuilles de plantules de blé et de *Pelargonium x hortorum*. *Agronomie* 2, 25-30.

Lancashire, P.D. and Jones, D.G. (1985) Components of partial resistance to *Septoria nodorum* in winter wheat. *Annals of Applied Biology* 106, 541-553.

Lehtinen, U. (1993) Plant cell wall degrading enzymes of *Septoria nodorum*. *Physiological and Molecular Plant Pathology* 43, 121-134.

Loughman, R., Wilson, R.E. and Thomas, G.J. (1996) Components of resistance to *Mycosphaerella graminicola* and *Phaeosphaeria nodorum* in spring wheats. *Euphytica* 89, 377-385.

Lovell, D.J., Parker, S.R., Hunter, T., Royle, D.J. and Coker, R.R. (1997) Influence of crop growth and structure on the risk of epidemics by *Mycosphaerella graminicola* (*Septoria tritici*) in winter wheat. *Plant Pathology* 46, 126-138.

Luthra, J.C., Sattar, A. and Ghani, M.A. (1938) A comparative study of species of *Septoria* occurring on wheat. *The Indian Journal of Agricultural Science* 7, 271-289.

Magboul, A.M., Geng, S., Gilchrist, D.G. and Jackson, L.F. (1992) Environmental influence on the infection of wheat by *Mycosphaerella graminicola*. *Phytopathology* 82, 1407-1413.

Magro, P. (1984) Production of polysaccharide-degrading enzymes by *Septoria nodorum* in culture and during pathogenesis. *Plant Science Letters* 37, 63-68.

Mäkelä, K. (1977) *Septoria* and *Selenophoma* species on Gramineae in Finland. *Annales Agriculturae Fenniae* 16, 256-276.

Martin, S.I. and Cooke, B.M. (1979) Effect of wheat and barley hosts on pathogenicity and cultural behaviour of barley and wheat isolates of *Septoria nodorum*. *Transactions of the British Mycological Society* 72, 219-224.

Mielke, H. (1995) Untersuchungen zur Anfälligkeit inländischer Weizen- und Triticalesorten gegenüber *Septoria tritici* Rob. ex Desm. *Nachrichtenblatt des Deutschen Pflanzenschutzdienstes* 47, 96-99.

Mullaney, E.J., Martin, J.M. and Scharen, A.L. (1982) Generation mean analysis to identify and partition the components of genetic resistance to *Septoria nodorum* in wheat. *Euphytica* 31, 539-545.

Murray, G.M., Martin, R.H. and Cullis, B.R. (1990) Relationship of the severity of Septoria tritici blotch of wheat to sowing time, rainfall at heading and average susceptibility of wheat cultivars in the area. *Australian Journal of Agricultural Research* 41, 307-315.

Nelson, L.R. (1980) Inheritance of resistance to *Septoria nodorum* in wheat. *Crop Science* 20, 447-449.

Newton, A.C. and Caten, C.E. (1991) Characteristics of strains of *Septoria nodorum* adapted to wheat or to barley. *Plant Pathology* 40, 546-553.

O'Reilly, P. and Downes, M.J. (1986) Form of survival of *Septoria nodorum* on symptomless winter wheat. *Transactions of the British Mycological Society* 86, 381-385.

Osbourn, A.E., Scott, P.R. and Caten, C.E. (1986) The effects of host passaging on the adaptation of *Septoria nodorum* to wheat or barley. *Plant Pathology* 35, 135-145.

Osbourn, A.E., Caten, C.E. and Scott, P.R. (1987) Adaptation of *Septoria nodorum* to wheat or barley on detached leaves. *Plant Pathology* 36, 565-576.

Perelló, A., Cordo, C.A. and Alippi, H.E. (1990) Características morfológicas y patogénicas de aislamientos de *Septoria tritici* Rob ex Desm. *Agronomie* 10, 641-648.

Rapilly, F., Auriau, P., Laborie, Y., Depatureaux, C. and Skajennikoff, M. (1981) Résistance partielle du blé, *Triticum aestivum* L., à *Septoria nodorum* Berk. - Etude du temps d'incubation. *Agronomie* 1, 771-782.

Rosielle, A.A. (1972) Sources of resistance in wheat to speckled leaf blotch caused by *Septoria tritici*. *Euphytica* 21, 152-161.

Rosielle, A.A. and Brown, A.G.P. (1979) Inheritance, heritability and breeding behaviour of three sources of resistance to *Septoria tritici* in wheat. *Euphytica* 28, 385-392.

Rubiales, D., Snijders, C.H.A., Nicholson, P. and Martín, A. (1996) Reaction of tritordeum to *Fusarium culmorum* and *Septoria nodorum*. *Euphytica* 88, 165-174.

Rufty, R.C., Herbert, T.T. and Murphy, C.F. (1981) Variation in virulence in isolates of *Septoria nodorum*. *Phytopathology* 71, 593-596.

Saadaoui, E.M. (1987) Physiologic specialization of *Septoria tritici* in Morocco. *Plant Disease* 71, 153-155.

Scharen, A.L. (1963) Effect of age of wheat tissues on susceptibility to *Septoria nodorum*. *Plant Disease Reporter* 47, 952-954.

Scharen, A.L. (1964) Environmental influences on development of glume blotch in wheat. *Phytopathology* 54, 300-303.

Scharen, A.L. (1966) Cyclic production of pycnidia and spores in dead wheat tissue by *Septoria nodorum*. *Phytopathology* 56, 580-581.

Scharen, A.L. and Eyal, Z. (1980) Measurement of quantitative resistance to *Septoria nodorum* in wheat. *Plant Disease* 64, 492-496.

Scharen, A.L. and Eyal, Z. (1983) Analysis of symptoms on spring and winter wheat cultivars inoculated with different isolates of *Septoria nodorum*. *Phytopathology* 73, 143-147.

Scharen, A.L. and Krupinsky, J.M. (1969) Effect of *Septoria nodorum* infection on CO_2 absorption and yield of wheat. *Phytopathology* 59, 1298-1301.

Scharen, A.L. and Krupinsky, J.M. (1970) Cultural and inoculation studies of *Septoria nodorum*, cause of glume blotch of wheat. *Phytopathology* 60, 1480-1485.

Scharen, A.L. and Taylor, J.M. (1968) CO_2 assimilation and yield of Little Club wheat infected by *Septoria nodorum*. *Phytopathology* 58, 447-451.

Scharen, A.L., Eyal, Z., Huffman, M.D. and Prescott, J.M. (1985) The distribution and frequency of virulence genes in geographically separated populations of *Leptosphaeria nodorum*. *Phytopathology* 75, 1463-1468.

Scott, P.R. (1973) Incidence and effects of *Septoria nodorum* on wheat cultivars. *Annals of Applied Biology* 75, 321-329.

Scott, P.R., Benedikz, P.W. and Cox, C.J. (1982) A genetic study of the relationship between height, time of ear emergence and resistance to *Septoria nodorum* in wheat. *Plant Pathology* 31, 45-60.

Shaner, G. and Finney, R.E. (1976) Weather and epidemics of Septoria leaf blotch of wheat. *Phytopathology* 66, 781-785.

Shaner, G., Finney, R.E. and Patterson, F.L. (1975) Expression and effectiveness of resistance in wheat to Septoria leaf blotch. *Phytopathology* 65, 761-766.

Sharma, H.S.S. and Brown, A.E. (1983a) Effect of plant and leaf age on the susceptibility of wheat and barley cultivars to *Septoria nodorum* Berk. *Record of Agricultural Research* 31, 59-62.

Sharma, H.S.S. and Brown, A.E. (1983b) Temperature enhanced sectoring in barley isolates of *Septoria nodorum* and the possible relationship with varying pathogenicity. *Transactions of the British Mycological Society* 81, 263-267.

Sharma, H.S.S., Brown, A.E. and Swinburne, T.R. (1982) Relationship of spore germination and appressoria formation in isolates of *Septoria nodorum* to pathogenicity on wheat and barley leaves. *Transactions of the British Mycological Society* 79, 519-526.

Shearer, B.L. and Zadoks, J.C. (1972a) Observations on the host range of an isolate of *Septoria nodorum* from wheat. *Netherlands Journal of Plant Pathology* 78, 153-159.

Shearer, B.L. and Zadoks, J.C. (1972b) The latent period of *Septoria nodorum* in wheat. 1. The effect of temperature and moisture treatments under controlled conditions. *Netherlands Journal of Plant Pathology* 78, 231-241.

Shipton, W.A., Boyd, W.R.J., Rosielle, A.A. and Shearer, B.I. (1971) The common *Septoria* diseases of wheat. *The Botanical Review* 37, 231-262.

Shtienberg, D. (1992) Effects of foliar diseases on gas exchange processes: a comparative study. *Phytopathology* 82, 760-765.

Sieber, T. (1989) Substratabbauvermögen endophytischer Pilze von Weizenkörnern. *Zeitschrift für Pflanzenkrankheiten und Pflanzenschutz* 96, 627-632.

Smedegård-Petersen, V. (1974) *Leptosphaeria nodorum* (*Septoria nodorum*), a new pathogen on barley in Denmark, and its physiologic specialization on barley and wheat. *Friesia* 10, 251-264.

Somasco, O.A., Qualset, C.O. and Gilchrist, D.G. (1996) Single-gene resistance to Septoria tritici blotch in the spring wheat cultivar 'Tadinia'. *Plant Breeding* 115, 261-267.

Sprague, R. (1950a) Some leafspot fungi on western Gramineae. V. *Mycologia* 42, 758-771.

Sprague, R. (1950b) *Diseases of Cereals and Grasses in North America (Fungi, Except Smuts and Rusts)*. The Ronald Press Company, New York, 538 pp.

Tavella, C.M. (1978) Date of heading and plant height of wheat varieties, as related to Septoria leaf blotch damage. *Euphytica* 27, 577-580.

Tiedemann, A.V. and Firsching, K.H. (1993) Effects of ozone exposure and leaf age of wheat on infection processes of *Septoria nodorum* Berk. *Plant Pathology* 42, 287-293.

Trottet, M., Dosba, F. and Tanguy, A.-M. (1983) Analyse cytogénétique et comportement vis-à-vis de *Septoria nodorum* d'hybrides *Triticum* sp. x *Aegilops squarrosa* et de leurs descendances. *Agronomie* 3, 659-664.

Tvaruzek, L. (1994) Hodnocení výnosové reakce odrud psenice ozimé lisících se v morfologii a produktivité klausu na napadení branicatkou plevovou (*Septoria nodorum* Berk.) *Ochrana Rostlin* 30, 49-58.

Ueng, P.P., Cunfer, B.M., Alano, A.S., Youmans, J.D. and Chen, W. (1995) Correlation between molecular and biological characters in identifying the wheat and barley biotypes of *Stagonospora nodorum*. *Phytopathology* 85, 44-52.

van der Plank, J.E. (1968) *Disease Resistance in Plants*. Academic Press, New York, 206pp.

van Ginkel, M. and Scharen, A.L. (1986) Genetics of resistance in durum wheat to *Septoria tritici*. *Phytopathology* 76, 1112.

van Ginkel, M. and Scharen, A.L. (1988a) Diallel analysis of resistance to *Septoria tritici* isolates in durum wheat. *Euphytica* 38, 31-37.

van Ginkel, M. and Scharen, A.L. (1988b) Host-pathogen relationships of wheat and *Septoria tritici*. *Phytopathology* 78, 762-766.

Verreet, J.A. and Hoffmann, G.M. (1986) The distribution of nitrogen in plant organs after infection of wheat with *Septoria nodorum*. *Mededelingen van de Faculteit Landbouwwetenschappen Rijksuniversiteit Gent* 51/2b, 581-588.

Verreet, J.A. and Hoffmann, G.M. (1990) Auswirkungen des Blatt- und Ährenbefalls durch *Septoria nodorum* in verschiedenen Entwicklungsstadien des Weizens auf den N-Gehalt der Pflanzen und die Zusammensetzung der Aminosäuren. *Zeitschrift für Pflanzenkrankheiten und Pflanzenschutz* 97, 1-12.

Verreet, J.A., Hoffmann, G.M. and Amberger, A. (1987) Der Einfluss des Blatt- und Ährenbefalles durch *Septoria nodorum* in verschiedenen Entwicklungsstadien des Weizens auf den Chlorophyll- und Kohlenhydrat-Gehalt. *Zeitschrift für Pflanzenkrankheiten und Pflanzenschutz* 94, 462-477.

Wainshilbaum, S.J. and Lipps, P.E. (1991) Effect of temperature and growth stage of wheat on development of leaf and glume blotch caused by *Septoria tritici* and *S. nodorum*. *Plant Disease* 75, 993-998.

Weber, G.F. (1922) II. *Septoria* diseases of wheat. *Phytopathology* 12, 537-585 with plates XXIII-XXXVI.

Welling, B., Nielsen, G.C. and Nielsen, B.J. (1984) Sygdomme i vinterhvede. I. Gråplet (*Septoria tritici*). *Tidsskrift for Planteavl* 88, 519-526.

Woś, H. and Maćkowiak, W. (1995) Odpornosc polskich odmian pszenóyta ozimego na *Stagonospora nodorum*. *Biuletyn Instytutu Hodowli i Aklimatyzacji Roslin* No. 195/196, 183-189.

Yechilevich-Auster, M., Levi, E. and Eyal, Z. (1983) Assessment of interactions between cultivated and wild wheats and *Septoria tritici*. *Phytopathology* 73, 1077-1083.

Zinkernagel, V. and Riess, F. (1987) Elektronenmikrosckopische Untersuchungen der Infektionsstrukturen von *Septoria nodorum* (Berk.) Berk. an Weizen. *Mededelingen van de Faculteit Landbouwwetenschappen Rijksuniversiteit Gent* 52/3a, 841-850.

Zinkernagel, V., Riess, F., Wendland, M. and Bartscherer, H.-C. (1988) Infektionsstrukturen von *Septoria nodorum* in Blättern anfälliger Weizensorten. *Zeitschrift für Pflanzenkrankheiten und Pflanzenschutz* 95, 169-175.

Zuckerman, E., Eshel, A. and Eyal, Z. (1997) Physiological aspects related to tolerance of spring wheat cultivars to Septoria tritici blotch. *Phytopathology* 87, 60-65.

Chapter ten:

Genetics of Biological and Molecular Markers in *Mycosphaerella graminicola*, the Cause of Septoria Tritici Leaf Blotch of Wheat

G.H.J. Kema[1], E.C.P. Verstappen[1], C. Waalwijk[1], P.J.M. Bonants[1], J.R.A. de Koning[1], M. Hagenaar-de Weerdt[1], S. Hamza[2], J.G.P. Koeken[1] and T.A.J. van der Lee[3]

[1]*DLO-Research Institute for Plant Protection (IPO-DLO), P.O.Box 9060, 6700 GW Wageningen, The Netherlands.* [2]*Institut National Agronomique de Tunisie (INAT), 43 Av. Charles Nicolle, 1082 Tunis El Mahrajène, Tunisia.* [3]*Department of Phytopathology, Wageningen Agricultural University, Wageningen, The Netherlands*

Introduction

More than 150 years ago Desmazières (1842) described *Septoria tritici* as the cause of Septoria tritici leaf blotch of wheat. Since then, a number of key papers have been published to elucidate the relationship between this fungus and its host. Early work (Weber, 1922; Sprague, 1944; 1950) described the host range, preservation, and the effects of differences in artificial media and illumination on the fungus. Hilu and Bever (1957) studied the oversummering, inoculation techniques and host-parasite interactions. These papers, that also included some histological observations on pathogenesis, were the foundation for further work on the *S. tritici*-wheat pathosystem. Since then, it took approximately another decade for further papers on the subject to be published. Two areas of research emerged, one dealing mainly with epidemiology and (micro-) environmental effects on spore germination and infection (Cooke and Jones, 1970; 1971; Jones and Cooke, 1970; Benedict, 1971a,b; Jones and Odebunmi, 1971), and the other concentrating mainly on host responses (Baker, 1970; Rosielle, 1972). Epidemiological work continued (Shaner and Finney, 1976; Hess and Shaner, 1987; Shaw, 1987; Shaw *et al.*, 1993; Shaner and Buechley, 1995) until Sanderson (1972; 1976) identified *Mycosphaerella graminicola* as the ascogenous state of *S. tritici*. The sexual stage was also identified in other wheat-producing areas (Madariaga, 1986; Scott *et al.*, 1988; Cordo *et al.*, 1990; Verreet *et al.*, 1990; Garcia and Marshall, 1992; Halama, 1996), and its role in epidemiology was recognized and emphasized (Brown *et al.*, 1978; Sanderson and Hampton, 1978; Shaw and Royle, 1989; Shaw *et al.*, 1993). A necessary complement to these epidemiological studies was the work of McDonald and co-workers on the population genetics of *M. graminicola* (McDonald and Martinez, 1990; Boeger *et al.*, 1993; Chen and McDonald, 1996; McDonald *et al.*, Chapter 3 this volume). Data suggested that sexual reproduction is a major factor in determining the structure of *M. graminicola* populations.

An important paper was published by Eyal *et al.* (1973) suggesting that the gene-for-gene hypothesis applied to the *M. graminicola*-wheat pathosystem. This was based on inoculation experiments on different host cultivars with different pathogen strains. Following this, a number of papers were published that either supported or disputed this finding (Rosielle, 1972; Yechilevich-Auster *et al.*, 1983; Eyal *et al.*, 1985; Eyal and Levy, 1987; Saadaoui, 1987; Van Ginkel and Scharen, 1988b; Van Beuningen and Kohli, 1990; Johnson, 1992; Van Ginkel and Rajaram, 1993; Ballantyne and Thomson, 1995). Whether observed specific interactions were due to the interaction of specific genes was not determined. Genetic studies of the host were necessary to explain these observations and were, importantly, indispensable for developing breeding strategies. A broad range of possibilities with respect to the number of genes involved and to their mode of action emerged from these reports. Cultivars like Veranopolis and Bulgaria 88 were reported to carry a single major gene for resistance (Wilson, 1979; 1983; Somasco *et al.*, 1996). In other cultivars, such as the winter bread wheats Kavkaz and Aurora, several genes were detected, whereas it was proposed that resistance in spring wheats from Latin America was controlled by several recessive genes with additive effects (Danon and Eyal, 1990). In durum wheats, there was little indication that specific genes for resistance were of importance (Van Ginkel, 1986; Van Ginkel and Scharen, 1987; 1988a; Jlibene *et al.*, 1994). In virtually all studies on host resistance, general combining ability and specific combining ability effects were significant, although the relative contribution of each component in the explanation of the results was interpreted differently. A complicating factor was that there seemed to be a relationship between disease severity, tallness and earliness (Rosielle and Brown, 1979; Danon *et al.*, 1982; Scott and Benedikz, 1983; Baltazar *et al.*, 1990; Eyal and Talpaz, 1990). Hence, although these studies contributed significantly to the understanding of the genetic control of resistance, they did not elucidate either the underlying mechanism or how genes for resistance would interact with genes for (a)virulence in the pathogen. Both these aspects are being studied at the DLO-Research Institute for Plant Protection [IPO-DLO]. In this chapter we will review the results obtained so far and identify future strategies for molecular and genetic research on *M. graminicola.*

Genetic Variation for Virulence and Resistance in the *Mycosphaerella graminicola*-Wheat Pathosystem

Seedling Tests

The obvious approach for estimating genetic variation for virulence in pathogen populations is to screen a number of host cultivars with a number of pathogen isolates. Data sets retrieved by such experiments can be subjected to different analyses. Person's analysis (Person, 1959) has been an appropriate method to identify possible differences in resistance or virulence in host cultivars and pathogen isolates, respectively. However, the method is best suited to situations where severities can be assessed qualitatively. This hampers its application to pathosystems where both qualitative and quantitative data are collected. Nevertheless, it has also been used in the *M. graminicola*-wheat pathosystem to detect specific resistance components in differential cultivars and specific virulence factors in pathogen isolates (Eyal *et al.*, 1985; Van Ginkel and Scharen, 1988b). One aspect that can be clearly identified in inoculation experiments is the apparent adaptation of *M. graminicola* isolates either to bread wheat or to durum wheat. This phenomenon was first observed by Eyal *et al.* (1973) and later discussed by Van Ginkel and Scharen (1988b). We recently provided more evidence for this adaptation by testing larger numbers of isolates originating from both host species (Kema *et al.*, 1996a). Figure 10.1 shows a dendrogram derived from a cluster analysis that applied the interaction mean-square values from an analysis of variance as a proximity measure. In this analysis, isolates or cultivars are considered to be genetically different if they do interact with each other. It is evident that the only durum wheat-derived isolate in the dendrogram (TK5) is separated from all the others that were derived from bread wheat. In addition, all durum wheat cultivars (Vo, Et, Ca and In) were merged into one cluster since they were resistant to the bread wheat-derived isolates. This relationship was confirmed in experiments where the majority of isolates were derived from durum wheat (Kema *et al.*, 1996a).

Here, we present new data from experiments where we inoculated 110 *M. graminicola* isolates on wheat cultivars that were highly susceptible in previous experiments (bread wheat cultivars: Gerek '79, Shafir, Ceeon and Lakhish; durum wheat cultivars: Volcani 447, Cakmak, Waha and Zenati Bouteille). The latter cultivar was replaced by cv. Cocorit due to seed shortage. The isolates originated from 36 locations in eight countries (Table 3.1). Special care was taken to include samples that were obtained from bread wheat and durum wheat at the same location, namely Berrahal in Algeria and Lattakhia in Syria. In virtually all cases, isolates sampled from durum wheat only induced symptoms on durum wheat cultivars, and isolated sampled from bread wheat only infected bread wheat cultivars, even if isolates originated from the same location. Interestingly, seven isolates were able to induce symptoms in both wheat species. Inoculation experiments with monospore cultures of these isolates should confirm this capability. In addition, it was evident that even by testing only four wheat cultivars of each host species, specificity among these isolates for some cultivars was observed, an observation that confirms the vast

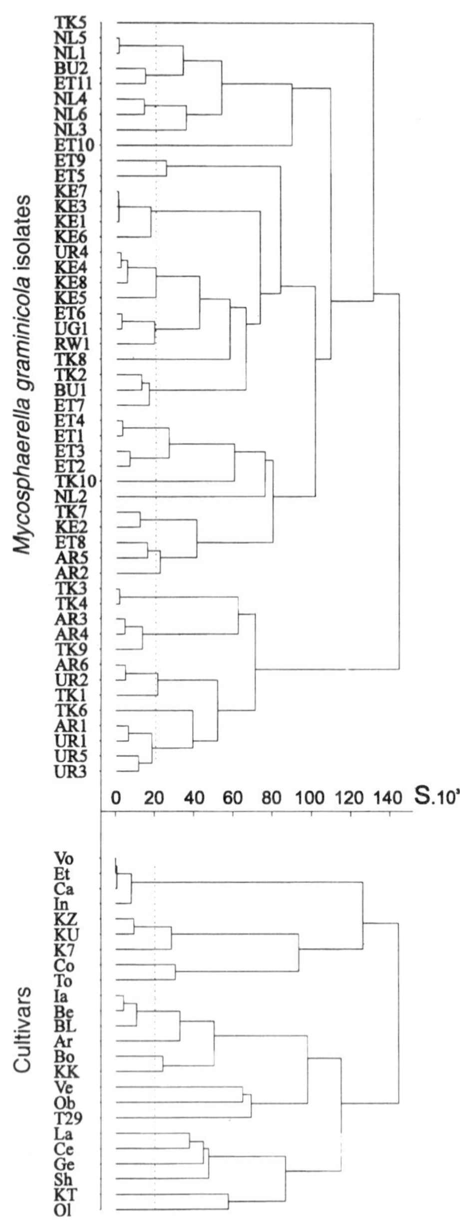

Fig. 10.1. Dendrograms of simultaneously clustered genotypes of wheat (23) and triticale (1), and *Mycosphaerella graminicola* isolates (50), based on *N* in experiment 1. The positions of the nodes correspond to the cumulative sum of squares for interactions between cultivars and isolates (*S*) on the horizontal axis. The area at the left of the vertical dotted line represents non-significant differences at R=0.05, S=28.12x10^3 (for R=0.01, S=29.78x10^3).

Table 10.1. Adaptation of *Mycosphaerella graminicola* isolates to bread wheat or to durum wheat. Results of inoculation experiments using 110 isolates originating from 36 locations in eight countries on four bread wheat and on four durum wheat cultivars.

Country	Location	Number of isolates adapted to		
		Bread wheat	Durum wheat	Both
Algeria	Ain Berda	1	-	-
	Azzaba	1	1	-
	Berrahal	12	12	2
	Bouchegouf	-	1	-
	Elkarma	-	1	-
	Gmelig	1	-	-
	Guelma	6	1	-
	Hazer	-	1	-
	Khroub	-	1	-
	Menzel el abtal	-	3	-
	Oum Bouachi	-	3	-
	Rahouia	-	2	-
	Skikda	-	1	-
	Tiaret	1	-	-
Israël	Tel Aviv	1	-	-
Morocco	Doukkala	-	1	-
	Fez	1	-	-
	Jenica Shaim	-	1	-
The Netherlands	Barendrecht	1	-	-
	Ulrum	1	-	-
	West Brabant	1	-	-
	Zuid Flevoland	1	-	-
Portugal	Casas Velhas	2	-	-
	Elvas	2	-	2
Syria	Derbasiyeh	-	1	-
	Khou el Zarar	-	1	-
	Lattakhia	10	10	2
	Minbeg	2	4	-
	Tel Aboid	-	2	-
Tunisia	Beja	1	2	-
	Nateur	-	1	-
	Sidi Ncir	-	1	-
Turkey	Adana	2	-	1
	Altinova	-	1	-
	Gönen	1	-	-
	Tasci	3	-	-
Totals		53	50	7

genetic variation for virulence in populations of *M. graminicola*.

Previously we performed phylogenetic analysis by amplification and digestion of nuclear and mitochondrial internally transcribed spacer ribosomal DNA (ITS rDNA) followed by sequencing of the nuclear amplicons (Kema *et al.*, 1996a). These analyses did not show any difference between a durum wheat-adapted strain and a bread wheat-adapted strain. The majority of the 110 isolates were characterized for mtDNA RFLPs (see McDonald *et al.*, Chapter 3 this volume), which revealed a novel mtDNA haplotype (type 4) that had a much higher frequency in durum wheat-derived isolates than in bread wheat-derived isolates. The ITS rDNA data suggest that both variants belong to a similar taxonomic rank, which, in principle, enables genetic exchange between them. If this occurs, an important question remains to be answered: why and how does the adaptation persist?

Adult Plant Tests

A valid argument against results obtained in seedling experiments (Yechilevich-Auster *et al.*, 1983; Eyal *et al.*, 1985; Van Ginkel and Scharen, 1988b; Ballantyne and Thomson, 1995; Kema *et al.*, 1996a,b) is what relevance they have for adult plants under field conditions. As discussed previously (Kema *et al.*, 1996d; Kema and Van Silfhout, 1997), resistance to *M. graminicola* usually deteriorates gradually. Data in Table 10.2 illustrate such a process on three wheat cultivars in naturally infected wheat disease observation nurseries in The Netherlands. Cultivars Ritmo, Vivant and Bercy have accounted for most of the Dutch winter wheat area (approximately 120 x 10^3 ha) over the last 5 years. Several field experiments have been done to study resistance in wheat cultivars (Cooke and Jones, 1971; Rosielle, 1972; Brokenshire, 1976; Shaner and Finney, 1982; Van Beuningen and Kohli, 1990). However, these studies did not consider possible genetic differences between pathogen isolates since they relied upon application of a single isolate, of isolate mixtures or natural infection. Kema and Van Silfhout (1997) only recently performed comparative experiments on seedlings and adult plants. They observed that, in general, adult plants were more susceptible than seedlings. The data provided evidence for the specific expression of resistance genes at the seedling, but not at the adult, stage. Whether genes that control resistance to *M. graminicola* at the seedling stage are also responsible for adult plant resistance is not yet known. These experiments showed that interactions between a range of wheat cultivars and *M. graminicola* isolates, as observed in seedling experiments, could be reproduced in field experiments on adult plants over the years. Recently, these observations have been confirmed in an international experiment where wheat cultivars and breeding lines were exposed to specific genotypes of the fungus at three locations over several years. An interesting observation in these experiments was the cultivar-specific virulence difference of two ascospore isolates that originated from the same ascus (Brown *et al.*, 1998). This emphasizes the existence of important qualitative variation in the relationship between wheat and *M. graminicola*. A gradual decrease of resistance, as usually observed in agronomic practice and illustrated in Table 10.2, is not necessarily contradictory to qualitative variation unless the particular epidemiological aspects of the pathosystem are considered.

Table 10.2. Decline of resistance to *Mycosphaerella graminicola* in winter wheat as observed in naturally infected annual disease observation trials (four in 1994 and three in 1997; data, on a 2-9 scale, were adjusted so that the worst plot was scored 2 and the best 9). Source: L. van den Brink, PAV, Lelystad, The Netherlands.

Cultivar	Year	
	1994	1997
Ritmo	7.2	6.5
Vivant	6.5	6.1
Bercy	5.7	3.6

The next challenge is to study the genetic basis of the observed variation. Recently protocols became available to cross and transform *M. graminicola* (Kema *et al.*, 1996c; Pnini-Cohen *et al.*, 1996; Payne *et al.*, 1998), which open up possibilities for fundamental genetic research to elucidate the interaction between this pathogen and wheat.

Genetics in *Mycosphaerella graminicola*

As mentioned before, *M. graminicola* was found to be the perfect state of *S. tritici* more than 20 years ago and its role in epidemiology was recognized (see above) and recently emphasized (Kema *et al.*, 1996c; Hunter *et al.*, 1998; Chapter 7 this volume). Despite the studies on specificity in this pathosystem, formal genetic studies were not conducted perhaps because it appeared to be difficult to cross specific isolates of the fungus. We developed an *in planta* protocol to generate segregating *M. graminicola* populations and will report on the results obtained so far.

Mating Types in *Mycosphaerella graminicola*

We initially screened a set of eight ascospores, that were thought to originate from a single ascus, with 33 RAPD primers. Eighteen of these primers revealed two polymorphic patterns among seven of the isolates, from which we concluded that the eighth ascospore did not belong to the same ascus. The patterns could be arranged in such a way that the individual tetrads could be identified (Table 10.3). These data suggested that *M. graminicola* was heterothallic, which was in agreement with the large variation observed by McDonald and co-workers (see Introduction). Crossing one of the ascospore isolates with the six others resulted in three successful crosses, which indicated a bipolar heterothallic mating system in *M. graminicola* (Kema *et al*., 1996c), confirmed by the generation of the above mentioned BC1 progenies. By generating over 100 (back) crosses with our mating type tester strains either for isolation of segregating populations or for mating type we have observed that one of the crosses was always successful. Hence, the segregation of the MAT locus suggests a single genetic locus with two alleles, MAT 1-1 and MAT 1-2. We unsuccessfully tried to isolate the MAT locus by heterologous hybridization with *Magnaporthe grisea* mating type probes (developed and kindly provided by Dr B.S. Valent, DuPont, USA). Also, a recently developed PCR strategy aimed at amplification of conserved regions of the MAT locus in ascomycetes was not successful either (Arie *et al*., 1997). In our mapping project (see below), we have identified linked AFLP markers. These will be used for map-based cloning strategies in addition to a continuation of heterologous hybridization experiments with previously cloned MAT genes from other fungi.

Genetics of Avirulence

A large collection of well-characterized *M. graminicola* isolates is available for genetic analysis. We have selected two contrasting genotypes; isolate IPO323 which is consistently avirulent on cvs Veranopolis, Kavkaz and Shafir in both seedling and adult plant experiments, and IPO94269 which is virulent on these cultivars. Moreover, both isolates are avirulent on cv. Kavkaz-K4500 and virulent on cv. Taichung 29. Generated F1, F2 and BC1 progenies were isolated and the isolates individually inoculated on the test cultivars in replicated experiments. Avirulence on the differential cultivars was, without exception, under monogenic control. An intriguing observation was that the segregating populations only showed the parental phenotypes. This suggests either that only one avirulence gene is involved or that several avirulence genes are tightly linked into a complex locus. Additional data (Kema *et al*., 1996a) suggest that the latter option is more likely. As expected, all populations were entirely virulent on cv. Taichung 29 and avirulent on cv. Kavkaz-K4500.

These data provide the first insight into the genetic control of avirulence in *M. graminicola*. They confirm the suggested specificity in the *M. graminicola*-wheat relationship as discussed previously (Kema *et al*., 1996a,b; Kema and Van Silfhout, 1997). There are currently no indications of polygenic control of avirulence. This was not expected as we selected fungi rated at the extremes of the disease rating scale. Most data, however, indicate that resistance to *M. graminicola* in wheat is a continuum, a range of levels between 0-100%. Our approach is to analyse the genetic control of large differences of virulence and fitness parameters. Subsequently, more complex interactions can be studied. Currently, it seems most likely that the *M. graminicola*-wheat pathosystem is characterized by both qualitative and quantitative genetic variation for virulence and resistance.

Table 10.3. Tetrad-analysis of *Mycosphaerella graminicola* isolates IPO94265-IPO94272, which were assumed to originate from the same ascus, based on 19 RAPD polymorphisms revealed with 18 primers.

Polymorphisms	Isolates						
	Tetrad 1		Tetrad 2		Tetrad 3		Tetrad 4
	IPO94265	IPO94267	IPO94270	IPO94271	IPO94266	IPO94269	IPO94268
I (N=6)[a]	A[d]	A	B	B	A	A	B
II (N=6)[b]	A	A	A	A	B	B	B
II (N=7)[c]	A	A	B	B	B	B	A

[a]Primers OPA-4, OPA-9, OPA-12, OPA-13, OPA-16 and OPG-14.
[b]Primers OPA-5, OPA-12, OPA-14, OPB-6, OPH-5 and OPH-14. Note that primer OPA-12 produced polymorphisms in class I as well as in class II.
[c]Primers OPA-7, OPA-8, OPA-11, OPA-17, OPB-8, OPG-6 and OPG-7.
[d]A and B symbolize polymorphic patterns obtained with each of the primers.

Mapping the *Mycosphaerella graminicola* Genome

The aforementioned F1 population originating from the cross between isolates IPO323 and IPO94269 was used to start a genomic mapping project. We have used the AFLP technique and an automated sequencer to characterize 68 progeny isolates with 11 primer combinations. These were selected for optimal polymorphism after screening the parents with 64 primer combinations. In total, 317 polymorphic AFLP markers were scored of which 303 have been placed on a preliminary map comprising 24 linkage groups. The aforementioned avirulence and MAT loci and several linked AFLP markers were mapped on different small linkage groups. In addition, pulsed field gel electrophoresis (PFGE) was used for electrophoretic karyotyping, which resulted in 16-18 chromosomes. The difference between the two approaches might be due to co-migrating chromosomes in the PFGE. Saturation of the map with additional markers is in progress and will probably result in a better agreement between both approaches. The AFLP markers linked to the avirulence and MAT loci are currently being used as probes to identify chromosomes that contain these loci in hybridization experiments.

Genome Plasticity and Fertility in *Mycosphaerella graminicola*

Chromosome length polymorphisms have been described among isolates that originated from a single population (McDonald and Martinez, 1991). We have studied such polymorphisms by PFGE using a CHEF system and found differences within a single ascus (Fig. 10.2A). Interestingly, a chromosome of similar size to the largest *Schizosaccharomyces pombe* size marker was identified, which had not been detected previously (McDonald and Martinez, 1991). Whether this was due to the different applied systems (TAFE vs. CHEF) is unclear, but the finding enlarges the total genome size by an additional 5.5 Mb to approximately 36 Mb. In addition, we have studied the electrophoretic karyotypes of isolates IPO323 and IPO94269, the parental strains of the aforementioned mapping populations (Fig. 10.2B). Despite the fact that we have identified major differences in the electrophoretic karyotypes, fertility does not seem to be affected by such polymorphisms. In general, crosses between any two isolates with opposite mating types result in sufficient viable ascospores.

Comparison of two progenies originating from crosses between isolates from the same ascus showed that a 850 bp RAPD fragment, present in both parents, was inherited independently in one of the progenies (Fig. 10.3). The fragments from both parents were sequenced and found to be identical. RFLP probes were developed and used to identify which parent contributed its 850 bp sequence to the individual progeny isolates. A large part of the 850 bp fragment was a multicopy sequence, hence, a 150 bp probe was developed from the original 850 bp product. Using this probe, diploidy for the fragment was determined in an expected 0.25 Mendelian fraction of the population. Hybridization of this probe to the parental DNA on PFGE filters identified that the independently segregating, though identical, 850 bp markers are located

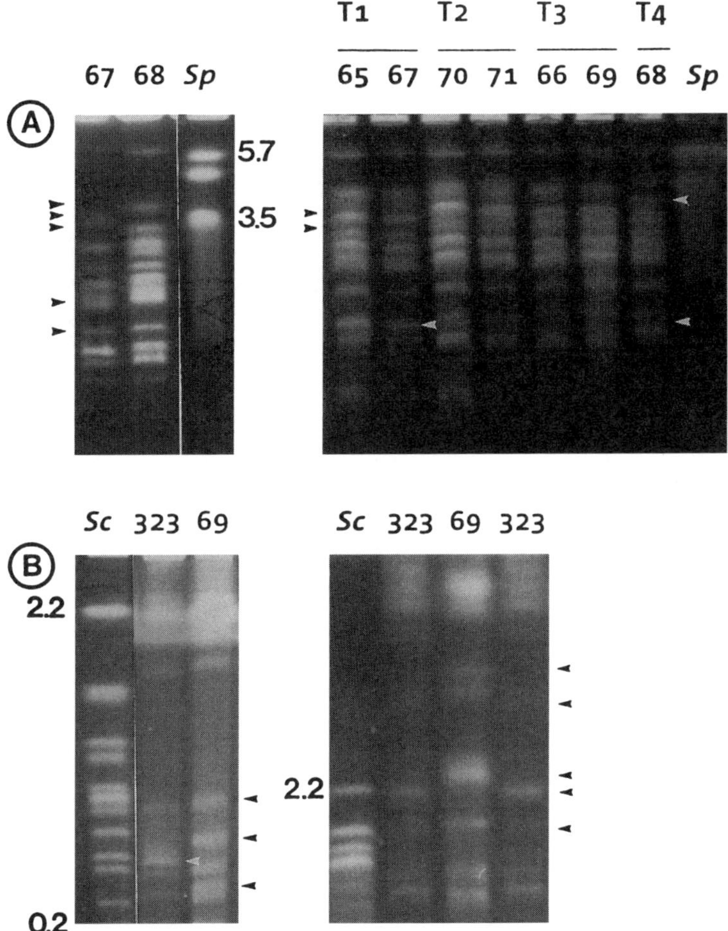

Fig. 10.2. Electrophoretic karyotype polymorphisms in *Mycosphaerella graminicola* as revealed by pulsed field gel electrophoresis using a CHEF system. Differences in size or number of chromosomes are indicated by arrowheads (‣). **A**. Polymorphism within a single ascus. Each lane is an individual isolate; '67'=IPO94267, '68'=IPO94268 etc., *Sp* is chromosome size marker (in Mb) *Schizosaccharomyces pombe*. Left panel: 45 V, 9C, 0.8% SeaKem Gold agarose, 24 h: pulse 600 s ➡ 1500 s, 48 h: pulse 1500 s ➡ 2700 s, 96 h: pulse 2700 s ➡ 3600 s, 15 h: 100 V, pulse 60 s ➡ 90 s. Right panel: isolates arranged by tetrads (T1-T4), conditions identical to left panel except the last step, 40 V, 48 h: pulse 400 s ➡ 600 s, 18 h: pulse 60 s ➡ 90 s. **B**. Polymorphism between the parental strains ('323'=IPO323 and '69'=IPO94269) of the mapping population, *Sc* is chromosome size marker (in Mb) *Saccharomyces cerevisiae*. Left panel: separation of small chromosomes, 200 V, 14C, 1% SeaKem Gold agarose, 24 h: pulse 50 s ➡ 90 s. Right panel: separation of large chromosomes, 100 V, 9C, 1% SeaKem Gold agarose, 8 h: pulse 3000 s, 64 h: pulse 500 s.

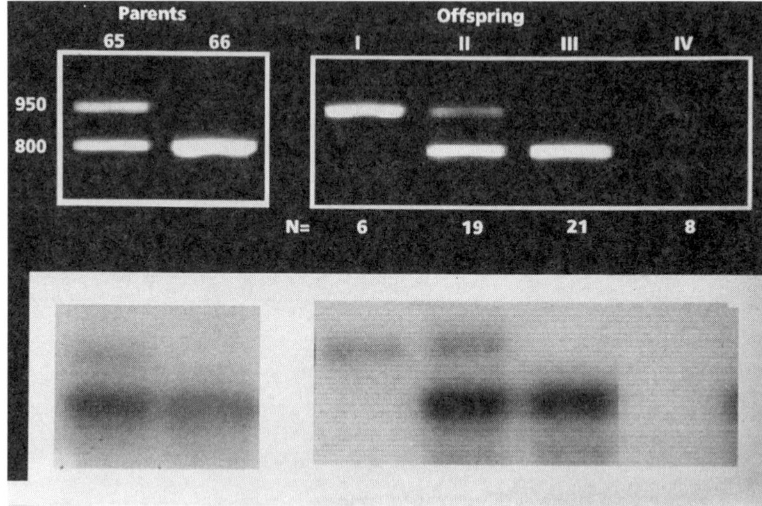

Fig. 10.3. Random ascospore progeny analysis (N=54) obtained from the cross between *Mycosphaerella graminicola* isolates IPO94265 x IPO94266. Upper panel: RAPD patterns of parental and offspring isolates with primer OPB-6. The number of each offspring genotype, categorized in parental patterns (II, III) or in recombinant patterns (I, IV), is shown underneath. Lower panel: Southern blot of the same gel with a 800 bp probe derived from isolate IPO94265.

on different chromosomes. Other isolates from the ascus carried three copies of the marker. It is not yet clear whether translocation or transposition was involved in the initial event that resulted in the chromosomal rearrangement. However, we have shown that once the rearrangement is established it can be considered as a regular genetic marker. Hence, duplication of the 850 bp sequence in progeny isolates results from conventional Mendelian genetics.

Pathology of Compatible and Incompatible Interactions

The aforementioned genetic approach to study the *M. graminicola*-wheat relationship needs to be accompanied by a continued thorough analysis of the pathogenesis. Previous work (Weber, 1922; Hilu and Bever, 1957; Cohen and Eyal, 1993; Kema *et al.*, 1996d) has largely elucidated how *M. graminicola* infects and colonizes its host. Major characteristics of the infection process are that pycnidiospores (Fig. 10.4A) and ascospores (Fig. 10.4G) have a very high germination frequency and penetrate the leaves through stomata (Fig.

10.4B, C, E and H). Recently, some emphasis was put on the capability of *M. graminicola* to penetrate the host directly through the anticlinal walls of epidermal cells (A.C. Payne, IACR-Long Ashton Research Station. 1998, personal communication). We have observed pycnidiospore germ tubes producing a branched structure on, but not penetrating, epidermal cells (Fig. 10.4E). When germ tubes of pycnidiospores and ascospores penetrate a stoma they may produce a specific structure (Fig. 10.4B, C and H) which might be necessary for penetration and which could be induced by thigmotropy, for it was also observed 'erroneously' on an accidental fold (Fig. 10.4D). The function of these structures is not yet understood. It is of interest, however, that pycnidiospores, as well as ascospores, produce them on stomatal slits (Fig. 10.4B, C and H). After penetration, further colonization of the host is intercellular until approximately 8-10 days post-inoculation (dpi). At that time, mesophyll cells collapse in compatible interactions, which suggests an active role of toxic compounds (Kema *et al.*, 1996d). Although at 8 dpi initials of pycnidia are already formed in substomatal cavities, closely located to the stomatal aperture, rapid colonization of the tissue and fructification occurs only after the mesophyll cell collapse. Possible effects of toxic compounds were clearly observed on chloroplasts in more intermediate responses, where the colonization process is far less aggressive. The resistance mechanism of wheat to *M. graminicola* is poorly understood. Conversely, there is no indication that a hypersensitive response is involved; it is more likely that resistance in this pathosystem results from processes that retard fungal colonization (Kema *et al.*, 1996d).

An exciting area to be developed is the combination of advanced histological techniques such as confocal laser scan microscopy, immuno-TEM, based on the suggested toxic compounds, and molecular genetic approaches of avirulence using reporter genes. This will result in an improved understanding of the specific components of the interaction between *M. graminicola* and wheat.

Discussion

The direction of research into the *M. graminicola*-wheat pathosystem is likely to undergo some major changes. The ability to cross and genetically transform isolates of the fungus will especially boost genetic approaches. New insights into the epidemiology (Kema *et al.*, 1996c; Hunter *et al.*, 1998) should lead to a better understanding of the relative contribution of ascospores and pycnidiospores to an epidemic. Since we can now inoculate field plots with molecularly well characterized strains of opposite mating type, such experiments are currently feasible. Hence, the mode of clonality in *M. graminicola* populations can be estimated much more precisely and the selective influence of wheat cultivars on pathogen populations will be better understood. Such information is indispensable for developing breeding strategies. The knowledge of the complexity of the population structure (see McDonald *et al.*, Chapter 3 this volume) for non-selectable markers, such as RFLPs, needs to be complemented with data of selectable markers such as avirulence factors. Since the development of molecular markers for these genes is underway, natural

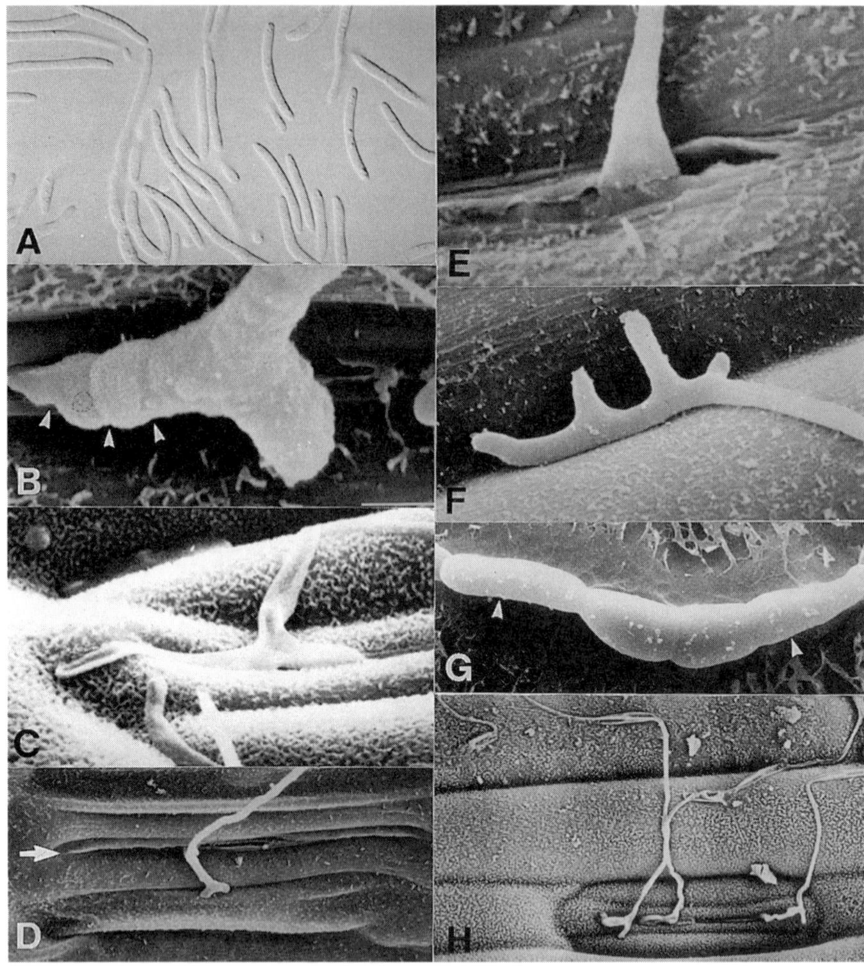

Fig. 10.4. Sexual and asexual propagules of *Mycosphaerella graminicola*, their mode of infection and their infection organs. **A**. Multicelled asexual pycidiospores. **B-C**. Infection structures penetrating a stoma, constrictions in the structure (B) are possible septae in the structure. **D**. Infection structure of a pycnidiospore attempting to penetrate on an accidental fold in the leaf of similar shape the stomatal cells (➡). **E**. Penetration of a stomatal slit by a pycnidiospore sperm tube without the formation of particular structures. **F**. Structure of a pycnidiospore germ tube formed on top of an epidermal cell. **G**. Sexual ascospore. **H**. Multiple penetration of a stoma by ascospore germ tubes. Note the similarity of infection structures with those produced by asexual spores (B-D).

populations can be screened for the distribution and dynamics of avirulence genes, with the possibility of developing a panel of selective wheat cultivars. The apparently contradictory observations of important qualitative variation and gradual decrease of resistance might be due largely to clonality rarely occurring, as *M. graminicola* is able to complete several sexual cycles throughout the growing season (Kema *et al.*, 1996c; Hunter *et al.*, 1998). An integration of molecular genetics with population genetics is a very powerful method to improve the understanding of the above-mentioned natural processes.

Due to the long-lasting discussion and contrasting interpretation of experimental evidence as to whether or not specificity in the *M. graminicola*-wheat pathosystem existed, effective breeding for resistance has been severely hampered. This has left the genetics of resistance to *M. graminicola* in its infancy. Today, however, we currently have isolates available that are well characterized for virulence towards certain cultivars and the inheritance of resistance to *M. graminicola* in wheat can be studied precisely so that markers for the genes controlling it can be identified.

In our laboratory, we will continue to study genome plasticity in *M. graminicola* and explore the possibilities to dissect the genetic factors that control the interaction between host and pathogen, as well as mating between pathogen strains. The completion of a high-density genetic linkage map, including AFLP, RFLP, RAPD and biological markers, is a high priority.

Acknowledgements

Part of this work is financially supported by EC grant EU-BIOTECH PL96-0352. EU-COST817 established a *M. graminicola* working group, which will follow-up on some of the experiments discussed in this chapter. We gratefully acknowledge the technical assistance of Maria Rosa Simon, La Plata University, Argentina.

References

Arie, T., Christiansen, S.K., Yoder, O.C. and Turgeon, G. (1997) Efficient cloning of ascomycete mating type genes by PCR amplification of the conserved MAT HMG box. *Fungal Genetics and Biology* 21, 118-130.

Baker, C.J. (1970) Varietal reaction of wheat to leaf infection by *Leptosphaeria nodorum* and *Septoria tritici*. *Transactions of the British Mycological Society* 54, 500-504.

Ballantyne, B. and Thomson, F. (1995) Pathogenic variation in Australian isolates of *Mycosphaerella graminicola*. *Australian Journal of Agricultural Research* 46, 921-934.

Baltazar, B.M., Scharen, A.L. and Kronstad, W. E. (1990) Association between dwarfing genes 'Rht1' and 'Rht2' and resistance to Septoria tritici blotch in winter wheat (*Triticum aestivum* L. em Thell). *Theoretical and Applied Genetics* 79, 422-426.

Benedict, W.G. (1971a) Differential effect of light intensity on the infection of wheat by *Septoria tritici* Desm. under controlled environmental conditions. *Physiological Plant Pathology* 1, 55-66.

Benedict, W.G. (1971b) Effect of light intensity on the infection of wheat by *Septoria tritici. American Journal of Botany* 58, 457.

Boeger, J.M., Chen, R.S. and McDonald, B.A. (1993) Gene flow between geographic populations of *Mycosphaerella graminicola* (anamorph *Septoria tritici*) detected with restriction fragment length polymorphism markers. *Phytopathology* 83, 1148-1154.

Brokenshire, T. (1976) The reaction of wheat genotypes to *Septoria tritici. Annals of Applied Biology* 82, 415-423.

Brown, J.K.M., Kema, G.H.J., Verstappen, E.C.P., Forrer, H.R., Arraiano, L.S., Brading, P.A., Foster, E.M., Hecker, A. and Jenny, E. (1998) Resistance of wheat varieties and breeding lines to specific genotypes of *Septoria tritici* (*Mycosphaerella graminicola*) under field conditions, (in preparation).

Brown, J.S., Kellock, A.W. and Paddick, R.G. (1978) Distribution and dissemination of *Mycosphaerella graminicola* (Fuckel) Schroeter in relation to the epidemiology of speckled leaf blotch of wheat. *Australian Journal of Agricultural Research* 29, 1139-1145.

Chen, R.S. and McDonald, B.A. (1996) Sexual reproduction plays a major role in the genetic structure of populations of the fungus *Mycosphaerella graminicola. Genetics* 142, 1119-1127.

Cohen, L. and Eyal, Z. (1993) The histology of processes associated with the infection of resistant and susceptible wheat cultivars with *Septoria tritici. Plant Pathology* 42, 737-743.

Cooke, B.M. and Jones, D.G. (1970) The epidemiology of *Septoria tritici* and *S. nodorum*. II. Comparative studies of head infection by *Septoria tritici* and *S. nodorum* on spring wheat. *Transactions of the British Mycological Society* 54, 395-404.

Cooke, B.M. and Jones, D.G. (1971) The epidemiology of *Septoria tritici* and *S. nodorum*. III. The reaction of spring and winter wheat varieties to infection by *Septoria tritici* and *S. nodorum. Transactions of the British Mycological Society* 56, 121-135.

Cordo, C.A., Perelló, A., Alippi, H.E. and Arriaga, H.O. (1990) Presencia de *Mycosphaerella graminicola* (Fuckel) Schroeter teleomorfo de *Septoria tritici* Rob apud Desm. en trigos maduros de la Argentina. *Revista de la Facultad de Agronomía* 66/67, 49-55.

Danon, T. and Eyal, Z. (1990) Inheritance of resistance to two *Septoria tritici* isolates in spring and winter bread wheat cultivars. *Euphytica* 47, 203-214.

Danon, T., Sacks, J.M. and Eyal, Z. (1982) The relationships among plant stature, maturity class, and susceptibility to Septoria leaf blotch of wheat. *Phytopathology* 72, 1037-1042.

Desmazières, J.B.H.J. (1842) Cryptogames nouvelles. *Annales des Sciences Naturelles* 17, 91-118.

Eyal, Z. and Levy, E. (1987) Variations in pathogenicity patterns of *Mycosphaerella graminicola* within *Triticum* spp. in Israel. *Euphytica* 36, 237-250.

Eyal, Z. and Talpaz, H. (1990) The combined effect of plant stature and maturity on the response of wheat and triticale accessions to *Septoria tritici*. *Euphytica* 46, 133-141.

Eyal, Z., Amiri, Z. and Wahl, I. (1973) Physiologic specialization of *Septoria tritici*. *Phytopathology* 63, 1087-1091.

Eyal, Z., Scharen, A.L., Huffman, M.D. and Prescott, J.M. (1985) Global insights into frequencies of *Mycosphaerella graminicola*. *Phytopathology* 75, 1456-1462.

Garcia, C. and Marshall, D. (1992) Observations on the ascogenous stage of *Septoria tritici* in Texas. *Mycological Research* 96, 65-70.

Halama, P. (1996) The occurrence of *Mycosphaerella graminicola*, teleomorph of *Septoria tritici* in France. *Plant Pathology* 45, 135-138.

Hess, D.E. and Shaner, G. (1987) Effect of moisture on Septoria tritici blotch development on wheat in the field. *Phytopathology* 77, 220-226.

Hilu, H.M. and Bever, W.M. (1957) Inoculation, oversummering, and suscept-pathogen relationship of *Septoria tritici* on *Triticum* species. *Phytopathology* 47, 474-480.

Hunter, T., Coker, R.R. and Royle, D.J. (1998) The sexual phase, *Mycosphaerella graminicola*, in epidemics of leaf blotch in UK winter wheat. *Plant Pathology* (in press).

Jlibene, M., Gustafson, J.P. and Rajaram, S. (1994) Inheritance of resistance to *Mycosphaerella graminicola* in hexaploid wheat. *Plant Breeding* 112, 301-310.

Johnson, R. (1992) Past, present and future opportunities in breeding for diseases resistance, with examples from wheat. *Euphytica* 63, 3-22.

Jones, D.G. and Cooke, B.M. (1970) *Septoria tritici* Rob. et Desm. on the heads of winter wheat in West Wales. *Plant Pathology* 19, 99-100.

Jones, D.G. and Odebunmi, K. (1971) The epidemiology of *Septoria tritici* and *S. nodorum*. V. Effect of mixed inocula on disease symptoms and yield in two spring wheat varieties. *Transactions of the British Mycological Society* 57, 153-159.

Kema, G.H.J. and Van Silfhout, C.H. (1997) Genetic variation for virulence and resistance in the wheat-*Mycosphaerella graminicola* pathosystem. III. Comparative seedling and adult plant experiments. *Phytopathology* 87, 266-272.

Kema, G.H.J., Annone, J.G., Sayoud, R., Van Silfhout, C.H., Van Ginkel, M. and De Bree, J. (1996a) Genetic variation for virulence and resistance in the wheat-*Mycosphaerella graminicola* pathosystem. I. Interactions between pathogen isolates and host cultivars. *Phytopathology* 86, 200-212.

Kema, G.H.J., Sayoud, R., Annone, J.G. and Van Silfhout, C.H. (1996b) Genetic variation for virulence and resistance in the wheat-*Mycosphaerella graminicola* pathosystem. II. Analysis of interactions between pathogen isolates and host cultivars. *Phytopathology* 86, 213-220.

Kema, G.H.J., Verstappen, E.C.P., Todorova, M. and Waalwijk, C. (1996c) Successful crosses and molecular tetrad and progeny analyses demonstrate heterothallism in *Mycosphaerella graminicola*. *Current Genetics* 30, 251-258.

Kema, G.H.J., Yu, D.Z., Rijkenberg, F.H.J., Shaw, M.W. and Baayen, R.P. (1996d) Histology of the pathogenesis of *Mycosphaerella graminicola* in wheat. *Phytopathology* 86, 777-786.

Madariaga, B.R. (1986) Presencia en Chile de *Mycosphaerella graminicola* (Fuckel) Schroeter, estado sexuado de *Septoria tritici* Rob. ex. Desm. *Agricultura Tecnica* 46, 209-211.

McDonald, B.A. and Martinez, J.P. (1990) DNA restriction fragment length polymorphisms among *Mycosphaerella graminicola* (anamorph *Septoria tritici*) isolates collected from a single wheat field. *Phytopathology* 80, 1368-1373.

McDonald, B.A. and Martinez, J.P. (1991) Chromosome length polymorphisms in a *Septoria tritici* population. *Current Genetics* 19, 265-271.

Payne, A.C., Grosjean-Cournoyer, M.-C. and Hollomon, D.W. (1998) Transformation of the phytopathogen *Mycosphaerella graminicola* to carbendazim and hygromycin B resistance. *Current Genetics* (in press).

Person, C. (1959) Gene-for-gene relationships in host-parasite systems. *Canadian Journal of Botany* 37, 1101-1130.

Pnini-Cohen, S., Ezrati, S., Zilberstain, A., Schuster, S. and Eyal, Z. (1996) Genetic transformation in the wheat pathogen *Septoria tritici*. *Phytopathology* 86, S40.

Rosielle, A.A. (1972) Sources of resistance in wheat to speckled leaf blotch caused by *Septoria tritici*. *Euphytica* 21, 152-161.

Rosielle, A.A. and Brown, A.G.P. (1979) Inheritance, heritability and breeding behaviour of three sources of resistance to *Septoria tritici* in wheat. *Euphytica* 28, 385-392.

Saadaoui, E.M. (1987) Physiologic specialization of *Septoria tritici* in Morocco. *Plant Disease* 71, 153-155.

Sanderson, F.R. (1972) A *Mycosphaerella* species as the ascogenous state of *Septoria tritici* Rob. and Desm. *New Zealand Journal of Botany* 10, 707-709.

Sanderson, F.R. (1976) *Mycosphaerella graminicola* (Fuckel) Sanderson comb. nov., the ascogenous state of *Septoria tritici* Rob. apud Desm. *New Zealand Journal of Botany* 14, 359-360.

Sanderson, F.R. and Hampton, J.G. (1978) Role of the perfect states in the epidemiology of the common *Septoria* diseases of wheat. *New Zealand Journal of Agricultural Research* 21, 277-281.

Scott, P.R. and Benedikz, P.W. (1983) The effect of Rht-2 and other height genes on resistance to *Septoria nodorum* and *Septoria tritici* in wheat. In: Scharen, A.L. (ed.) *Septoria of Cereals.* Bozeman, Montana, USA, pp. 18-21.

Scott, P.R., Sanderson, F.R. and Benedikz, P.W. (1988) Occurrence of *Mycosphaerella graminicola*, teleomorph of *Septoria tritici*, on wheat debris in the UK. *Plant Pathology* 37, 285-290.

Shaner, G. and Buechley, G. (1995) Epidemiology of leaf blotch of soft red winter wheat caused by *Septoria tritici* and *Stagonospora nodorum*. *Plant Disease* 79, 928-938.

Shaner, G. and Finney, R.E. (1976) Weather and epidemics of Septoria leaf blotch of wheat. *Phytopathology* 66, 781-785.
Shaner, G. and Finney, R.E. (1982) Resistance in soft red winter wheat to *Mycosphaerella graminicola. Phytopathology* 72, 154-158.
Shaw, M.W. (1987) Assessment of upward movement of rain splash using a fluorescent tracer method and its application to the epidemiology of cereal pathogens. *Plant Pathology* 36, 201-213.
Shaw, M.W. and Royle, D.J. (1989) Airborne inoculum as a major source of *Septoria tritici* (*Mycosphaerella graminicola*) infections in winter wheat crops in the UK. *Plant Pathology* 38, 35-43.
Shaw, M.W., Royle, D.J. and Shtienberg, D. (1993) Factors determining the severity of epidemics of *Mycosphaerella graminicola* (*Septoria tritici*) on winter wheat in the UK. *Plant Pathology* 42, 882-899.
Somasco, O.A., Qualset, C.O. and Gilchrist, D.G. (1996) Single-gene resistance to Septoria tritici blotch in the spring wheat cultivar 'Tadinia'. *Plant Breeding* 115, 261-267.
Sprague, R. (1944) Septoria diseases of Gramineae in western United States. *Botany* 6, 151.
Sprague, R. (1950) Some leafspot fungi on western Gramineae V. *Mycologia* 42, 758-771.
Van Beuningen, L.T. and Kohli, M.M. (1990) Deviation from the regression of infection on heading and height as a measure of resistance to Septoria tritici blotch in wheat. *Plant Disease* 74, 488-493.
Van Ginkel, M. (1986) Inheritance of resistance in wheat to *Septoria tritici.* In: Scharen, A.L. (ed.) *Septoria of Cereals.* Bozeman, Montana, USA, pp. 93-102.
Van Ginkel, M. and Rajaram, S. (1993) Breeding for durable resistance in wheat: An international prespective. In: Jacobs, Th. and Parlevliet, J.E. (eds) *Durability of Disease Resistance.* Kluwer Academic Publishers, Dordrecht, pp. 259-272.
Van Ginkel, M. and Scharen, A.L. (1987) Generation mean analysis and heritabilities of resistance to *Septoria tritici* in durum wheat. *Phytopathology* 77, 1629-1633.
Van Ginkel, M. and Scharen, A.L. (1988a) Diallel analysis of resistance to *Septoria tritici* isolates in durum wheat. *Euphytica* 38, 31-37.
Van Ginkel, M. and Scharen, A.L. (1988b) Host-pathogen relationships of wheat and *Septoria tritici. Phytopathology* 78, 762-766.
Verreet, J.A., Hoffman, G.M. and Portner, J. (1990) Nachweis des Teleomorph *Mycosphaerella graminicola* (Fuckel) Schroeter (Anamorph: *Septoria tritici* Rob. apud Desm.) in der Bundesrepublik Deutschland. *Journal of Phytopathology* 130, 105-113.
Weber, G.F. (1922) II. Septoria diseases of wheat. *Phytopathology* 12, 537-588.
Wilson, R.E. (1979) Resistance to *Septoria tritici* in two wheat cultivars, determined by independent, single dominant genes. *Australasian Plant Pathology* 8, 16-18.

Wilson, R.E. (1983) Inheritance of resistance to *Septoria tritici* in wheat. In: Scharen, A.L. (ed.) *Septoria of Cereals.* Bozeman, Montana, USA, pp. 33-35.

Yechilevich-Auster, M., Levi, E. and Eyal, Z. (1983) Assessment of interactions between cultivated and wild wheats by *Septoria tritici. Phytopathology* 73, 1077-1083.

Chapter eleven:

Host Responses Controlling Basic Incompatibility to *Stagonospora nodorum* in Cereals

C. Stevens[1], E. Titarenko[2], J. Keon[3], S. Gurr[4] and J. Hargreaves[3]
[1]*Department of Biological Sciences, Wye College, University of London, Wye, Ashford, Kent TN25 5AH, UK.* [2]*Centro Nacional de Biotechnologia, Consejo Superior de Investigaciones Cientificas, Campus Universidad Autonoma, Cantoblanco, Madrid, Spain.* [3]*IACR-Long Ashton Research Station, Department of Agricultural Sciences, University of Bristol, Long Ashton, Bristol BS41 9AF, UK.* [4]*Department of Plant Sciences, University of Oxford, South Parks Road, Oxford OX1 3RB, UK*

Introduction

The majority of interactions between plants and fungal pathogens are unsuccessful as a result of a **basic incompatibility** between the organisms involved. In other words, plants are resistant to the majority of fungi they encounter in everyday life, including those that are pathogenic to other plant species. This type of incompatibility is often referred to as **non-host resistance** and it is the mechanisms underlying this type of defence response in cereals against *Stagonospora nodorum* (Berk) Castellani & E.G. Germano (teleomorph *Phaeosphaeria nodorum*) that will be the subject of this chapter. Where appropriate, examples from other cereal host-fungal pathogen interactions will be drawn upon to illustrate some of the more salient points about this type of disease resistance mechanism.

However, before considering the processes operating at the level of basic incompatibility in cereals it is worth noting that not all individual attempts at infection by a potential pathogen are successful on cultivars of the host which are considered to be susceptible to the pathogen in question (Keon and Hargreaves, 1983). For example, during interactions between a wheat-biotype of *S. nodorum* and wheat cultivars which succumb to infection by this pathogen, a significant number of attempted infections failed during the initial penetration event (Bird and Ride, 1981). A unique feature of the infection process by *S. nodorum* is the ability of hyphae growing on the surface of the leaf to attempt several penetrations via rudimentary undifferentiated appressoria; the inability to form a successful association with the host leads to resumption of hyphal growth and further attempts to infect thus leading to multiple penetration events (Fig. 11.1). Understanding the mechanisms underlying this type of defence response and why it is overcome under certain circumstances is important and should give a clearer insight into the basis of pathogenicity and virulence in this and other fungal pathogens of cereals.

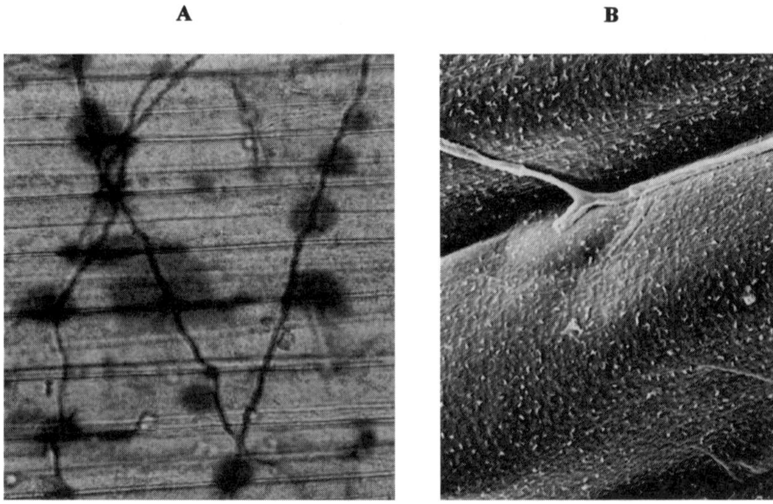

Fig. 11.1. Penetration of leaf epidermal cells by *Stagonospora nodorum*. A: Light micrograph illustrating the nature of multiple penetration events from single *S. nodorum* hypha. Each penetration event is marked by the formation of a reaction site by the host cell stained with Coomassie Brilliant Blue R250 (x250). B: SEM of a hyphae growing on the leaf surface (x380). The hypha has produced a rudimentary appressorium which has unsuccessfully attempted to penetrate an epidermal cell. Growth over the leaf surface then resumed as a hyphal branch formed just behind the penetration site.

Structural Aspects of Basic Incompatibility in Cereals

The inability of non-pathogenic fungi to infect cereals and other members of the Gramineae is often associated with rapid, highly localized structural changes to the cell wall surrounding the initial penetration event (Sherwood and Vance, 1980). These changes involve the deposition of appositional material on the inner surface of the wall beneath the penetration site as a **papilla** and an alteration to the cell wall surrounding the site of penetration to form a **halo** or **disc** (Fig. 11.2). Collectively, these modifications to the cell wall at the site of attempted infections have been termed **reaction sites** (Hargreaves and Keon, 1986b) and are characterized by their autofluorescence and altered histochemical staining properties (Ride and Pearce, 1979; Hargreaves, 1982). Most papillae are composed of a callose (β-1,4 glucans) matrix, within which a diverse array of materials, including lignin-like polymers and other phenolic derivatives, are deposited. Similarly, modified cell walls surrounding the penetration site are encrusted with the same type of material which is deposited in the intermicrofibrillar spaces (Ride, 1978; Aist, 1983). Some of the components of reaction sites are thought to be covalently linked to polymers in papillae or to existing components of the cell wall through the oxidative cross-linking of proteins and esterified phenolic acids (Hargreaves and Keon, 1986a). In this context, the identification of putative guanidine compounds as the 'basic staining material' bound to cell walls is interesting because this type of material may provide additional functional groups for the covalent linking of induced polymeric substances to the cell wall (Wei *et al.*, 1994). On the basis of histochemical staining properties, reaction sites induced by *S. nodorum* in a range of cereals appear to be composed of different materials (Hargreaves and Keon, 1986b). For example, although lignin-like polymers containing syringyl groups have been shown to accumulate in reaction sites formed by wheat leaves in response to infection by non-pathogenic fungi (Ride and Pearce, 1979), material of a similar composition cannot be detected in reaction sites formed by either barley or oats (Hargreaves and Keon, 1986b). Indeed, papillae and modified walls formed by oats are stained most strongly by Nile Blue sulphate, a stain with affinity for lipids and for basic material (Hargreaves, 1982). That reaction sites formed by different cereals, although structurally remarkably similar, are composed of chemically different materials raises the possible importance of components of these structures in determining host range and specificity of a pathogen to a given species.

In the case of *S. nodorum*-challenged barley leaf epidermal cells, penetrating hyphae are invariably restricted to the outer layer of the cell wall just beneath the cuticle and appear unable to grow within or through the cell wall (Keon and Hargreaves, 1984). This observation raises two important points. First, the reaction of the host cells to fungal invasion is triggered as, or soon after, the cuticle has been breached, and second, cessation of fungal growth occurs as a result of changes occurring within the modified plant cell wall surrounding the infection site. However, it should be noted that in some cereal-fungal pathogen interactions inhibition of fungal growth can occur at a later stage after the cell wall has been breached. In these cases, infection structures formed in the invaded cell are often encased in material similar to that deposited in reaction sites (Hargreaves, 1982).

It should also be stressed that the initial reaction of the host cell to attempted penetration is remarkably rapid and highly localized and is likely to be the result of a number of highly coordinated cellular and biochemical processes. However, unlike the hypersensitive reaction to fungal infection, the responding host cells do not die during this defence reaction. On the contrary, evidence suggests that the metabolic activity of pathogen-challenged cells is greatly enhanced during this response (Fig. 11.3). For example, cytoplasmic aggregates often rapidly appear within the cell beneath the site of penetration just prior to the deposition of appositional material (Bushnell and Bergquist, 1975). More recently, it has been shown that reorientation of microtubules and actin filaments occurs during the initial perception of the pathogen (Kobayashi *et al.*, 1992) and these changes may serve to redirect cell organelles and the nucleus towards the vicinity of the penetration site (Kobayashi *et al.*, 1996). How the cellular activities leading to the formation of reaction sites are activated and orchestrated still remains to be resolved, although materials as diverse as chitin and related compounds (Pearce and Ride, 1982) and potassium phosphate (Inoue *et al.*, 1994) are known to elicit components of this defence reaction. Furthermore, besides understanding the mechanisms by which the invading pathogen is perceived and unravelling the signalling transduction pathways involved in this process, information on how the responses invoked following fungal ingress are organized within the cell and directed to specific locations beneath the site of penetration is required. A clearer knowledge of the components of and the regulation of secretory pathways in plant cells would obviously help address this latter issue.

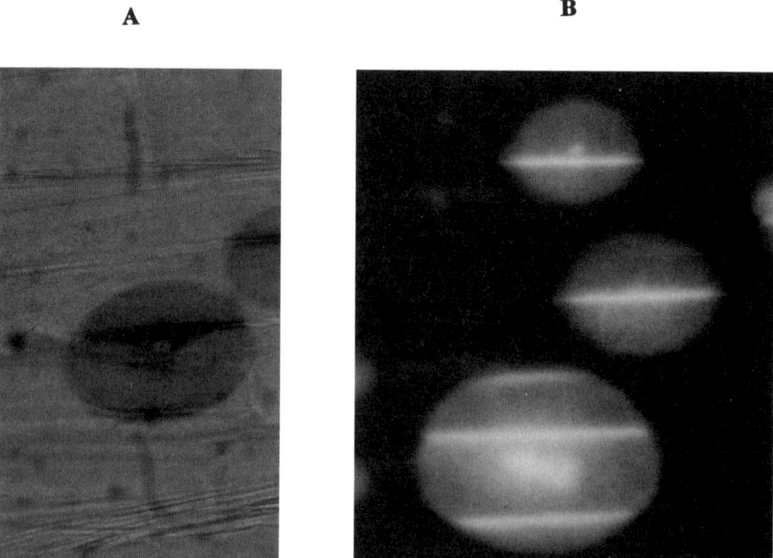

Fig. 11.2. Light micrographs of reaction sites formed in response to infection by *Stagonospora nodorum*. A: Reaction site stained with ammoniacal silver nitrate; the site of penetration is clearly visible in the centre of the modified cell wall (x640). B: Autofluorescence of papillae and cell walls surrounding the penetration site (x700).

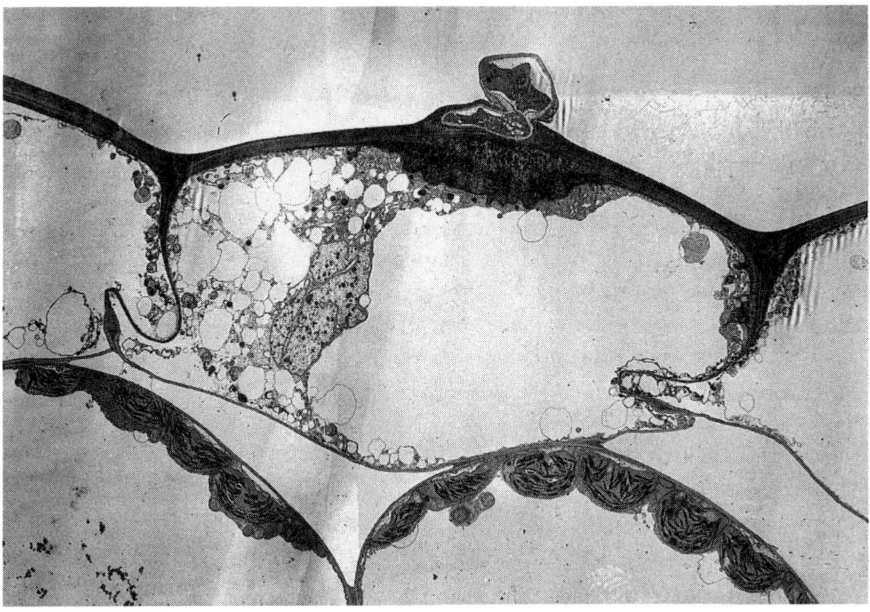

Fig. 11.3. TEM through an abortive penetration of a barley leaf epidermal cell by *Pyrenophora teres,* illustrating cytoplasmic aggregates associated with the deposition of a papilla (x2230).

An additional point that should also be considered concerns the function of these structures in defence against fungal attack and the contribution that each individual component of this response makes to restricting fungal growth. Material deposited in papillae and in the cell wall surrounding attempted penetration sites appears to serve at least two functions. First, it represents a wound healing response of the cell which seals off and repairs the physical damage caused by the penetrating fungus and, thus, prevents the loss of water through the breached cuticle (Aist, 1976). Second, these alterations to the cell wall seem to result in a mechanical strengthening (Israel *et al.*, 1980), in a manner similar to secondary thickenings, which may serve as an impermeable barrier that restricts the movement of nutrients and metabolites to and from the infection court. In this way, the pathogen may be starved of host-derived nutrients necessary for growth and secreted fungal-derived metabolites, such as enzymes and toxins, may be unable to reach their target in the cell (Aist, 1983). Furthermore, it has been suggested that changes similar to those that occur in the plant cell wall (e.g. lignification) might also occur within the walls of the invading hyphae and that this could lead to a reduction in the plasticity of the hyphal tip and, hence, restrict extension growth of the infection hypha (Ride, 1978). The resistance of reaction sites to commercial fungal hydrolases (Ride and Pearce, 1979; Keon and Hargreaves, 1984) also implies that the material deposited in these modified cell walls may act by protecting polymeric components of the wall from degradation by hydrolytic enzymes emanating from the fungal hyphae. These induced cell wall changes could, therefore, serve at least two distinct functions in protecting the host against fungal attack. Furthermore, the accumulation of antimicrobial peptides, such as leaf-specific thionins, within papillae and cell walls surrounding penetration sites (Ebrahim-Nesbat *et al.*, 1993) raises the possibility that other factors may also contribute to the development of a hostile antifungal environment in the cell wall surrounding the invading hypha. In conclusion, what appears to be emerging about this defence response is that it is a highly complex and highly coordinated multicomponent reaction of the plant to fungal attack, in which the components of the response need to be assembled at an appropriate time and place to ward off fungal infection.

Defence-related Gene Expression During an Incompatible Interaction Between *Stagonospora nodorum* and Barley Epidermal Cells

In order to investigate the role of specific sets or subsets of defence-related genes in the resistance of barley to *S. nodorum,* a cDNA library was constructed from *S. nodorum*-challenged coleoptile cells and then screened by differential hybridization (Titarenko *et al.*, 1992). Two induced genes were identified by this procedure (Stevens *et al.*, 1996). The first gene, which was called *Bsi1* (Barley-*Stagonospora* Induced), was highly induced following inoculation and was found to encode a cysteine-rich protein containing 89 amino acids with a M_r of 9405. The deduced amino acid sequence of this protein contained a putative N-terminal secretory sequence (von Heijne, 1986), with a predicted cleavage after the serine residue at position 22. This implied that Bsi1 was secreted and, therefore, this protein could be a good candidate as a protein that accumulates in cell wall reaction sites in response to attempted penetration. The product of the *Bsi1* gene was found to exhibit a high degree of homology to the products of three wheat genes of unknown function (Wali5a, 86.5%; Wali6, 64.4% and Wali3, 63.3%) that are induced in roots after treatment with aluminium (Snowden and Gardner, 1993; Richards *et al.*, 1994) and limited homology (40%) to the product of a maize wound-induced gene, which showed similarities with Bowman-Birk proteinase inhibitors (Rohrmeier and Lehle, 1993). Of particular interest, was the highly conserved position of the cysteine residues with in these sequences (Fig. 11.4), since this would imply that the secondary structure of these proteins is conserved and that they have a similar

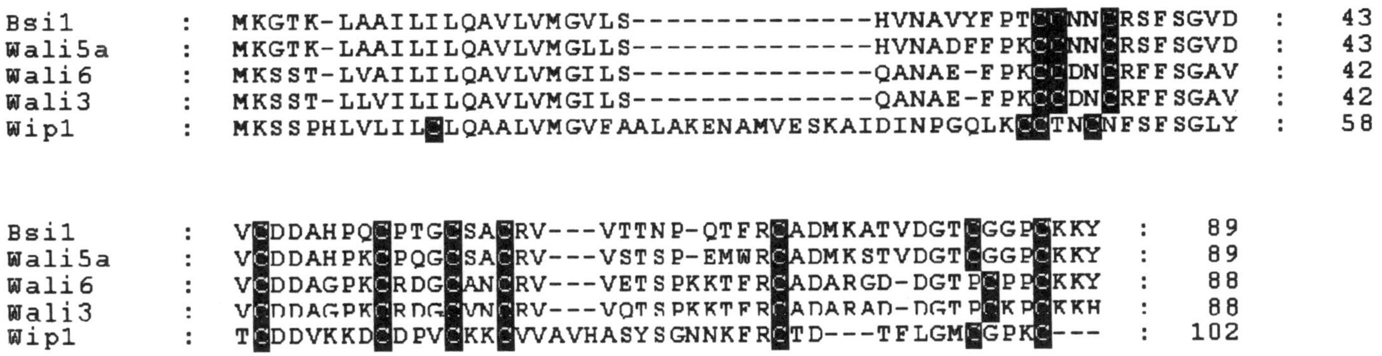

```
Bsi1    :  MKGTK-LAAILILQAVLVMGVLS--------------HVNAVYFPTCCNNCRSFSGVD  :  43
Wali5a  :  MKGTK-LAAILILQAVLVMGLLS--------------HVNADFFPKCCNNCRSFSGVD  :  43
Wali6   :  MKSST-LVAILILQAVLVMGILS--------------QANAE-FPKCCDNCRFFSGAV  :  42
Wali3   :  MKSST-LLVILILQAVLVMGILS--------------QANAE-FPKCCDNCRFFSGAV  :  42
Wip1    :  MKSSPHLVLILCLQAALVMGVFAALAKENAMVESKAIDINPGQLKCCTNCNFSFSGLY  :  58

Bsi1    :  VCDDAHPQCPTGCSACRV---VTTNP-QTFRCADMKATVDGTCGGPCKKY  :   89
Wali5a  :  VCDDAHPKCPQGCSACRV---VSTSP-EMWRCADMKSTVDGTCGGPCKKY  :   89
Wali6   :  VCDDAGPKCRDGCANCRV---VETSPKKTFRCADARGD-DGTPCPPCKKY  :   88
Wali3   :  VCDDAGPKCRDGCVNCRV---VQTSPKKTFRCADARAD-DGTPCKPCKKH  :   88
Wip1    :  TCDDVKKDCDPVCKKCVVAVHASYSGNNKFRCTD---TFLGMCGPKC---  :  102
```

Fig. 11.4. Alignment of the deduced amino acid sequence of Bsi1 with three wheat aluminium-induced (*Wali5a, Wali6* and *Wali3*) gene products and with a maize wound-induced proteinase inhibitor (*Wip1*). The conserved positions of the cysteine residues are marked by a solid box.

function. However, Bsi1, as perhaps expected from its relationships with these other proteins, was found to be induced by a variety of stresses, including wounding, UV light and water stress (C. Stevens, unpublished results). This finding perhaps indicates that Bsi1 may not have a specific role to play in resistance to fungi but may instead be involved in the wound repair response of the cell.

The second induced gene recovered, which was called *Bpr1*, encoded a basic protein (predicted pI of 8.28) containing 174 amino acids with a M_r of 18,859. The amino acid sequence deduced from this gene sequence was highly homologous to type-1 pathogenesis-related (PR-1) proteins, including two other PR-1 proteins previously isolated from barley (HvPR-1a, 62.8% and HvPR, 62.8%; Bryngelsson *et al.*, 1994) and PR-1 proteins from rice (61.7%; Bhargava and Hamer, 1997), from maize (57.7%; Gilllikin *et al.*, 1991) and from tobacco (60.4%; Payne *et al.*, 1989). This protein also had a putative N-terminal signal sequence for translocation to the endoplasmic reticulum and, like the tobacco PR-1 protein, it contained a charged 10 amino acid C-terminal extension that may represent a vacuolar targeting sequence (Neuhaus *et al.*, 1991). Although the function of PR-1 proteins is still unknown, some have been shown to possess antifungal properties (Niderman *et al.*, 1995) and may, therefore, have a direct role to play in resistance. Of more significance, perhaps, was the finding that the sequence of a genomic clone of *Bpr1* contained the 10 bp nucleotide sequence TCATCTTCTT within its promoter region. This motif has been found in a number of stress-induced genes and it has recently been shown that the binding of a tobacco nuclear protein to this sequence is induced by salicylic acid (Goldsbrough *et al.*, 1993). This finding raises the possibility that activation of *Bpr1* is mediated through a salicylic acid signal transduction pathway, even though it is not systemically induced following infection (C. Stevens, unpublished results).

A number of other defence-related genes were also found to be activated during this non-host resistance reaction of barley cells to infection by *S. nodorum* (Stevens *et al.*, 1996). These include genes encoding β-1,3 glucanases, hydroxymethylglutaryl CoA reductase, a homologue of a wheat wound-induced peroxidase and leaf-specific thionins. Interestingly, transcripts of leaf-specific thionins appear later in the interaction than transcripts of the other genes. This group of proteins has been shown to possess antifungal activity and to accumulate in papillae and modified cell walls formed in barley epidermal cells following penetration by *Erysiphe graminis* (Apel *et al.*, 1990; Ebrahim-Nesbat *et al.*, 1993). It is, therefore, possible that these proteins could contribute to the creation of a hostile environment within modified cell walls surrounding penetration sites. Indeed, these, and other antifungal proteins, may function to consolidate the resistance process once the initial response to wounding and infection has served to repair and seal off the damaged part of the cell.

Acknowledgements

C. Stevens and E. Titarenko were funded through BBSRC Link research grant No. LR24/567. IACR-Long Ashton Research Station receives grant-aided support from the Biotechnology and Biological Research Council of the United Kingdom.

References

Aist, J.R. (1976) Papillae and related wound plugs of plant cells. *Annual Review of Phytopathology* 14, 145-163.

Aist, J.R. (1983) Structural responses as resistance mechanisms. In: Bailey, J.A. and Deverall, B.J. (eds) *The Dynamics of Host Defence*. Academic Press, London, pp. 33-70.

Apel, K., Bohlmann, H. and Reimann-Philipp, U. (1990) Leaf thionins, a novel class of putative defence factors. *Physiologia Plantarum* 80, 315-321.

Bhargava, T. and Hamer, J.E. (1997) DNA sequence submitted to GenBank in February 1997 (Accession number U89895).

Bird, P.M. and Ride, J.P. (1981) The resistance of wheat to *Seportia nodorum*: fungal development in relation to host lignification. *Physiological Plant Pathology* 19, 289-299.

Bryngelsson, T., Sommer-Knudsen, J., Gregersen, P.L., Collinge, D.B., Ek, B. and Thordal-Christensen, H. (1994) Purification, characterisation and molecular cloning of basic PR-1-type pathogenesis-related proteins from barley. *Molecular Plant-Microbe Interactions* 7, 267-275.

Bushnell, W.R. and Bergquist, S.E. (1975) Aggregation of host cytoplasm and formation of papillae and haustoria in powdery mildew of barley. *Phytopathology* 65, 310-318.

Ebrahim-Nesbat, F., Bohl, S., Heitefuss, R. and Apel, K. (1993) Thionin in cell walls and papillae of barley in compatible and incompatible interactions with *Erysiphe graminis* f. sp. *hordei*. *Physiological and Molecular Plant Pathology* 43, 343-352.

Gillikin, J., Burkart, W. and Graham, J.S. (1991) Protein sequence submitted to PIR in February 1991 (Accession number A33155).

Goldsbrough, A.P., Albrecht, H. and Stratford, R. (1993) Salicylic acid-inducible binding of a tobacco nuclear protein to a 10 bp sequence which is highly conserved amongst stress-inducible genes. *The Plant Journal* 3, 563-571.

Hargreaves, J.A. (1982) The nature of the resistance of oat leaves to infection by *Pyrenophora teres*. *Physiological Plant Pathology* 20, 165-171.

Hargreaves, J.A. and Keon, J.P.R. (1986a) Cell wall modifications associated with the resistance of cereals to fungal pathogens. In: Bailey, J. (ed.) *Biology and Molecular Biology of Plant-Pathogen Interactions*. Springer-Verlag, Berlin, pp. 133-140.

Hargreaves, J.A. and Keon, J.P.R. (1986b) A comparison of the reaction sites associated with the resistance of coleoptiles of barley, oats and wheat to infection by *Septoria nodorum*. *Journal of Phytopathology* 115, 72-82.

Inoue, S., Macko, V. and Aist, J.R. (1994) Identification of the active component in the papilla-regulating extract from barley leaves. *Physiological and Molecular Plant Pathology* 44, 441-453.

Israel, H., Wilson, R., Aist, J. and Kunoh, H. (1980) Cell wall appositions and plant disease resistance: Acoustic microscopy of papillae that block fungal ingress. *Proceedings of the National Academy of Sciences, USA* 77, 2046-2049.

Keon, J.P.R. and Hargreaves, J.A. (1983) A cytological study of the net blotch disease of barley caused by *Pyrenophora teres*. *Physiological Plant Pathology* 22, 321-329.

Keon, J.P.R. and Hargreaves, J.A. (1984) The response of barley leaf epidermal cells to infection by *Septoria nodorum*. *New Phytologist* 98, 387-398.

Kobayashi, I., Kobayashi, Y., Yamaoka, N. and Kunoh, H. (1992) Recognition of a pathogen and non-pathogen by barley coleoptile cells. III. Responses of microtubules and actin filaments in barley coleoptile cells to penetration attempts. *Canadian Journal of Botany* 70, 1815-1823.

Kobayashi, I., Kobayashi, Y., Yamada, M. and Kunoh, H. (1996) The involvement of the cytoskeleton in the expression of nonhost resistance in plants. In: Mills, D., Kunoh, H., Keen, N.T. and Mayama, S. (eds) *Molecular Aspects of Pathogenicity and Resistance: Requirement for Signal Transduction.* APS Press, Minnesota, pp. 185-195.

Neuhaus, J.M., Sticher, L., Meins, F. and Boller, T. (1991) A short C-terminal sequence is necessary and sufficient for the targeting of chitinases to the plant vacuole. *Proceedings of the National Academy of Sciences, USA* 88, 10362-10366.

Niderman, T., Genetet, I., Bruyere, T., Gees, R., Stinzi, A., Legrand, M., Fritig, B. and Mosinger, E. (1995) Pathogenesis-related proteins are antifungal: isolation and characterisation of 3 14-kilodalton proteins of tomato and a basic PR-1 of tobacco with inhibitory activity against *Phytophthora infestans*. *Plant Physiology* 108, 17-27.

Payne, G., Middlesteadt, W., Desai, N., Williams, S., Dincher, S., Carnes, M. and Ryals, J. (1989) Isolation and sequence of a genomic clone encoding the basic form of pathogenesis-related protein-1 from *Nicotiana tabacum*. *Plant Molecular Biology* 12, 595-596.

Pearce, R.B. and Ride, J.P. (1982) Chitin and related compounds as elicitors of the lignification response in wounded wheat leaves. *Physiological Plant Pathology* 20, 119-123.

Richards, K.D., Snowden, K.C. and Gardner, R.C. (1994) *Wali6* and *Wali7* - Genes induced by aluminium in wheat (*Triticum aestivum* L.) roots. *Plant Physiology* 105, 1455-1456.

Ride, J.P. (1978) The role of cell wall alterations in resistance to fungi. *Annals of Applied Biology* 89, 302-306.

Ride, J.P. and Pearce, R.B. (1979) Lignification and papilla formation at sites of attempted penetration of wheat leaves by non-pathogenic fungi. *Physiological Plant Pathology* 15, 79-92.

Rohrmeier, T. and Lehle, L. (1993) WIP1, a wound-inducible gene from maize with homology to Bouwman-Birk proteinase-inhibitors. *Plant Molecular Biology* 22, 783-792.

Sherwood, R.T. and Vance, C.P. (1980) Resistance to fungal penetration in Gramineae. *Phytopathology* 70, 273-279.

Snowden, K.C. and Gardner, R.C. (1993) Five genes induced by aluminium in wheat (*Triticum aestivum* L.) roots. *Plant Physiology* 103, 855-861.

Stevens, C., Titarenko, E., Hargreaves, J.A. and Gurr, S.J. (1996) Defence-related gene activation during an incompatible interaction between *Stagonospora (Septoria) nodorum* and barley (*Hordeum vulgare* L.) coleoptile cells. *Plant Molecular Biology* 31, 741-749.

Titarenko, E., Hargreaves, J., Keon, J. and Gurr, S. (1992) Defence-related gene expression in barley coleoptile cells following infection by *Septoria nodorum*. In: Fritig, B. and Legrand, M. (eds) *Mechanisms of Plant Defence Responses*. Kluwer Academic Publishers, Dordrecht, pp. 308-311.

von Heijne, G. (1986) A new method for predicting signal sequence cleavage sites. *Nucleic Acid Research* 14, 4683-4692.

Wei, C.D., de Neergaard, E., Thordal-Christensen, H., Collinge, D.B. and Smedegaard-Petersen, V. (1994) Accumulation of a putative guanidine compound in relation to other early defence reactions in epidermal cells of barley and wheat exhibiting resistance to *Erysiphe graminis* f. sp. *hordei*. *Physiological and Molecular Plant Pathology* 45, 469-484.

Chapter twelve:

Restriction Enzyme-mediated Integration in the Rice Blast Fungus

J.A. Sweigard, A.M. Carroll and B. Valent
DuPont Experimental Station, Wilmington, DE 19880, USA

Introduction

Magnaporthe grisea, the rice blast fungus, causes a devastating and economically significant disease of rice. The experimental advantages of this fungus have led to its development as a model system for the exploration of molecular genetic aspects of the fungal pathogenesis of plants (Valent and Chumley, 1991). We have explored insertional mutagenesis via transformation as a means of tagging genes important for pathogenicity (Sweigard *et al.*, 1998). Hygromycin-resistant transformants (5538), most generated by using restriction enzyme-mediated integration (REMI) of transforming DNA, were screened in a hunt for non-pathogenic mutants. Twenty-seven mutants with an altered pathogenic phenotype were obtained, in 18 of which the non-pathogenic defect was tagged by the transforming DNA. Genes have been cloned from 12 of these tagged mutants. These genes have been designated *PTH* for pathogenicity. This chapter provides details of additional experiments involving REMI transformation of *M. grisea* that will not be provided in the mutant hunt paper (Sweigard *et al.*, 1998). A short review of REMI in filamentous fungi has already been published (Sweigard, 1996).

REMI, as initially established in *Saccharomyces cerevisiae,* was based on the powerful homologous recombination system in yeast (Schiestl and Petes, 1991). Transforming DNA was cut with a restriction endonuclease (e.g. *Eco*RI) and transformed into yeast cells in the presence of additional *Eco*RI. The *Eco*RI-cut transforming DNA integrated into *Eco*RI sites in the genome that were apparently generated by the restriction enzyme added to the transformation. *Eco*RI sites were regenerated during the integration. This result implied that the 4 basepair (bp) single-stranded overhang generated by the *Eco*RI digestion was sufficient to stimulate homologous recombination. It further suggested that integration events that had previously been termed illegitimate recombination (as opposed to homologous recombination) might be homologous recombination events that were catalysed by very short regions (as small as 4 bp) of homology.

REMI in *Magnaporthe grisea*

Table 12.1 shows data from two independent REMI transformation experiments using *M. grisea* strain 4091-5-8. The transformation vectors, pCB1173 and pCB1004, contain the hygromycin resistance gene. pCB1173 (top portion of Table 12.1) does not contain sites for any of the restriction enzymes used in the experiment, except *Hae*III. *Hae*III has a 4 bp recognition sequence while the other enzymes have a 6 bp recognition sequence.

Table 12.1. The effect of restriction enzymes on transformation efficiency in *Magnaporthe grisea*.

Enzyme added	Enzyme units added	Number of hygromycin-resistant transformants[a]	
		Uncut pCB1173	*Eco*RV-cut pCB1173
None		2	9
*Avr*II	4	32	50
*Bcl*I	10	32	60
*Hae*III	6	84	79
*Xho*I	50	22	47
		*Bam*HI-cut pCB1004[b]	*Xho*I-cut pCB1004
None		28	23
Bcl	20	315	303

[a]Average of two independent transformations. 1 μg DNA per transformation.
[b]*Bcl*I and *Bam*HI have identical cohesive ends.

These data illustrate several points that are typical of results we have obtained with REMI transformations in *M. grisea*. First, REMI transformation can increase the frequency of transformation, and many different enzymes have this effect. Increases in transformation efficiency have been previously reported for *M. grisea* (Shi *et al.*, 1995), for *Colletotrichum lindemuthianum* (Redman and Rodriguez, 1994), and for *Cochiobolus heterostrophus* (Lu *et al.*, 1994). In *Ustilago maydis* (Bölker *et al.*, 1995), REMI did not affect transformation efficiency. Thus, in addition to the application of REMI as a tool for insertional mutagenesis, a practical benefit of the addition of restriction enzymes to normal transformations is the improvement of transformation efficiency. Although we have primarily used REMI in generating transformants for the non-pathogenic mutant hunt, REMI is not necessarily more mutagenic than transformation without restriction enzymes. Integrative transformation, by definition, is mutagenic because insertions occur in the chromosome. Additionally, integrative transformation has been associated with increased chromosome rearrangements in *Neurospora crassa* (Perkins *et al.*, 1993). Restriction enzymes simply provide sites for integration of the transforming DNA in addition to those sites normally present in a transformation-competent nucleus. We have not found that the addition of restriction enzymes to standard transformations (e.g. for complementing a specific mutant with a wild-type gene) is any more likely to produce abnormal transformants than a transformation without enzymes. Even if a REMI transformation is slightly more mutagenic than a transformation without restriction enzymes, the increase in transformation efficiency outweighs this possible drawback.

Data in Table 12.1 also show that enzymes that can cut vector sequences can still increase transformation efficiency. *Hae*III cuts pCB1173 22 times, including eight times in the hygromycin resistance gene, and yet addition of this enzyme still increases transformation efficiency of this vector about tenfold. This result suggests that sufficient undamaged vector remains intact for integration and further suggests that restriction sites within transforming DNA should not limit choices of restriction enzymes used in REMI transformation.

Finally, restriction enzymes can increase transformation efficiency even without correspondence between the restriction enzyme used to cut the transformation vector and the enzyme used in the transformation. Indeed, restriction enzymes increase transformation efficiency even with uncut vector. Furthermore, since pCB1173 has no sites for *Avr*II, *Bcl*I or *Xho*I, these enzymes could not have modified the transforming DNA during the transformation procedure. Not only is enzyme correspondence not necessary to increase transformation efficiency, the extent of the increase is the same with and without correspondence (e.g. bottom portion of Table 12.1). We speculate that the single-stranded overhangs on transforming DNA are destroyed by nuclease activity in the cell (see below). This result highlights a key difference in the results we have obtained with REMI both in *M. grisea* and in *S. cerevisiae*. In yeast, the 4 bp of homology between the transforming vector and the restricted genomic DNA catalyse the integration event and the restriction site is regenerated after transformation. The increases in transformation efficiency seen with *Avr*II, *Bcl*I or *Xho*I cannot be due to this 4 bp homology since sites for these restriction enzymes do not occur in the transforming vector. Thus, for REMI transformation in *M. grisea* there need not be any correspondence between the restriction enzyme used in transformation and the restriction enzyme used to cut the transforming DNA (if the transforming DNA is cut at all).

Restriction Enzymes Cut Genomic DNA

Despite this difference between REMI both in *M. grisea* and in *S. cerevisiae* important similarities do remain. First, restriction enzymes cut genomic DNA at their cognate recognition site, and second, small regions of homology catalyse integration. Table 12.2 shows the alignment of sequences from three integration sites for three of the mutants that we have analysed. These sequences are from one side of the integration. Three sequences are aligned: the wild-type genomic sequence; the sequence of the mutant genomic DNA; and the sequence from the vector in the region where the integration event occurred. These three integration events were from transformations where the vector was digested with the same enzyme that was added to the transformation. Two key points emerge from these data. First, the integration event occurred at a genomic site with a restriction site corresponding to the enzyme added to the transformation. (This site is in lowercase in the sequence of the wild-type DNA.) Analysis of seven additional integration events showed that a restriction site corresponding to the enzyme used in the transformation is present in the genomic DNA at, or near, the integration site (data not shown). These data clearly indicate that in REMI transformations of *M. grisea* the restriction enzymes enter the cell, make their way to the nucleus and restrict genomic DNA. Second, small regions of homology, 3-5 bp (microhomologies), exist at the junction of the genomic DNA and the vector DNA. (These basepairs are in lowercase in the vector line for each insertion.) In yeast, small regions of homology are provided by the sticky ends provided by the restriction sites in the genome and the vector. In *M. grisea*, these sticky ends apparently do not survive cellular nuclease activity and do not provide the homology that catalyses the integration of the transforming DNA. (Although a *Xho*I site was regenerated in the integration event that led to a mutation in *PTH2*, this *Xho*I site was not the result of recombination of the *Xho*I site in the genome and the vector *Xho*I site.) In all three integration events shown, the recombination site in the vector occurred a considerable distance from the restriction enzyme site. For the integration events that led to mutations in *PTH2*, *PTH3* and *PTH4* this distance was 730 bp, 89 bp and 2025 bp, respectively. Nonetheless, the presence of microhomologies at the integration sites that we have analysed suggests that small regions of homology are necessary for recombination.

REMI Mutagenesis is Not Random in *Magnaporthe grisea*

The inclusion of restriction enzymes in a mutagenesis procedure should, theoretically, be beneficial in randomizing the sites of integration. Use of a large number of different enzymes, or an enzyme with a frequent recognition site (e.g. *Sau*3A, an enzyme with a 4 bp recognition site), should help direct the integration events to nearly random sites in the genome. This rationale assumes both that restriction of genomic DNA actually occurs (and our data indicate that genomic restriction does occur) and that genomic restriction happens in a random fashion (i.e. that each restriction enzyme site in the genome has an equal likelihood of being cut). However, two types of information suggest that integration of transforming DNA is not random in the REMI that we have performed with *M. grisea*.

Table 12.2. Sequence at the integration sites of restriction enzyme-mediated transformants in *Magnaporthe grisea*.

Gene and restriction enzyme	Sequence of wild-type, mutant, and transforming vector[a]	
PTH2		
*Xho*I	ACCAACGctcgagGAGACTGCC	wild-type
	ACCAACGCTCGAGGTCGCCAAC	mutant
	TTCCCAATAcgaggTCGCCAAC	vector
PTH3		
*Hind*III	CCGATGTCaagcttCATTGGAG	wild-type
	CCGATGTCAAGCTGCAGGCGGC	mutant
	GATGGCGCCagctGCAGGCGGC	vector
PTH4		
*Hind*III	TTGGGGTAaagcttACTCATCAT	wild-type
	TTGGGGTAAAGCTGCGCTCGGTC	mutant
	TCACTGACTCgctGCGCTCGGTC	vector

[a]The six basepairs in lowercase in the wild-type sequences indicate the recognition sequence for the enzyme added to the transformation. The basepairs in lowercase in the vector sequences indicate the region of microhomology at the site of recombination.

Among the 13 integration events that we have explored in detail (with one exception, the tagged genes have been cloned and identified), three pairs of related integration events have occurred. In two of the pairs, two independent insertions occurred in the same gene. In the other pair, although integration did not occur in the same gene, the integration event occurred in DNA within a 40 kilobase (kb) region. Since it will probably take 100,000 insertions to saturate the *M. grisea* genome (Kang and Metzenberg, 1993), these related integration events in the first 5538 transformants suggest that the sites of integration in REMI transformation are not random.

We also attempted to evaluate the randomness of integration by screening for auxotrophs. Transformants (13,285) were selected for hygromycin resistance on a rich complete medium and then transferred to minimal medium. Twenty different restriction enzymes, including enzymes with 4 bp and 6 bp recognition sequences, were used to produce these transformants in REMI transformations. Sixteen auxotrophs were found with the following nutritional requirements: cysteine or methionine (11), choline (2), methionine (but could not be supplemented by cysteine) (1), arginine (1) and inositol (1). Our results from the non-pathogenic mutant hunt showed that one-third of the mutations obtained were not tagged. Since no attempt was made to determine if these auxotrophic requirements were due to the transformation integration event (i.e. if the mutations were tagged), the number of tagged auxotrophic requirements is likely to be less than 16. An accurate estimate of the number of auxotrophs that should be expected is difficult. We estimate (conservatively) that 150 kb of DNA are dedicated to biosynthesis of amino acids, nucleotides and vitamins where an insertion would lead to an auxotroph. The *M. grisea* genome is about 45,000 kb (Skinner *et al.*, 1993). Thus, if REMI is random, then about one auxotroph should be found in every 300 transformants and among 13,285 transformants there should be 44 tagged auxotrophs, while less than one-third this number was obtained. The skewed distribution of nutritional requirements more directly challenges the notion of random integration. A similar hunt for *M. grisea* auxotrophs based on UV mutagenesis yielded a much wider spectrum of auxotrophs (Crawford *et al.*, 1986). The most straightforward explanation for these results, especially in light of the three pairs of related integration events in the non-pathogenic mutant hunt, is that REMI integration is not random in *M. grisea*.

Despite the lack of randomness in the integration patterns obtained in *M. grisea* and the differences in the integration events compared to *S. cerevisiae*, we have found gene tagging in REMI transformations to be a valuable tool in exploring pathogenesis. The gene tagging approach has allowed us to discover and clone new and interesting genes involved in the ability of the fungus to be a pathogen. Chemical, UV or radiation mutagenesis would no doubt provide a more random spectrum of mutants, but the downstream molecular genetic analysis via cloning by complementation would be considerably more laborious. Thus, REMI is a useful, though flawed, tool for molecular analysis of *M. grisea.*

References

Bölker, M., Böhnert, H.U., Braun, K.H., Görl, J. and Kahmann, R. (1995) Tagging pathogenicity genes in *Ustilago maydis* by restriction enzyme-mediated integration (REMI). *Molecular and General Genetics* 248, 547-552.

Crawford, M.S., Chumley, F.G., Weaver, C.G. and Valent, B. (1986) Characterization of the heterokaryotic and vegetative diploid phases of *Magnaporthe grisea. Genetics* 114, 1111-1129.

Kang, S. and Metzenberg, R.L. (1993) Insertional mutagenesis in *Neurospora crassa*: cloning and molecular analysis of the preg+ gene controlling the activity of the transcriptional activator NUC-1. *Genetics* 133, 193-202.

Lu, S., Lyngholm, L., Yang, G., Bronson, C., Yoder, O.C. and Turgeon, B.G. (1994) Tagged mutations at the *Tox*1 locus of *Cochliobolus heterostrophus* by restriction enzyme-mediated integration. *Proceedings of the National Academy of Sciences, USA* 91, 12649-12653.

Perkins, D.D., Kinsey, J.A., Asch, D.K. and Frederick, G.D. (1993) Chromosome rearrangements recovered following transformation of *Neurospora crassa*. *Genetics* 134, 729-736.

Redman, R.S. and Rodriguez, R.J. (1994) Factors affecting the efficient transformation of *Colletotrichum* species. *Experimental Mycology* 18, 230-246.

Schiestl, R.H. and Petes, T.D. (1991) Integration of DNA fragments by illegitimate recombination in *Saccharomyces cerevisiae*. *Proceedings of the National Academy of Sciences, USA* 88, 7585-7589.

Shi, Z., Christian, D. and Leung, H. (1995) Enhanced transformation in *Magnaporthe grisea* by restriction enzyme mediated integration of plasmid DNA. *Phytopathology* 85, 329-333.

Skinner, D.Z., Budde, A.D., Farman, M.L., Smith, J.R., Leung, H. and Leong, S.A. (1993) Genome organization of *Magnaporthe grisea*: genetic map, electrophoretic karyotype, and occurrence of repeated DNAs. *Theoretical and Applied Genetics* 83, 545-557.

Sweigard, J.A. (1996) A REMI primer for filamentous fungi. *International Society for Molecular Plant-Microbe Interactions Reporter*, Spring 1996, 3-5.

Sweigard, J.A., Carroll, A.M., Farrall, L., Chumley, F.G. and Valent, B. (1998) *Magnaporthe grisea* pathogenicity genes obtained through insertional mutagenesis. *Molecular Plant-Microbe Interactions* 11, 404-412.

Valent, B. and Chumley, F.G. (1991) Molecular genetic analysis of the rice blast fungus, *Magnaporthe grisea*. *Annual Review of Phytopathology* 29, 443-467.

Chapter thirteen:

Host-Pathogen Recognition in the Barley-*Rhynchosporium* Pathosystem

W. Knogge, M. Fiegen, A. Gierlich, D. Kruse, V. Li, H. Liedgens, A. van 't Slot, S. Steiner and R. Vöcking
Department of Biochemistry, Max-Planck-Institut für Züchtungsforschung, Carl-von-Linné-Weg 10, D-50829 Köln, Germany

The imperfect fungus *Rhynchosporium secalis* (Oudem.) J.J. Davis is a foliar pathogen of barley causing scald, a disease that is widespread in Europe, North America, Australia and other semi-humid barley-growing areas (Shipton *et al.*, 1974). The infection cycle starts with the penetration of the leaf cuticle by the fungus which stays confined to the subcuticular region throughout most of its life cycle. The early stages of pathogenesis are characterized by the collapse of a few epidermal cells (Jones and Ayres, 1974; Hosemans and Branchard, 1985; Lehnackers and Knogge, 1990). Subsequently, mesophyll cells underlying the affected epidermal cells collapse. As a perthotrophic fungus, *R. secalis* appears, therefore, to get access to the host nutrient supply by killing plant cells. At later stages of infection, the subcuticular mycelium forms a dense stroma, causes the formation of necrotic lesions and finally sporulates (Hosemans and Branchard, 1985; Lehnackers and Knogge, 1990; Lyngs Jørgensen *et al.*, 1993). In incompatible interactions, the early infection stages are similar to those found in compatible interactions. However, after the collapse of a few epidermal cells fungal growth is arrested, probably as a result of the induction of plant defence genes. No symptoms are macroscopically visible on resistant barley leaves. Although a hypersensitive response does not occur, small local wall appositions, as well as a weak autofluorescence in the vicinity of fungal hyphae, can be detected under the microscope (Jones and Ayres, 1974; Lehnackers and Knogge, 1990; Lyngs Jorgensen *et al.*, 1993).

Resistance of barley to *R. secalis* is governed by several major resistance (R) genes (Habggod and Hayes, 1971; Shipton *et al.*, 1974; Abbott *et al.*, 1992). Among these genes is the codominant gene *Rrs1* (Hahn *et al.*, 1993) that originates from cv. Turk and that was introgressed into cv. Atlas to yield the near-isogenic cv. Atlas 46 (Riddle and Briggs, 1950; Arias *et al.*, 1983). Several fungal races that are avirulent on cvs Turk and Atlas 46 (*Rrs1*) but virulent on cv. Atlas (*rrs1*) were isolated (Lehnackers and Knogge, 1990; Rohe *et al.*, 1995). Assuming that the interaction of *R. secalis* and barley shows gene-for-gene complementarity (Flor, 1955; 1971), the observed avirulence specificity suggested that these races harbour the avirulence (Avr) gene *AvrRrs1*, matching the host R gene *Rrs1*.

The gene-for-gene hypothesis is the genetic description of a recognition process in which a pathogen-encoded signal, whose generation is controlled by the complementary Avr gene, is directly or indirectly recognized by the R gene product. This signal perception is subsequently transduced into the activation of plant defence reactions (Keen, 1982; 1990). Therefore, the product of an Avr gene can be presumed to be a cultivar-specific elicitor that is only active in plants carrying the corresponding R gene. Identification and characterization of Avr genes and their products, as well as of R genes and receptors of Avr gene products, are, therefore, expected to yield insight into and to finally unravel the defence strategies of a plant towards a given pathogen and the respective plant signalling processes.

From an imperfect fungus such as *R. secalis*, an Avr gene cannot be isolated using genetic techniques. Either transformation-based molecular biological or biochemical strategies need to be followed instead. The earliest symptoms during pathogenesis of *R. secalis* on barley consist of collapsing epidermal cells, although the fungus does not significantly degrade the host cell walls. Consequently, fungal toxins were proposed to be involved in the killing of plant cells with the purpose of accessing the host's nutrient reservoir (Ayesu-Offei and Clare, 1971). It was, therefore, tempting to speculate that toxic compounds, which are secreted very early by the invading pathogen, may also be the factors that are recognized by the plant (Knogge, 1991).

The involvement of low molecular mass toxins, the rhynchosporosides (Auriol *et al.*, 1978), as well as of high molecular mass glycoprotein toxins (Mazars *et al.*, 1984; 1987; 1989) has been described. However, none of these compounds could be detected in culture filtrates of the fungal races used in the present study. Instead, a family of small necrosis-inducing proteins, termed NIP1, NIP2 and NIP3, was identified that shows non-specific toxic activity in all barley cultivars, in other cereals, and in a few tested dicotyledonous species following injection into leaf tissues (Wevelsiep *et al.*, 1991). NIP1 and NIP3, although structurally unrelated, were found to stimulate the plant plasma membrane-localized H^+-ATPase, whereas nothing is known about the mode of action of NIP2 (Wevelsiep *et al.*, 1993).

When these proteins were analysed for elicitor activity, it turned out that NIP1 triggered the rapid and transient synthesis of pathogenesis-related (PR) proteins exclusively in barley cultivars carrying the R gene, *Rrs1* (Hahn *et al.*, 1993). Furthermore, mixing a suspension of spores from a virulent race with purified NIP1 prior to inoculation shifted the interaction to incompatibility. This protein was, therefore, the prime candidate for the product of the fungal *AvrRrs1* gene. After N-terminal sequencing of the protein, the *Nip1* gene was cloned using a PCR-assisted approach. The deduced amino acid sequence consisted of 82 amino acids. Cleavage of a 22 amino acid signal peptide yielded the mature protein of 60 amino acids, including ten cysteine residues (Rohe *et al.*, 1995). Electrospray mass spectrometry detected a single peak with a mass corresponding to a protein structure with five disulphide bridges (V. Li and W. Knogge, unpublished results). No amino acid sequence similarities were found with sequences in the databases. However, the characteristic cysteine pattern C-CC-C-C-CC-C-(C-C) is identical with that of a group of fungal proteins, the hydrophobins (Wessels, 1996; 1997). An additional similarity with these types of proteins became obvious in a hydropathy plot of NIP1 (Talbot *et al.*, 1993; V. Li and W. Knogge, unpublished results).

After adapting a transformation protocol for *R. secalis* (Rohe *et al.*, 1996), genetic complementation (Rohe *et al.*, 1995) and gene replacement (W. Knogge, unpublished results) demonstrated that NIP1 is sufficient and necessary to determine incompatibility on *Rrs1* plants. Furthermore, leaves of susceptible barley plants (*rrs1*) inoculated with spores of a *Nip1* disruption mutant showed reduced disease symptoms as compared to the wild-type race indicating that NIP1 is a fungal virulence factor (W. Knogge, unpublished results). These finding thus confirmed the original hypothesis of a fungal offensive weapon to be utilized by the plant to recognize the invader (Knogge, 1991; 1996a).

Analysis of DNA from 12 fungal isolates originating from different regions of the world revealed that the *Nip1* gene is present in a single copy in the genome of all fungal races ($Nip1^+$) that are avirulent on *Rrs1* plants. Races lacking the gene ($Nip1^-$) are virulent. However, in addition to a race whose interaction with barley is not controlled by the *Rrs1* gene, the *Nip1* gene was also present in a virulent race from Australia (Rohe *et al.*, 1995). Sequence comparison of 1.6 kb of this virulent allele exposed only eight nucleotide alterations in comparison with the originally isolated avirulent allele from a Californian race, indicating that the *Nip1* gene is highly conserved. Three of the alterations were found in a 1 kb stretch 5' of the coding region, one in a 280 bp downstream region, and four in the coding region. Sequencing of the *Nip1* coding region from the six additional $Nip1^+$ races revealed four types of alleles. Type I was present in a Californian and a British race. Type II was found in three Australian races and a German race, and differred from type I in three nucleotide positions that all translate into amino acid alterations. Both types are characteristic for avirulent races. Types III and IV were from Australian races and deviated in two different nucleotide and amino acid positions from type II. Type III was present in the race whose avirulence was not due to the *Rrs1* gene, whereas type IV was found in the virulent $Nip1^+$ race (Rohe *et al.*, 1995).

The products of all four types of *Nip1* alleles were isolated from culture filtrates of a respective race. Elicitor assays revealed that the type I protein has higher activity than the type II. In contrast, the type III and the type IV proteins are inactive. Surprisingly, the same result was obtained when the necrosis-inducing activity and the stimulation of the H^+-ATPase activity were tested. Type II is a weaker toxin and H^+-ATPase stimulator than type I, while types III and IV are inactive (Rohe *et al.*, 1995; Knogge, 1996a,b). The type III- and type IV-specific nucleotides were introduced into the type I sequence. After expression of the resulting proteins in a heterologous system their elicitor inactivity was demonstrated, while the wild-type I protein was active (A. Gierlich and W. Knogge, unpublished results). This indicates that the two amino acid positions alone and independent of each other are essential for the activity of the protein.

NIP1 activity is associated with two adverse phenotypes, host susceptibility through its toxicity and resistance through its elicitor function. This may result from two independent signal transduction pathways involving distinct plant receptors. Alternatively, NIP1 binding to a single receptor may lead to the different phenotypes via branched signal transduction pathways. However, the parallel effects the different amino acid alterations exert on both activities of NIP1 are compatible with the assumption of a single receptor mediating both functions of the protein. In this scenario, the receptor may or may not be encoded by the R gene, *Rrs1*. The product of the *Rrs1* gene may, however, be involved in downstream signalling.

The resistance response of barley to *R. secalis* infection includes a rapid and transient biosynthesis of PR proteins such as PR1, PR5 and peroxidase. Furthermore, mRNA *in situ* hybridization studies revealed that epidermis and mesophyll exhibit different functions in the defence response (E. Schmelzer and W. Knogge, unpublished results). In early stages of the interaction, when the fungus has barely passed the cuticle, high levels of PR5 and peroxidase mRNAs could already be detected in extended regions of the mesophyll of resistant leaves. In contrast, in the epidermis other genes are activated very locally at infection sites of resistant and susceptible leaves (S. Steiner and W. Knogge, unpublished results). It is not clear yet whether or not the epidermal genes are also activated by NIP1.

Using the mRNA differential display technique, more epidermis-specifically regulated clones were isolated and are currently being characterized. However, preliminary results indicate that non-specificity is not characteristic for the activation of all genes that are activated in the epidermis, but that some genes are induced in this tissue in a resistance-specific manner (S. Steiner and W. Knogge, unpublished results). Furthermore, increased cross-linking was observed in cell walls between epidermal and mesophyll cells of resistant leaves (W. Knogge, unpublished results). These data are interpreted such that a primary defence response occurs in the epidermis which also includes the generation of a second, endogenous signal. Amplification of this signal is thought to initiate the secondary resistance reactions observed in the mesophyll. Gene activation in the epidermis may be triggered by NIP1 but part of the response may also by due to other fungal signal molecules.

The present work aims primarily at the identification, isolation, and characterization of the NIP1 receptor(s). However, future research will be directed towards the isolation of additional components of the signalling pathways leading both to the activation of defence reactions in barley and to the stimulation of the plasma membrane-localized H^+-ATPase. This will also include the identification of the putative endogenous signal that is involved in cell-to-cell signalling. In addition, the *Nip2* and *Nip3* genes were cloned to analyse their role in virulence expression on barley.

References

Abbott, D.C., Brown, A.H.D. and Burdon, J.J. (1992) Genes for scald resistance from wild barley (*Hordeum vulgare* ssp *spontaneum*) and their linkage to isozyme markers. *Euphytica* 61, 225-232.

Arias, G., Reiner, L., Penger, A. and Mangstl, A. (1983) *Directory of Barley Cultivars and Lines*. Verlag E. Ulmer, Stuttgart, 391pp.

Auriol, P., Strobel, G., Pio Beltran, J. and Gray, G. (1978) Rhynchosporoside, a host-selective toxin produced by *Rhynchosporium secalis*, the causal agent of scald disease of barley. *Proceedings of the National Academy of Sciences, USA* 75, 4339-4343.

Ayesu-Offei, E.N. and Clare, B.G. (1971) Symptoms of scald disease induced by toxic metabolites of *Rhynchosporium secalis*. *Australian Journal of Biological Sciences* 24, 169-174.

Flor, H.H. (1955) Host-parasite interactions in flax rust - Its genetics and other implications. *Phytopathology* 45, 680-685.

Flor, H.H. (1971) Current status of the gene-for-gene concept. *Annual Review of Phytopathololgy* 9, 275-296.

Habggod, R.M. and Hayes, J.D. (1971) The inheritance of resistance to *Rhynchosporium secalis* in barley. *Heredity* 27, 25-37.

Hahn, M., Jüngling, S. and Knogge, W. (1993) Cultivar-specific elicitation of barley defense reactions by the phytotoxic peptide NIP1 from *Rhynchosporium secalis*. *Molecular Plant-Microbe Interactions* 6, 745-754.

Hosemans, D. and Branchard, M. (1985) Étude *in vitro* de la rhynchosporiose de l'orge: cycle de reproduction du parasite - histopathologie de l'hôte. *Phytopathologische Zeitschrift* 112, 127-142.

Jones, P. and Ayres, P. G. (1974) *Rhychosporium* leaf blotch of barley studied during the subcuticular phase by electron microscopy. *Physiological Plant Pathology* 4, 229-233.

Keen, N.T. (1982) Specific recognition in gene-for-gene host-parasite systems. *Advances in Plant Pathology* 1, 35-82.

Keen, N.T. (1990) Gene-for-gene complementarity in plant-pathogen interactions. *Annual Review of Genetics* 24, 447-463.

Knogge, W. (1991) Plant resistance genes for fungal pathogens - physiological models and identification in cereal crops. *Zeitschrift für Naturforschung* 46c, 969-981.

Knogge, W. (1996a) Fungal infection of plants. *The Plant Cell* 8, 1711-1722.

Knogge, W. (1996b) Molecular basis of specificity in plant/fungus interactions. *European Journal of Plant Pathology* 102, 807-816.

Lehnackers, H. and Knogge, W. (1990) Cytological studies on the infection of barley cultivars with known resistance genotypes by *Rhynchosporium secalis*. *Canadian Journal of Botany* 68, 1953-1961.

Lyngs Jørgensen, H.J., de Neergaard, E. and Smedegaard-Petersen, V. (1993) Histological examination of the interaction between *Rhynchosporium secalis* and susceptible and resistant cultivars of barley. *Physiological and Molecular Plant Pathology* 42, 345-358.

Mazars, C., Hapner, K.D. and Strobel, G.A. (1984) Isolation and partial characterization of a phytotoxic glycoprotein from culture filtrates of *Rhynchosporium secalis*. *Experientia* 40, 1244-1247.

Mazars, C., Rossignol, M. and Auriol, P. (1987) Purification par HPLC de glycoproteines phytotoxic produites par *Rhynchosporium secalis* (Oud.) Davis. *Bio-Sciences* 6, 63-68.

Mazars, C., Rossignol, M., Marquet, P.Y. and Auriol, P. (1989) Reassessment of the toxic glycoprotein isolated from *Rhynchosporium secalis* (Oud.) Davis culture filtrates. Physicochemical properties and evidence of its presence in infected barley plants. *Plant Science* 62, 165-174.

Riddle, O.C. and Briggs, F.N. (1950) Inheritance of resistance to scald in barley. *Hilgardia* 20, 19-27.

Rohe, M., Gierlich, A., Hermann, H., Hahn, M., Schmidt, B., Rosahl, S. and Knogge, W. (1995) The race-specific elicitor, NIP1, from the barley pathogen, *Rhynchosporium secalis*, determines avirulence on host plants of the *Rrs1* resistance genotype. *The EMBO Journal* 14, 4168-4177.

Rohe, M., Searle, J., Newton, A.C. and Knogge, W. (1996) Transformation of the plant pathogenic fungus, *Rhynchosporium secalis*. *Current Genetics* 29, 587-590.

Shipton, W.A., Boyd, W.J.R. and Ali, S.M. (1974) Scald of barley. *Review of Plant Patholology* 53, 839-861.

Talbot, N.J., Ebbole, D.J. and Hamer, J.E. (1993) Identification and characterization of *MPG1*, a gene involved in pathogenicity from the rice blast fungus *Magnaporthe grisea*. *The Plant Cell* 5, 1575-1590.

Wessels, J.G.H. (1996) Fungal hydrophobins: proteins that function at an interface. *Trends in Plant Science* 1, 9-15.

Wessels, J.G.H. (1997) Hydrophobins: proteins that change the nature of the fungal surface. *Advances in Microbial Physiology* 38, 1-45.

Wevelsiep, L., Kogel, K.H. and Knogge, W. (1991) Purification and characterization of peptides from *Rhynchosporium secalis* inducing necrosis in barley. *Physiological and Molecular Plant Pathology* 39, 471-482.

Wevelsiep, L., Rüpping, E. and Knogge, W. (1993) Stimulation of barley plasmalemma H^+-ATPase by phytotoxic peptides from the fungal pathogen *Rhynchosporium secalis*. *Plant Physiology* 101, 297-301.

Chapter fourteen:

Host-Pathogen Recognition in the Cereal-*Pyrenophora* Pathosystem

G.M. Ballance, L. Lamari and C.C. Bernier
Department of Plant Science, University of Manitoba, Winnipeg, Manitoba, Canada R3T 2N2

Introduction

The genus *Pyrenophora* represents a large group of species which cause disease on grasses and economically important cereals. Taxonomically, *Pyrenophora* belongs to the class Loculoascomycetes within the order Pleosporales (Luttrell, 1973). Septoria species belong to the order Dothideales which is a parallel order in the same class. Some of the genera from these two orders are shown in Table 14.1. Most, if not all, of these genera have species that are responsible for leaf spot diseases and many are pathogens of cereals.

Pyrenophora species are important pathogens of a number of cereal crops, the most significant being *P. tritici-repentis* on wheat, *P. graminea*, *P. hordei* and *P. teres* on barley, *P. avenea* on oats, and *P. bromi* on brome grass. Most of these species have a fairly narrow host range or are restricted to their main host. *Pyrenophora tritici-repentis* has the broadest host range and will infect and reproduce on barley, *Agropyron* spp., and brome grass, from which it is often isolated (Morall and Howard, 1975; Krupinsky, 1982; S. Ali and L. Lamari, unpublished results). The focus of our work at the University of Manitoba has been on *P. tritici-repentis* and the foliar disease of wheat that it produces known as tan spot.

Pyrenophora tritici-repentis has been reported to occur worldwide. However, its significance as a pathogen has only become apparent in the last two decades and only in certain wheat-growing areas. The recent increases in disease incidence have been generally attributed to changes in cultural practices (Rees and Platz, 1979; Rees, 1982). The shift from stubble burning to its retention, along with conservation and zero-tillage practices, are believed to be the most important causes of the increase in disease incidence in Australia (Rees, 1982) and elsewhere. *Pyrenophora tritici-repentis,* as well as many other leaf spot pathogens, overwinter on the stubble and surface residues of infected host plants from the previous crop. Thus, high levels of inoculum can build up on repeated cropping with cereals. The considerable success in breeding wheats for rust resistance could also have contributed to this problem, as the leaf surface remained open to invasion by other organisms which may previously have been out-competed or insignificant by comparison.

Table 14.1. Genera taxonomically related to *Pyrenophora.*

Class: Filamentous Ascomycetes

Subclass: Loculoascomycetes

Order: Dothideales

Genus: *Mycosphaerella*, anamorphs may be *Cercospora*, *Septoria* and others

Order: Pleosporales

Genera: *Cochliobolus*, anamorph *Bipolaris*
Pyrenophora, anamorph *Drechslera*
Setosphaera, anamorph *Exserophilum*
Pleospora, anamorph *Stemphylium*
Leptosphaeria, anamorph *Phoma*
Venturia, anamorph *Spiloceae*
Guignardia, anamorph *Phyllosticta*

In Canada, yield losses due to tan spot vary from 0-50%. Because tan spot was not acknowledged, until recently, as a serious problem from a breeding priority standpoint, identification and incorporation of resistance into Canadian cultivars is only now occurring. In the USA, there are a number of resistant cultivars, but resistance to tan spot was introduced 'accidentally' into them. In Europe, the problems with fungal leaf diseases are more difficult to analyse because of the more intensive use of fungicides for control of fungal diseases.

Much of the early work to identify resistance to *P. tritici-repentis* was done from the mid 1970s to mid 1980s. The expression of wheat resistance to tan spot was reported (Hosford, 1982; Gilchrist *et al.,* 1984; Raymond *et al.*, 1985; Cox and Hosford, 1987) and found to be influenced by the duration of the post-inoculation leaf wetness period (Hosford, 1982). It was generally recognized that a leaf wetness period of 24 hours was adequate for discrimination between susceptible and resistant wheat genotypes (Hosford, 1982; Gilchrist *et al.*, 1984). However, the nature of resistance was not clear as it was believed that resistance broke down following wetness periods longer than 54 hours. Several rating systems were used to describe host reaction to *P. tritici-repentis*. These included percentage infection (Nagle *et al.*, 1982), lesion size and percentage infection (da Luz and Hosford, 1980), number of lesions per square centimetre (Nagle *et al.,* 1982), an index combining lesion size, percentage leaf area infected and leaf location (Raymond *et al.,* 1985), lesion type and number of lesions (Gilchrist *et al.*, 1984) or lesion size alone (Cox and Hosford, 1987). The multitude of rating scales used to describe a single disease within the same decade, and even within the same institution, was a testimony to the initial complexity of tan spot of wheat and reflected the confusion in the literature on how best to describe wheat reaction to *P. tritici-repentis*.

One of the difficulties in evaluating resistance and susceptibility in this disease is that both chlorotic and necrotic symptoms can be present. In the mid 1980s, Lamari and Bernier took a different approach and opted to re-examine this host-pathogen system and to develop a qualitative rating system, based on lesion type only. The idea of such a system was borrowed and adapted from the infection type concept used in rust pathology. The outcome of that work has been the development of a system involving well defined relationships between the pathogen and its wheat host. The significant stages in the development of our understanding of the wheat-*P. tritici-repentis* pathosystem can be summarized as follows:

(a) development of a qualitative rating system
(b) assessment of host for reaction to the pathogen
(c) identification of pathotypes (races) in *P. tritici-repentis*
(d) detailed study of host and pathogen components in the system
- genetics
- cytology
- biochemistry, physiology, molecular biology

Development of a Qualitative Rating System Based on Lesion Type

A difficulty which may have confounded early studies was that there is a range of host reactions. Lesion types vary from small dark brown to black spots only to coalescing chlorotic or necrotic lesions which cover most of the leaf (Hosford, 1971). The range of lesion types is illustrated in Fig. 14.1. Based on these recognized lesion types, a qualitative rating system was established and used to evaluate inoculated plants maintained under controlled environmental conditions.

Assessment of Host for Reaction to the Pathogen

All reactions were assessed following inoculation with a spore suspension of each field isolate, followed by 24 hours of leaf wetness and then routine plant maintenance in a growth room under controlled conditions (22°C/18°C; 16 hour

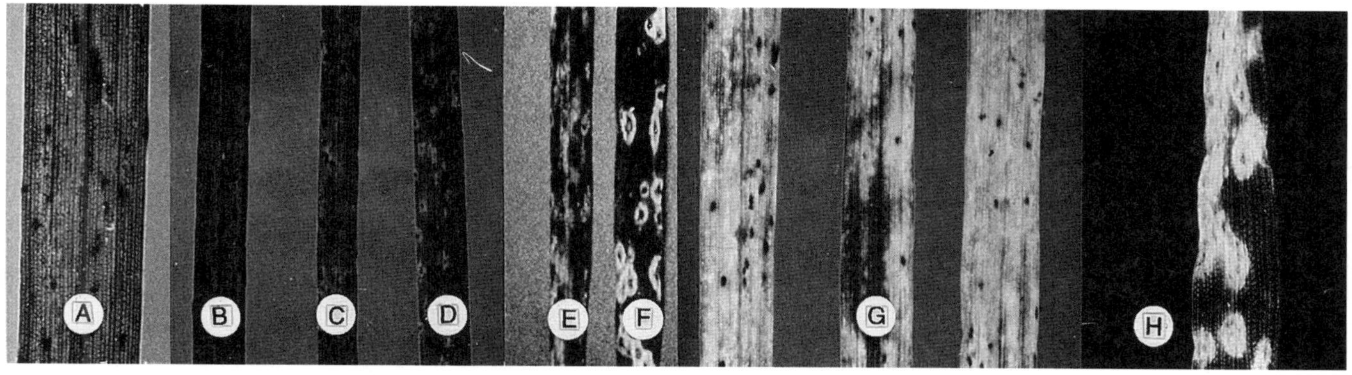

Fig. 14.1. Range of reactions (lesion types) induced in wheat by isolate ASC1 of *Pyrenophora tritici repentis*. A: lesion type 1: note the presence of small brown to black spots without chlorosis or tan necrosis; B: lesion type 2: small brown to black spots with little tan necrosis; C: lesion type 3: small brown to black spots surrounded by a thin ring of chlorosis; D and E lesion type 4 with chlorosis (D) and tan necrosis (E); F: lesion type 5, extensive chlorosis covering most of the leaves. Note the absence of sharp contours of the lesions; G: lesion type 5 consisting of tan necrosis. Note the well defined lesion borders; H: lesion type 5 (tan necrosis) not showing the small brown to black spots.

photoperiod) for 8 days prior to rating. Plants were originally tested at the 4-6-leaf stage but subsequently the 2-leaf stage was used because the results were similar.

Over 1200 accessions from diploid, tetraploid (durum), hexaploid (bread) and octoploid wheats were evaluated, on the basis of lesion type, against a single virulent isolate of *P. tritici-repentis* for their reaction to tan spot in the growth room. A range of reactions was observed and high levels of resistance were identified in all ploidy groups (Lamari and Bernier, 1989a). Resistance was not affected by leaf wetness duration. The most important contribution of this study was the demonstration that tan spot of wheat was a complex disease consisting of two distinct symptoms: tan necrosis and chlorosis, which can be very severe in some genotypes (extensive chlorosis) (Fig. 14.1). The necrosis and chlorosis symptoms were exhibited by plants from diploid, tetraploid and hexaploid wheat accessions, making them a general phenomenon in wheat. Susceptible reactions for a given genotype were either a chlorotic or a necrotic lesion and in a few host genotypes, both chlorotic and necrotic reactions were observed to occur simultaneously (Lamari *et al.*, 1991).

Identification of Pathotypes (Races) in *Pyrenophora tritici-repentis*

A differential set of host wheat lines, representing the reactions observed in the previous study, was created and used to assess the variation in virulence of the pathogen population from western Canada. In a survey of 92 isolates, mainly from Manitoba, Lamari and Bernier (1989b) demonstrated that the isolates of the pathogen exhibited different abilities to induce necrosis and chlorosis in the same way as the host plant. Isolates were consequently grouped into pathotypes based on their ability to induce necrosis and chlorosis in specific wheat differential genotypes (Table 14.2). Pathotype 1 isolates have the ability to induce necrosis and chlorosis and were designated ($nec^{+}chl^{+}$). Pathotype 2 isolates induce only necrotic lesions ($nec^{+}chl^{-}$) and pathotype 3 isolates induce only chlorotic lesions ($nec^{-}chl^{+}$) on the corresponding susceptible host lines. Pathotype 4 isolates are avirulent and produce a resistant reaction on all the test host lines ($nec^{-}chl^{-}$). Subsequent to the description of this pathotype classification, it was discovered that there were two different groups of chlorosis-inducing isolates and that these groups were differentiated by two different host genotypes (i.e. cause chlorosis only, but on two different wheat lines). Thus, the classification of isolates was refined into races based on genotypic differences for a given symptom (Lamari *et al.*, 1995b). Isolates classified in pathotypes 1-4 are now in races 1-4 and the second subset of chlorosis-inducing isolates has been designated as race 5. From these studies, it became clear that the necrosis and chlorosis symptoms observed initially were the result of specific interactions between individual wheat genotypes and specific races of the pathogen. The fact that the capacity of a host genotype to develop necrosis and/or chlorosis was complemented by the ability of the pathogen to induce these symptoms in a race-specific manner suggested that the independence of necrosis and chlorosis represented a coherent working model. A subset of this interaction model, demonstrating the independence of these symptoms, is shown in Fig. 14.2.

Table 14.2. Pathotypes of *Pyrenophora tritici-repentis* based on host symptom reactions.

Pathotype		Host symptom[a]	
Number	Designation	Necrosis	Chlorosis
1	nec^{+}chl^{+}	+	+
2	nec^{+}chl^{-}	+	-
3	nec^{-} chl^{+}	-	-
4	nec^{-} chl^{-}	-	-

[a]+ and - refer to the presence or absence of the indicated symptom.

Detailed Study of Host and Pathogen Components of the System

Host Genetics

Several genetic studies were conducted to understand the relationship between necrosis and chlorosis in the host (Lamari and Bernier, 1991; Duguid, 1995; Gamba, 1996). In all studies, the inheritance of wheat reaction to *P. tritici-repentis* was found to be Mendelian in nature. Furthermore, necrosis and chlorosis were always found to be inherited independently as single loci. With the exception of chlorosis in one cross involving line 6B365, susceptibility was always dominant. In a recent study, we demonstrated (Gamba, 1996) that necrosis to race 2 and chlorosis to races 3 and 5 were inherited independently as three loci: one locus determining the reaction to each race (Table 14.3). In all three cases, susceptibility was dominant. This is in sharp contrast with previous findings that resistance to tan spot was quantitatively inherited (Nagle *et al.*, 1982; Elias *et al.*, 1989). We believe that the discrepancies between our results and those previously reported by various research groups are largely due to the use by our group of well defined host genotypes and specific races of the pathogen. It is important to note that independent genetic studies conducted after the necrosis-chlorosis model was proposed supported our finding that resistance to tan spot was inherited in a Mendelian fashion. Furthermore, the single locus controlling reaction to the necrosis caused by the Ptr necrosis toxin has now been mapped to chromosome arm 5BL by two independent groups and was designated tsn1 (Faris *et al.*, 1996; Stock *et al.*, 1996). Work to map the loci controlling chlorosis to races 3 and 5 to specific chromosomes is in progress.

Host Reaction

Microscopy studies of the infection process have shown that penetration occurs primarily through the anticlinal cell walls of the epidermis. No differences were observed between the degree of penetration of epidermal cells in resistant and susceptible reactions, indicating that discrimination is not at the leaf surface (Dushnicky *et al.*, 1996; Lamari and Bernier, 1989b). After development of secondary hyphae which move into the mesophyll tissue, the hyphae begins to ramify through the tissue intercellularly - hyphae do not invade mesophyll cells. Based on these microscopy studies, it was concluded that it is in the mesophyll tissue that resistance/susceptibility is determined (Larez *et al.*, 1986; Loughman and Deverall, 1986; Lamari and Bernier, 1989b; Dushnicky, 1993).

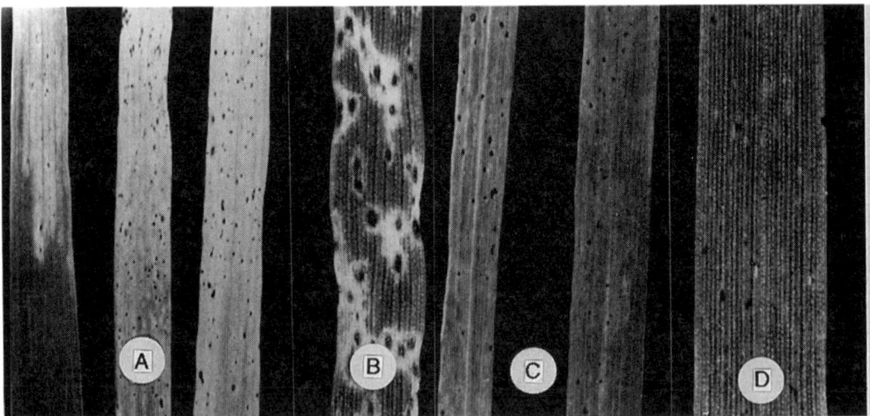

Fig. 14.2. Reaction of wheat lines BH1146 and 6B365 to three pathotypes of *Pyrenophora tritici-repentis*. A: extensive chlorosis induced by isolate ASC1 (nec^+chl^+) in line 6B365, B: tan necrosis induced by isolate ASC1 in line BH1146, C: reaction of line 6B365 to isolate 86-124 (nec^+chl^-); note the absence of extensive chlorosis, D: reaction of line BH1146 to isolate D308 (nec^-chl^+); note the absence of tan necrosis. Seedlings were inoculated at the 2-leaf stage with a conidial suspension (3000 spores ml^{-1}), incubated for 24 hours under continuous leaf wetness and photographed 8 days later.

Table 14.3. F_2 combined segregation ratios for necrosis to race 2 and chlorosis to races 3 and 5 of *Pyrenophora tritici-repentis*.

Cross	Ratio[a]		Probability[c]
	Observed	Expected[b]	
6B365/Katepwa	312:102:100:95:29:45:30:13	27:9:9:9:3:3:3:1	0.70-0.50
ST15/Katepwa	196:66:68:72:25:26:23:6	27:9:9:9:3:3:3:1	> 0.95

Seedlings were inoculated at the 2-leaf stage with a conidial suspension of isolate Hy331-9 (race 3) and infiltrated with the Ptr necrosis and Ptr chlorosis toxins as surrogates for, respectively, races 2 and 5. Each seedling was simultaneously rated for reaction to the three races.
[a]Phenotypic classes.
27 $S_{race2}/S_{race3}/S_{race5}$
9 $S_{race2}/S_{race3}/R_{race5}$: 9 $S_{race2}/R_{race3}/S_{race5}$: 9 $R_{race2}/S_{race3}/S_{race5}$
3 $S_{race2}/R_{race3}/R_{race5}$: 3 $R_{race2}/S_{race3}/R_{race5}$: 3 $R_{race2}/R_{race3}/S_{race5}$
1 $R_{race2}/R_{race3}/R_{race5}$
where R= resistant, S= susceptible.
[b]this ratio is consistent with the involvement of three independently inherited loci.
[c]A fit to the expected ratio is accepted if P >0.05 (chi-square test).

Pathogen-Produced Host-specific Toxins as Pathogenicity Factors

Tomas and Bockus (1987) reported the presence of a cultivar-specific toxin in culture filtrates of field isolates of *P. tritici-repentis* that could produce typical necrotic lesions when infiltrated into the leaves of susceptible cultivars. Lamari and Bernier (1989c) confirmed the presence of this toxin and showed that it occurred in culture filtrates of all nec$^+$ (pathotypes 1 and 2) isolates but in none of the nec$^-$ isolates tested. They further demonstrated that host sensitivity to the necrosis-inducing toxin followed necrotic susceptibility to the corresponding isolates exactly (Lamari and Bernier, 1989c) and that sensitivity and susceptibility in the host segregated in an identical manner (Lamari and Bernier, 1991).

The Ptr necrosis toxin has now been purified to homogeneity by several groups (Ballance *et al.*, 1989; Tomas *et al.*, 1990; Tuori *et al.*, 1995; Zhang *et al.*, 1997) and is the first protein host-specific toxin to be characterized. Using antiserum raised against the Ptr necrosis toxin protein, it was possible to demonstrate that the toxin is produced *in planta* in wheat plants but only when they are infected with nec$^+$ isolates (Lamari *et al.,* 1995a; Fig. 14.3). Because the toxin can be recovered from intercellular washing fluids of such infected plants, it suggests that the toxin is being secreted by the hypha in the intercellular space. The perfect correlation between toxin-producing isolates and susceptible reactions involving necrosis strongly suggests that the toxin is a pathogenicity factor for these isolates.

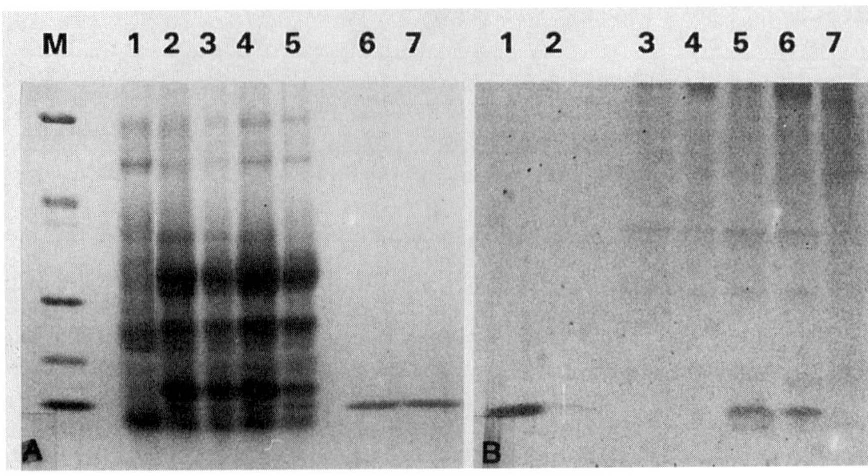

Fig. 14.3. SDS-polyacrylamide gel and Western blot of intercellular washing fluids from leaves of cv. Glenlea infected with *Pyrenophora tritici repentis* isolates. **A.** Coomassie blue-stained proteins: Lane M, protein size markers 68, 45, 30, 20.1 and 14.4 kDa; lanes 1-4, IWFs from leaves infected with isolates ASC1 (nec^+chl^+), 86-124 (nec^+chl^-), D308 (nec^- chl^+), and 88-1 (nec^- chl^-); lane 5, IWF from uninfected leaves; lanes 6 and 7, purified Ptr necrosis toxin, 1 μg and 2.5 μg respectively **B.** Western blot using Ptr necrosis toxin antiserum: Lanes 1 and 2, purified Ptr necrosis toxin, 100 ng and 10 ng respectively; lanes 3-6, IWFs from leaves infected with isolates 88-1, D308, 86-124 and ASC1; lane 7, IWF from uninfected cv. Glenlea leaves.

MRSILVLLFSAAAVLAAPTPEADPGYEIVKLFEAANSSELDARGLSLDWT 50

LKPRGLLQERQGSCMSITINPSRPSVNNIGQVDIDSVILGRPGAIGSWEL 100

▲

NNFITIGLNRVNADTVRVNIRNTGRTNRLIITQWDNTVTRGDVYELFGDY 150

ALIQGRGSFCLNIRSDTGRENWRMQLEN 178

Fig. 14.4. Amino acid sequence of Ptr necrosis toxin proprotein predicted from nucleotide sequence of cDNA clone. The underlined sequence indicates a putative signal peptide (based on von Heijne, 1986) and the triangle the likely N-terminal of the mature protein (Ciuffetti *et al.*, 1997).

The gene encoding a proprotein form of the toxin has been cloned (Ballance *et al.*, 1996; Ciuffetti *et al.*, 1997) from which the amino acid sequence of the toxin has been determined (Fig. 14.4). Direct protein sequencing as well as physiochemical studies on the toxin have also been carried out (Zhang *et al.*, 1997). The proprotein form of the toxin possesses an N-terminal amino acid sequence that is typical of a putative signal peptide (von Heijne, 1986). The signal peptide which serves as a signal for protein export is consistent with the appearance of the toxin in the intercellular space of the leaf, and with the toxin protein serving as a pathogenicity factor for nec^{+} isolates. Ciuffetti *et al.* (1997) have confirmed that the toxin can serve as a pathogenicity factor by transforming a non-toxin-producing *P. tritici-repentis* isolate with the necrosis toxin gene and demonstrating that the resulting transformant is pathogenic on wheat plants which are sensitive to the toxin.

A second toxin, which produces chlorosis in a host-specific manner, has been identified in culture filtrates, spore germination fluids and intercellular washing fluids from race 5 isolates, and designated Ptr chlorosis toxin (Orolaza *et al.*, 1995). This toxin, which appears to be a protein, has been purified, and the characterization work is in progress. The presence of another host-specific toxin which produces chlorosis has also been reported from a race 1 isolate (S.W. Meinhardt, North Dakota State University, 1997, personal communication). The genetics of host susceptibility for chlorotic reactions involving both race 1 and race 3 isolates appear to be the same, which might imply that the same pathogenicity factor is involved in the chlorotic reaction for both of these races. Thus, if the race 1 chlorosis toxin is confirmed as a pathogenicity factor for both race 1 and race 3 isolates, the basis of specificity for the four currently recognized virulent races of *P. tritici-repentis* could be explained. The currently identified specific host-pathogen interactions in the wheat-*P. tritici-repentis* system are summarized in Figure 14.5.

Summary

Within the past decade, our understanding of the relationship between *P. tritici-repentis* and its wheat host has changed considerably. We have been able to demonstrate that the tan spot 'syndrome' consisted of two distinct symptoms, tan necrosis and chlorosis. These symptoms are the phenotypic expression of highly specific interactions between individual wheat genotypes and races of the pathogen. We established conclusively that host reaction to the pathogen was a qualitative trait, inherited in a Mendelian fashion. Similarly, pathogen virulence was found to be specific. Physiological and molecular studies of individual interactions revealed, in each case, that compatibility was associated with the production by the pathogen of a host-specific toxin, which acts as a determinant of pathogenicity. It would appear that *P. tritici-repentis* produces multiple host-specific toxins to extend its range of virulence on its wheat host.

Race / Host[1]	2	1	3	5
Glenlea	Ptr necrosis toxin[2]			
6B365		reported chlorosis toxin[3]		
6B662				Ptr chlorosis toxin[4]

[1]Host: Glenlea = susceptible necrotic to races 1 and 2; 6B365 = susceptible chlorotic to races 1 and 3 and 6B662 = susceptible chlorotic to race 5

[2] Ptr necrosis toxin: 13.2 kDa protein, confirmed pathogenicity factor

[3] reported host specific chlorosis toxin. Awaiting confirmation

[4] Ptr chlorosis toxin: 5-6 kDa protein, confirmed pathogenicity factor

Fig. 14.5. Relationship between the virulent races of *Pyrenophora tritici-repentis* for production of host-specific toxins as pathogenicity factors.

Toxins have also been widely reported within other host-*Pyrenophora* pathosystems, but these were less well characterized. An observation at a recent meeting on host-specific toxins at Tottori, Japan (24-29 August 1997) indicated that most reported host-specific toxins are clustered into species which are taxonomically closely related. Although the toxins vary significantly in structure, similarities across genera do occur.

References

Ballance, G.M., Lamari, L. and Bernier, C.C. (1989) Purification and characterization of a host-selective necrosis toxin from *Pyrenophora tritici-repentis. Physiological and Molecular Plant Pathology* 35, 203-213.

Ballance, G.M., Lamari, L., Kowatsch, R. and Bernier, C.C. (1996) Cloning, expression and occurrence of the gene encoding the Ptr necrosis toxin from *Pyrenophora tritici-repentis. Molecular Plant Pathology On-Line* [http://www.bspp.org.uk/mppol/1996/ 1209ballance/].

Ciuffetti, L.M., Tuori, R.P. and Gaventa, J.M. (1997) A single gene encodes a selective toxin causal to the development of tan spot of wheat. *The Plant Cell* 9, 135-144.

Cox, D.J. and Hosford Jr, R.M. (1987) Resistant winter wheats compared at differing growth stages and leaf positions for tan spot severity. *Plant Disease* 71, 883-886.

da Luz, W.C. and Hosford Jr, R.M. (1980) Twelve *Pyronophora trichostoma* races for virulence to wheat in the Central Plains of North America. *Phytopathology* 70, 1193-1196.

Duguid, S. (1995) Genetics of resistance to *Pyrenophora tritici-repentis* (Died.) Drechs. in hexaploid wheat (*Triticum aestivum* L.). PhD thesis, The University of Manitoba, Winnipeg, Canada.

Dushnicky, L. (1993) A microscopy study of the infection process of *Pyrenophora tritici-repentis* in susceptible and resistant cultivars. MSc thesis, The University of Manitoba, Winnipeg, Canada.

Dushnicky, L., Ballance, G.M., Sumner, M. and MacGregor, A.W. (1996) Penetration and infection of susceptible and resistant wheat cultivars by a necrosis-producing isolate of *Pyrenophora tritici-repentis*. *Canadian Journal of Plant Pathology* 18, 392-402.

Elias, E., Cantrell, R.G. and Hosford Jr, R.M. (1989) Heritability of resistance to tan spot in durum wheat and its association with other agronomic traits. *Crop Science* 29, 299-304.

Faris, J.D., Anderson, J.A., Francl, L.J. and Jordhal, J.G. (1996) Molecular mapping in wheat to a necrosis-inducing protein produced by *Pyrenophora tritici-repentis*. *Phytopathology* 86, 459-463.

Gamba, F.M. (1996) Genetics of host-parasite interactions in tan spot of wheat. MSc thesis, The University of Manitoba, Winnipeg, Canada.

Gilchrist, L.S.L., Fuentes, S.F. and de la Isla de Bauer, M. de L. (1984) Determinacion de fuentes de resistencia contra *Helminthosporium tritici-repentis* bajo condiciones de campo e invernadero. *Agrociencia* 56, 95-105.

Hosford Jr, R.M. (1971) A form of *Pyrenophora trichostoma* pathogenic to wheat and other grasses. *Phytopathology* 61, 28-32.

Hosford Jr, R.M. (1982) Tan spot. In: Hosford Jr, R.M. (ed.) *Tan Spot of Wheat and Related Diseases.* North Dakota State University, pp. 1-24.

Krupinsky, J.M. (1982) Observations on the host range of isolates of *Pyrenophora trichostoma. Canadian Journal of Plant Pathology* 4, 42-46.

Lamari, L. and Bernier, C.C. (1989a) Evaluation of wheat lines and cultivars to tan spot [*Pyrenophora tritici-repentis*] based on lesion type. *Canadian Journal of Plant Pathology* 11, 49-56.

Lamari, L. and Bernier, C.C. (1989b) Virulence of isolates of *Pyrenophora tritici-repentis* on 11 wheat cultivars and cytology of the differential host reactions. *Canadian Journal of Plant Pathology* 11, 284-290.

Lamari, L. and Bernier, C.C. (1989c) Toxin of *Pyrenophora tritici-repentis*: host-specificity, significance in disease, and inheritance of host reaction. *Phytopathology* 79, 740-744.

Lamari, L. and Bernier, C.C. (1991) Genetics of tan necrosis and extensive chlorosis in tan spot of wheat caused by *Pyrenophora tritici-repentis. Phytopathology* 81, 1092-1095.

Lamari, L., Bernier, C.C. and Smith, R.B. (1991) Wheat genotypes that develop both tan necrosis and extensive chlorosis in response to isolates of *Pyrenophora tritici-repentis*. *Plant Disease* 75, 121-122.

Lamari, L., Ballance, G.M., Orolaza, N.P. and Kowatsch, R. (1995a) In planta production and antibody neutralization of the Ptr necrosis toxin from *Pyrenophora tritici-repentis*. *Phytopathology* 85, 333-338.

Lamari, L., Sayoud, R., Boulif, M. and Bernier, C.C. (1995b) Identification of a new race in *Pyrenophora tritici-repentis*: implications for the current pathotype classification system. *Canadian Journal of Plant Pathology* 17, 312-318.

Larez, C.R., Hosford Jr, R.M. and Freeman, T.P. (1986) Infection of wheat and oats by *Pyrenophora tritici-repentis* and initial characterization of resistance. *Phytopathology* 76, 931-938.

Loughman, R. and Deverall, B.J. (1986) Infection of resistant and susceptible cultivars of wheat by *Pyrenophora tritici-repentis*. *Plant Pathology* 35, 443-450.

Luttrell, E.S. (1973) Loculoascomycetes. In: Ainsworth, G.C., Sparrow, F.K. and Sussman, A.S. (eds) *The Fungi, An Advanced Treatise*. Academic Press, New York, pp. 135-219.

Morall, R.A.A. and Howard, R.G. (1975) The epidemiology of leaf spot diseases in a native prairie. II. Airborne spore populations of *Pyrenophora tritici-repentis*. *Canadian Journal of Botany* 53, 2345-2353.

Nagle, B.J., Frohberg, R.C. and Hosford Jr, R.M. (1982) Inheritance of resistance to tan spot of wheat. In: Hosford Jr, R.M. (ed.) *Tan Spot of Wheat and Related Diseases*. North Dakota State University, pp. 40-45.

Orolaza, N.P., Lamari, L. and Ballance, G.M. (1995) Evidence of a host-specific chlorosis toxin from *Pyrenophora tritici-repentis*, the causal agent of tan spot of wheat. *Phytopathology* 85, 1282-1287.

Raymond, P.J., Brockus, W.W. and Norman, B.L. (1985) Tan spot of winter wheat: procedures to determine host response. *Phytopathology* 75, 686-690.

Rees, R.G. (1982) Yellow spot, an important problem in the North-Eastern wheat areas of Australia. In: Hosford Jr, R.M. (ed.) *Tan Spot of Wheat and Related Diseases*. North Dakota State University, pp. 68-70.

Rees, R.G. and Platz, G.J. (1979) The occurrence and control of yellow spot of wheat in North Eastern Australia. *Australian Journal of Experimental Agriculture and Animal Husbandry* 19, 369-372.

Stock, W.S., Brule-Babel, A.L. and Penner, G.A. (1996) A gene for resistance to a necrosis-inducing isolate of *Pyrenophora tritici-repentis* located on 5BL of *Triticum aestivum* cv. Chinese Spring. *Genome* 39, 598-604.

Tomas, A. and Bockus, W.W. (1987) Cultivar-specific toxicity of culture filtrate of *Pyrenophora tritici-repentis*. *Phytopathology* 77, 1337-1366.

Tomas, A., Feng, G.H., Reeck, G.R., Bockus, W.W. and Leach, J.E. (1990) Purification of a cultivar specific toxin from *Pyrenophora tritici-repentis*, causal agent of tan spot of wheat. *Molecular Plant-Microbe Interactions* 3, 221-224.

Tuori, R.P., Wolpert, T.J. and Ciuffetti, L.M. (1995) Purification and immunological characterization of toxic components from cultures of *Pyrenophora tritici-repentis. Molecular Plant-Microbe Interactions* 8, 41-48.

von Heijne, G. (1986) A new method for predicting signal sequence cleavage sites. *Nucleic Acids Research* 14, 4683-4690.

Zhang, H-Z., Francl, L.J., Jordahl, J.G. and Meinhardt, S.W. (1997) Structure and physical properties of a necrosis-inducing toxin from *Pyrenophora tritici-repentis. Phytopathology* 87, 154-160.

Chapter fifteen:

Saponins and Septoria Pathogens

A.E. Osbourn[1], J.P. Wubben[2], R.E. Melton[1], L.M. Flegg[3], R.P. Oliver[4] and M.J. Daniels[1]

[1]*The Sainsbury Laboratory, John Innes Centre, Norwich Research Park, Colney Lane, Norwich NR4 7UH, UK.* [2]*Section of Molecular Genetics of Industrial Micro-organisms, Agricultural University Wageningen, Wageningen, The Netherlands.* [3]*University of East Anglia, Norwich NR4 7TJ, UK.* [4]*Carlsberg Laboratory, Copenhagen, Denmark*

Introduction

Saponins are glycosylated steroidal, steroidal glycoalkaloid or triterpenoid compounds, many of which are antifungal (Price *et al.*, 1987; Fenwick *et al.*, 1992; Hostettmann and Marston, 1995; Osbourn, 1996). These compounds have been found in many different plant species, sometimes representing over 20% of the dry weight (Hostettmann and Marston, 1995). Consequently they may be anticipated to contribute to the resistance of plants to attack by fungal pathogens (Price *et al.*, 1987; Hostettmann and Marston, 1995; Osbourn, 1996). The antifungal activity of saponins has been attributed to their ability to disrupt sterol-containing membranes (Roddick and Drysdale, 1984; Steel and Drysdale, 1988; Keukens *et al.*, 1995).

A number of phytopathogenic fungi are known to produce saponin-detoxifying enzymes, suggesting that detoxification of these compounds may be an important strategy for successful infection (Osbourn, 1996). We have confirmed that this is the case for the root-infecting pathogen of oats, *Gaeumannomyces graminis* var. *avenae.* This fungus produces a β-glucosyl hydrolase known as avenacinase, which detoxifies the triterpenoid avenacin saponins found in oat roots (Turner, 1961; Crombie *et al.*, 1986; Osbourn *et al.*, 1991). Fungal mutants lacking avenacinase are non-pathogenic to oats, but retain full pathogenicity to the alternative host wheat, which does not contain saponins (Bowyer *et al.*, 1995). In order to test whether the ability to detoxify saponins is a requirement for pathogenicity of other fungi to their host plants, we have extended our studies to include two *Septoria* species, *S. avenae* and *S. lycopersici*, which are pathogens of oat and tomato leaves respectively.

Detoxification of Oat Leaf Saponins by *Septoria avenae*

While the triterpenoid avenacin saponins are restricted to the roots of oats, foliar pathogens of oats encounter a different family of saponins, the steroidal avenacosides A and B (Tschesche *et al.*, 1969; Tschesche and Lauven, 1971). The avenacins are referred to as monodesmosidic, because they have a single sugar chain attached at the carbon 3 position. These saponins exist in healthy plants in an active form. In contrast, the avenacosides are bisdesmosidic and, as well as having a sugar chain attached to carbon 3, have a D-glucose molecule at carbon 26 (Fig. 15.1A). Avenacosides A and B are biologically inactive, but in response to tissue damage or pathogen attack are converted into the antifungal 26-desglucoavenacosides A and B (26-DGAs A and B) by removal of the C-26 D-glucose molecule (Lüning and Schlösser, 1975; 1976; Nisius, 1988). This activation is carried out by the oat enzyme avenacosidase, which is specific for the C-26 glucose and does not remove sugars from the C-3 sugar chain (Gus-Mayer *et al.*, 1994).

The toxicity of the 26-DGAs to a number of plant pathogenic fungi has been tested by measuring the effects of these compounds on amino acid leakage from mycelium (Lüning and Schlösser, 1976). These experiments indicated that the oat leaf pathogens *Drechslera avenacea* and *S. avenae* were highly resistant to these saponins, while the oat root pathogens *Fusarium avenaceum* and *G. graminis* var. *avenae*, and pathogens of other cereals (such as *Pseudocercosporella herpotrichoides* and *S. nodorum*), which would not be expected to encounter the 26-DGAs, were sensitive. Pathogens of other plants such as cyclamen and tomato were also insensitive, however. Further study of *D. avenacea* revealed that this pathogen was apparently capable of sequential hydrolysis of sugars from the 26-DGAs to give the aglycone, nuatigenin.

We have investigated the ability of *S. avenae* to detoxify the foliar saponins of oats. Isolates of *S. avenae* can be subdivided in two *formae speciales* , viz. f. sp. *avenae*, which infects oats but not wheat, and f. sp. *triticea*, which infects wheat but not oats (Weber, 1922; Shaw, 1957). The pathogenicity of a number of isolates of *Septoria* to oats and wheat was determined using the detached leaf assay of Johnston and Scott (1988), and a positive correlation was found between pathogenicity to oats and ability to grow on artificial medium supplemented with oat leaf extract (Wubben *et al.*, 1996). From TLC analysis of oat leaf extracts coupled with bioassay tests, the only antifungal compounds which could be detected in crude extracts were 26-DGA A and B. In addition, it was shown that culture filtrates of oat-attacking (but not wheat-attacking) isolates of *S. avenae* were able to detoxify 26-DGAs A and B by sequential hydrolysis of L-rhamnose and D-glucose molecules from the sugar chain attached to carbon 3 (Fig. 15.1).

An enzyme (*S. avenae* avenacosidase) capable of carrying out all the stages of this hydrolysis was purified from culture filtrates of the oat-infecting *S. avenae* WAC 1293 by ammonium sulphate precipitation followed by free flow

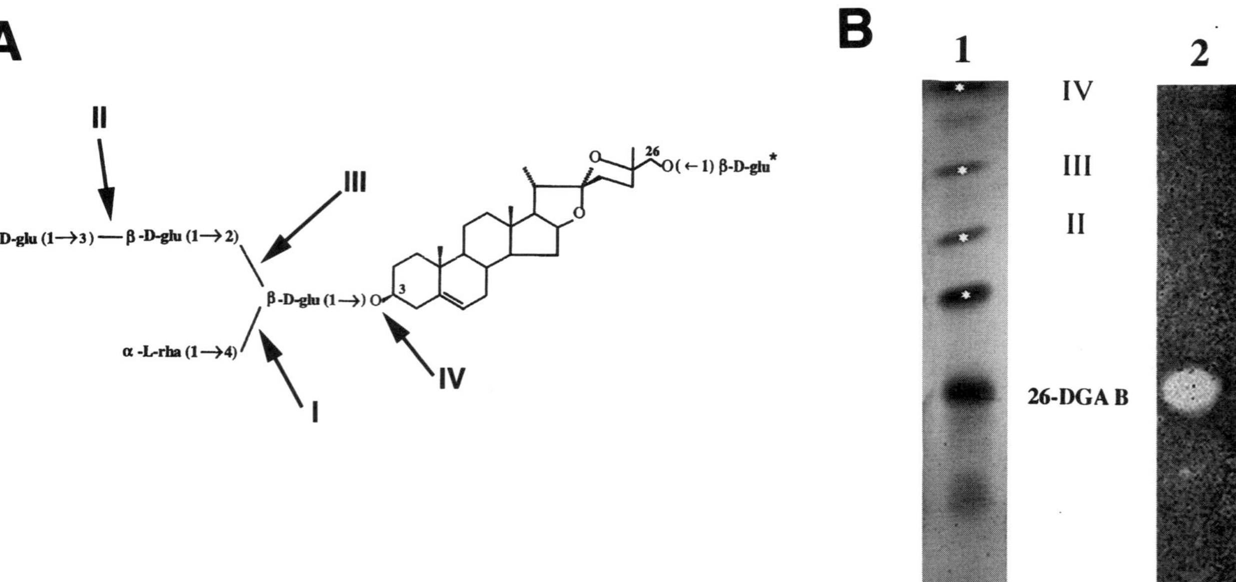

Fig. 15.1. A: Structure of the bisdesmosidic foliar oat saponin, avenacoside B. Removal of the C-26 D-glucose (indicated with an asterisk) by a specific oat glycosyl hydrolase converts avenacoside B to its biologically active monodesmosidic form, 26-desglucoavenacoside B (26-DGA B). Avenacoside A is not shown but differs from avenacoside B only in that it lacks the β,1-3-linked D-glucose molecule attached to the C-3 sugar chain. *Septoria avenae* f sp. *avenae* isolates remove sugars from the C-3 sugar chain by sequential hydrolysis, beginning with L-rhamnose (step I) followed by the remaining D-glucose molecules (steps II-IV). B: TLC separation of 26-DGA B and the deglycosylated hydrolysis products from steps I-IV (see above): lane 1, after visualization with the chromogenic reagent *p*-anisaldehyde/sulphuric acid; lane 2, after spraying with spores of a saponin-sensitive *Septoria* isolate and incubating. Fungal growth is inhibited by 26-DGA B, but not by the hydrolysis products.

isoelectric focusing, anion exchange HPLC, and size exclusion HPLC (Wubben *et al.*, 1996). N-terminal amino acid sequence information was obtained from peptide fragments after cyanogen bromide cleavage of the purified protein, and degenerate oligonucleotide primers corresponding to these peptide sequences were synthesized and used in PCR to amplify a part of the gene encoding avenacosidase. The cloned fragment was then used to isolate a full-length clone from a genomic library of *S. avenae* isolate WAC 1293. DNA sequence analysis identified an open reading frame predicted to encode an extracellular protein of 870 amino acids, with a calculated molecular weight of 94 kDa and an isoelectric point of pH 4.6, and which included the amino acid sequences derived from the purified avenacosidase protein. Southern blot analysis indicated that the gene was present as a single copy in WAC 1293.

A plasmid for targeted gene disruption was constructed by replacing part of the coding sequence of the cloned gene with the gene encoding hygromycin phosphotransferase, which confers resistance to hygromycin B. This plasmid was linearized and introduced into WAC 1293 by transformation in order to generate mutants by homologous recombination. After screening 400 hygromycin-resistant transformants, one mutant was identified which had undergone homologous recombination at the target gene. Two-dimensional polyacrylamide gel electrophoresis confirmed that this mutant was no longer able to produce the avenacosidase enzyme which had been characterized in this study. However, loss of this enzyme had no effect on growth in the presence of oat leaf extract, or on the overall ability to detoxify 26-DGAs *in vitro*, and pathogenicity to oats (as assessed in detached leaf assays) remained unaffected. Further biochemical analysis revealed that the avenacosidase activity remaining in the mutant fractionated in a similar way to the avenacosidase which had already been purified. Future work will focus on the characterization of this other avenacosidase activity and on the generation of mutants of *S. avenae* which completely lack the ability to detoxify 26-DGAs, in order to assess the importance of saponin detoxification for pathogenicity of *S. avenae* to oats.

Heterologous Expression of *Septoria lycopersici* Tomatinase in *Cladsporium fulvum*

Tomatoes contain the steroidal glycoalkaloid saponin, α-tomatine. In general, fungal pathogens of tomato are more resistant to α-tomatine *in vitro* than are fungi which do not infect this plant (Arneson and Durbin, 1968b; Schlösser, 1975; Steel and Drysdale, 1988; Suleman *et al.*, 1996). A number of tomato pathogens produce saponin-detoxifying enzymes known as tomatinases, which act on the tomato steroidal glycoalkaloid α-tomatine (Arneson and Durbin, 1967; Verhoeff and Liem, 1975; Schönbeck and Schlösser, 1976; Ford *et al.*, 1977; Pegg and Woodward, 1986; Osbourn *et al.*, 1995; Sandrock *et al.*, 1995; Lairini *et al.*, 1996; Quidde *et al.*, 1998). The contribution of these tomatinases to pathogenicity has not yet been established, although experiments are in progress in several laboratories to test this.

While most tomato pathogens are relatively resistant to α-tomatine, *Cladosporium fulvum* is highly sensitive to this saponin *in vitro* (Dow and Callow, 1978), and does not appear to be able to degrade it. Although the α-tomatine content of green leaf tissue of tomato is estimated to be approximately 1 mM (Arneson and Durbin, 1968a), it is not known how the saponin is distributed and compartmentalized within the leaf tissue. Since *C. fulvum* is a biotrophic pathogen and restricts its growth to the intercellular spaces of the mesophyll tissue, it is possible that the fungus may simply avoid coming into contact with α-tomatine during the infection process. In order to test whether α-tomatine is likely to restrict the growth of *C. fulvum* during infection of resistant and susceptible tomato lines, we transformed the cDNA encoding the tomatinase enzyme of the necrotrophic tomato pathogen *S. lycopersici* into two different races of this fungus (Osbourn *et al.*, 1995; Melton *et al.*, 1998). *Septoria lycopersici* tomatinase is extracellular, and detoxifies α-tomatine by the removal of a single terminal D-glucose molecule from the tetrasaccharide moiety of the saponin (Arneson and Durbin, 1967). Tomatinase-producing *C. fulvum* transformants showed increased sporulation on cotyledons of susceptible tomato lines (Table 15.1). They also caused more extensive infection of seedlings of resistant tomato lines (Fig. 15.2), but did not go on to sporulate. Thus, these experiments are supportive of a role for α-tomatine in inhibiting the growth of *C. fulvum* both in compatible and in incompatible interactions. The isolation of mutants of tomato which lack α-tomatine is currently underway, and should provide a direct genetic test of the importance of this saponin in protection of tomato plants against attack by *C. fulvum* and other pathogens.

A Family of Saponin-detoxifying Enzymes - Potential for Disease Control

The mechanisms of action of *G. graminis* avenacinase and *S. lycopersici* tomatinase are similar, since both enzymes remove a terminal β, 1-2-linked D-glucose molecule from their respective saponin substrates by hydrolysis. The two enzymes have been shown to be 68% similar at the amino acid level, despite having clearly different substrate specificities (Osbourn *et al.*, 1995). BLAST protein sequence similarity searches indicate that the *S. avenae* avenacosidase described here represents a third member of this family of saponin-detoxifying enzymes, and shares 62% amino acid similarity with avenacinase. Comparison of the sequences of these three saponin-detoxifying enzymes with other sequences in the databases indicates that, as might be expected from their biochemical function, these enzymes are related to a family of glucosyl hydrolases (defined as family 3 by Henrissat), which includes fungal and bacterial cellobiose-degrading enzymes (Henrissat, 1991; Henrissat and Bairoch, 1993). The effects of these other enzymes on saponins have not been

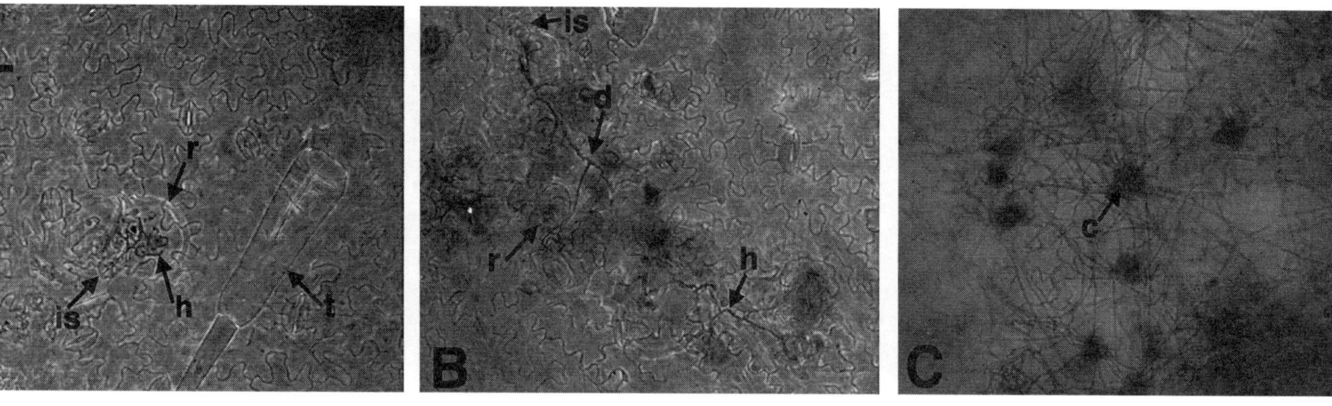

Fig. 15.2. Microscopic analysis of cotyledons of Moneymaker *Cf5* infected with *Cladosporium fulvum* (13 days after inoculation) showing: A: Limited hyphal growth of a control race 4 strain. A rounded host mesophyll cell is visible. B: Dichotomous branching and considerable hyphal growth of a tomatinase producing race 4 transformant within the plant tissue. Extensive host cell rounding and necrosis (the latter indicated by blue staining) are also evident. C: Sporulation of a compatible race (race 5) on the susceptible host, Moneymaker *Cf5*. h, hyphae; is, infection site; r, plant cell rounding; t, trichome; d, dichotomous branch; c, conidiophore. Intact cotyledons were stained with lactophenol trypan blue. The scale bar is equivalent to 20μM and applies to all three panels.

described. Currently there is no structure available for this major family of enzymes, although the recent crystallization of *S. lycopersici* tomatinase now offers an opportunity to investigate the relationship between substrate specificity and enzyme structure (V. Bamford, A. Osbourn and A. Hemmings, unpublished results).

Table 15.1. The effect of expression of *Septoria lycopersici* tomatinase on the ability of *Cladosporium fulvum* race 4 to sporulate on seedlings of the susceptible tomato line Moneymaker *Cf4*.

	Level of sporulation (% cotyledons)		
	0	+	++
Mock-inoculated	100	0	0
R4-C1[a]	53	47	0
R4-C2[a]	88	12	0
R4-C3[a]	69	31	0
R4-T1[b]	43	43	14
R4-T2[c]	0	25	75
R4-T3[c]	0	33	67

R4-T1, -T2 and -T3 are three independent transformants which express tomatinase activity; R4-C1, -C2 and -C3 are control transformants. Sporulation on the upper surface of 20 cotyledons per treatment was scored 11 days after inoculation as follows: 0 = no sporulation; + and ++ = sparse and heavy sporulation respectively. Strains marked with the same superscript letter had sporulation levels which were not significantly different from one another (P=5%).

In the future the continued characterization of saponin-detoxifying enzymes will determine whether or not saponin detoxification is a general pathogenicity mechanism employed widely by plant pathogenic fungi, and will also tell us how many different families of saponin-detoxifying enzymes exist. The extracellular nature of these enzymes makes them attractive targets for inhibitor-based control strategies, and searches for inhibitors are currently underway.

References

Arneson, P.A. and Durbin, R.D. (1967) Hydrolysis of tomatine by *Septoria lycopersici*; a detoxification mechanism. *Phytopathology* 57, 1358-1360.

Arneson, P.A. and Durbin, R.D. (1968a) The sensitivity of fungi to α-tomatine. *Phytopathology* 58, 536-537.

Arneson, P.A. and Durbin, R.D. (1968b) Studies on the mode of action of tomatine as a fungitoxic agent. *Plant Physiology* 43, 683-686.

Bowyer, P., Clarke, B.R., Lunness, P., Daniels, M.J. and Osbourn, A.E. (1995) Host range of a plant pathogenic fungus determined by a saponin-detoxifying enzyme. *Science* 267, 371-374.

Crombie, M.L., Crombie, L., Green, J.B. and Lucas, J.A. (1986) Pathogenicity of take-all fungus to oats: its relationship to the concentration and detoxification of the four avenacins. *Phytochemistry* 25, 2075-2083.

Dow, J.M. and Callow, J.A. (1978) A possible role for α-tomatine in the varietal-specific resistance of tomato to *Cladosporium fulvum*. *Phytopathologische Zeitschrift* 92, 211-216.

Fenwick, G.R., Price, K.R., Tsukamota, C. and Okubo, K. (1992) Saponins. In: D'Mello, J.P., Duffus, C.M. and Duffus, J.H. (eds) *Toxic Substances in Crop Plants*. Royal Society of Chemistry, Cambridge, pp. 285-327.

Ford, J.E., McCance, D.J. and Drysdale, R.B. (1977) The detoxification of α-tomatine by *Fusarium oxysporum* f.sp. *lycopersici*. *Phytochemistry* 16, 545-546.

Gus-Mayer, S., Brunner, H., Schneider-Poetsch, H.A.W. and Rüdiger, W. (1994) Avenacosidase from oat: purification, sequence analysis and biochemical characterization of a new member of the BGA family of β-glucosidases. *Plant Molecular Biology* 26, 909-921.

Henrissat, B. (1991) A classification of glycosyl hydrolases based on amino acid sequence similarities. *Biochemical Journal* 280, 309-316.

Henrissat, B. and Bairoch, A. (1993) New families in the classification of glycosyl hydrolases based on amino acid sequence similarities. *Biochemical Journal* 293, 781-788.

Hostettmann, K.A. and Marston, A. (1995) *Saponins. Chemistry and Pharmacology of Natural Products*, 1st edn. Cambridge University Press, Cambridge, 548 pp.

Johnston, H.W. and Scott, P.R. (1988) Identification of oat-adapted isolates of cereal *Septoria* in the UK using a detached leaf technique. *Plant Pathology* 37, 148-151.

Keukens, E.A.J., De Vrije, T., Van den Boom, C., De Waard, P., Plasman, H.H., Thiel, F., Chupin, V., Jongen, W.M.F. and De Kruijf, B. (1995) Molecular basis of glycoalkaloid induced membrane disruption. *Biochimica Biophysica Acta* 1240, 216-228.

Lairini, K., Perez-Espinosa, A., Pineda, M. and Ruiz-Rubio, M. (1996) Purification and characterization of tomatinase from *Fusarium oxysporum*. f. sp. *lycopersici*. *Applied and Environmental Microbiology* 62, 1604-1609.

Lüning, H.U. and Schlösser, E. (1975) Role of saponins in antifungal resistance. V. Enzymatic activation of avenacosides. *Zeitschrift für Pflanzenkrankheiten und Pflanzenschutz* 82, 699-703.

Lüning, H.U. and Schlösser, E. (1976) Role of saponins in antifungal resistance VI. Interactions *Avena sativa-Drechslera avenacea. Journal of Plant Disease Protection* 83, 317-327.

Melton, R.E., Flegg, L.M., Brown, J.K.M., Oliver, R.P., Daniels, M.J. and Osbourn, A.E. (1998) Heterologous expression of *Septoria lycopersici* tomatinase in *Cladosporium fulvum*: Effects on compatible and incompatible interactions with tomato seedlings. *Molecular Plant-Microbe Interactions* 11, 228-236.

Nisius, H. (1988) The stromacentre in Avena plastids: an aggregation of β-glucosidase responsible for the activation of oat-leaf saponins. *Planta* 173, 474-481.

Osbourn, A.E. (1996) Saponins and plant defence - a soap story. *Trends in Plant Science* 1, 4-9.

Osbourn, A.E., Clarke, B.R., Dow, J.M. and Daniels, M.J. (1991) Partial characterization of avenacinase from *Gaeumannomyces graminis* var. *avenae. Physiological and Molecular Plant Pathology* 38, 301-312.

Osbourn, A.E., Bowyer, P., Lunness, P., Clarke, B. and Daniels, M. (1995) Fungal pathogens of oat roots and tomato leaves employ closely related enzymes to detoxify host plant saponins. *Molecular Plant-Microbe Interactions* 8, 971-978.

Pegg, G.F. and Woodward, S. (1986) Synthesis and metabolism of α-tomatine in tomato isolines in relation to resistance to *Verticillium albo-atrum. Physiological and Molecular Plant Pathology* 28, 187-201.

Price, K.R., Johnson, I.T. and Fenwick, G.R. (1987) The chemistry and biological significance of saponins in food and feedingstuffs. *CRC Critical Reviews in Food Science and Nutrition* 26, 27-133.

Quidde, T., Osbourn, A.E. and Tudzynski, P. (1998) Detoxification of α-tomatine by *Botrytis cinerea. Physiological and Molecular Plant Pathology* (in press).

Roddick, J.J. and Drysdale, R.B. (1984) Destabilization of liposome membranes by the steroidal glycoalkaloid α-tomatine. *Phytochemistry* 23, 543-547.

Sandrock, R.W., Della Penna, D. and Van Etten, H.D. (1995) Purification and characterization of tomatinase, an enzyme involved in the degradation of α-tomatine and isolation of the gene encoding tomatinase from *Septoria lycopersici. Molecular Plant-Microbe Interactions* 8, 960-970.

Schlösser, E. (1975) Role of saponins in antifungal resistance III. Tomatine dependent development of fruit rot organisms of tomato fruits. *Zeitschrift für Pflanzenkrankheiten und Pflanzenschutz* 82, 476-484.

Schönbeck, F. and Schlösser, E. (1976) Preformed substances as potential protectants. In: Heitefuss, R. and Williams, P.H. (eds) *Physiological Plant Pathology*. Springer-Verlag, Berlin, pp. 653-678.

Shaw, D. (1957) Studies on *Leptosphaeria avenaria* f. sp. *triticea* on cereals and grasses. *Canadian Journal of Botany* 35, 113-118.

Steel, C.S. and Drysdale, R.B. (1988) Electrolyte leakage from plant and fungal tissues and disruption of liposome membranes by α-tomatine. *Phytochemistry* 27, 1025-1030.

Suleman, P., Tohamy, A.M., Saleh, A.A., Madkour, M.A. and Straney, D.C. (1996) Variation in sensitivity to tomatine and rishitin among isolates of *Fusarium oxysporum* f. sp. *lycopersici*, and strains not pathogenic on tomato. *Physiological and Molecular Plant Pathology* 48, 131-144.

Tschesche, R. and Lauven, P. (1971) Avenacosid B, ein zweites bisdesmosidisches Steroidsaponin aus *Avena sativa*. *Chemische Berichte* 104, 3549-3555.

Tschesche, R., Tauscher, M., Fehlhaber, H.W. and Wulff, G. (1969) Avenacosid A, ein bisdesmosidisches Steroidsaponin aus *Avena sativa*. *Chemische Berichte* 102, 2072-2082.

Turner, E.M. (1961) An enzymic basis for pathogen specificity in *Ophiobolus graminis*. *Journal of Experimental Botany* 12, 169-175.

Verhoeff, K. and Liem, J.I. (1975) Toxicity of tomatine to *Botrytis cinerea*, in relation to latency. *Phytopathologische Zeitschrift* 82, 333-338.

Weber, G.F. (1922) Septoria diseases of cereals. *Phytopathology* 12, 449-470.

Wubben, J.P., Price, K.R., Daniels, M.J. and Osbourn, A.E. (1996) Detoxification of oat leaf saponins by *Septoria avenae*. *Phytopathology* 86, 986-992.

Chapter sixteen:

Integrating Septoria Risk Variables

N.D. Paveley
ADAS High Mowthorpe, Duggleby, Malton, North Yorkshire YO17 8BP, UK

Introduction

Good crop managers are generally skilled observers who acquire, during their apprenticeship, a collection of empirically derived 'rules of thumb' that guide their decisions. Passioura (1996) points out that such rules should not be denigrated, as those that stand the test of time do so because they are generally (but importantly not always) robust. Although some recent decision support systems show promise (see Jørgensen *et al.*, Chapter17 this volume), attempts to replace or supplement subjective judgement with systems based on understanding of the population dynamics of the pest have generally failed to show a sustained economic benefit (Forrer, 1992; Hearn and Brook, 1994). There are numerous technical, political and psychological reasons for ineffective transfer of benefits from science to practice (Sylvester-Bradley, 1991). This chapter will focus on one: poor predictive precision.

Precision is improved by reducing error. Reynolds and Acock (1985) described the relationship between error and complexity in relation to plant growth models. Many of the same principles might equally apply to pathology. They divided the notional total error into two components, one arising from errors in estimating parameters, the other from systematic bias resulting from oversimplifying. The latter error, arising from the structure of the model, has the potential to decrease as complexity is increased, as the model better represents the complexity of the biological system. However, if the structure is flawed, the asymptote, below which error cannot be reduced, is raised and attempts to improve performance by adding parameters increase the total error. More importantly, because the reduction in error with increasing structural complexity follows a law of diminishing returns, oversimplification or neglect of the structure in one area cannot be compensated for by additional complexity elsewhere.

As structural form should follow function, it is important to be clear about the function - what is the model designed to predict? Wheat producers select cultivars predominantly for yield potential and grain quality. Disease resistance is only a secondary consideration (Smith and Webster, 1986). As the primary inoculum for *Septoria tritici* (anamorph of *Mycosphaerella graminicola*) epidemics arises from widespread and prolonged dispersal of ascospores (Scott *et al.*, 1988; Shaw and Royle, 1989a), practical opportunities for avoidance by cultural means are limited. It has, therefore, become important to ensure that the appropriate amount of the best fungicide active ingredient is applied at the right time. The optimum treatment can be judged by comparing input costs and output value. As input costs are known and grain value can be estimated, treatment optimization requires prediction of **yield response to treatment.**

Study of the *S. tritici* pathosystem was stimulated by the upsurgence of leaf blotch in the UK during the early 1980s (Polley and Thomas, 1991). Initial work to identify sources of host resistance (Bayles *et al.*, 1985) and fungicides with protectant or curative activity (Jordan *et al.*, 1986) was soon followed by quantitative epidemiological studies. Some of the latter work combined the interpretation of data from sequential measurements of disease severity with knowledge of mechanisms, obtained from independent experiments, in order to link cause and effect (Royle, 1994).

This progression, from the qualitative to the quantitative, followed the movement led by Van der Plank (1963) in other pathosystems and developed, often with increasing mathematical rigour and complexity, during the following three decades (Teng, 1985; Waggoner, 1986; Jeger, 1987; Campbell and Madden, 1990). However, improvements in quantitative epidemiology have not been matched by improvements in the understanding of the dynamics of fungicidal control, or the impact of disease on the processes of crop dry matter accumulation and partitioning to vegetative or reproductive growth. Precision in prediction of yield response to treatment has been limited by neglect of important structural components.

This chapter is an attempt to describe the crop-pathogen-fungicide-environment system, in such a way that there is a balance in the level of complexity adopted in each of the components. Successive sections describe risk variables associated with variation in disease progress; the effect of fungicide treatment on disease progress; and sources of variation in the relationship between disease and yield.

Variation in Disease Progress Curves

If sequential measurements of *S. tritici* severity in the upper canopy of a winter wheat crop are plotted, a disease progress curve similar to the example shown in Fig. 16.1 generally results. Where disease becomes severe, progress curves often show signs of inflection. If disease progress curves are plotted from crops across a range of sites and seasons, they tend to conform to an exponential or logistic form, but show a wide range of variation in time of onset of visual symptoms and in its rate of increase.

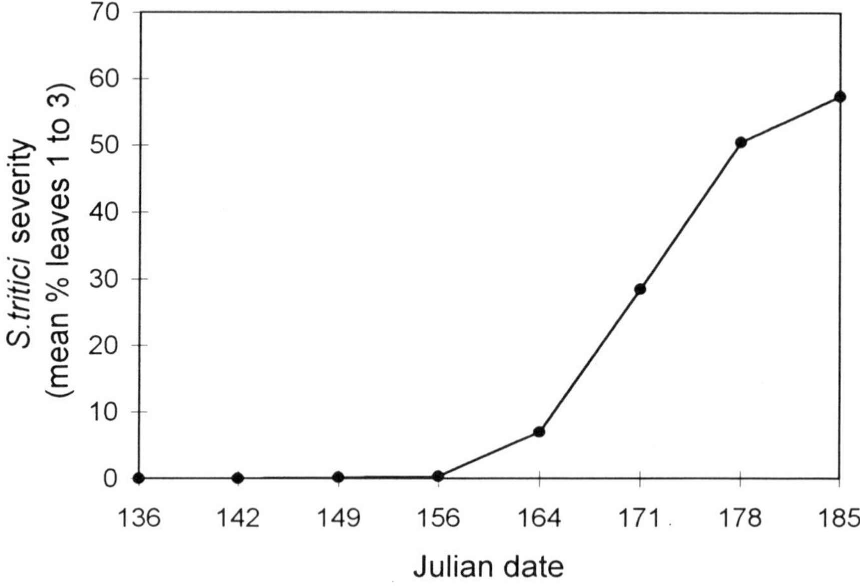

Fig. 16.1. Example disease progress curve for *Septoria tritici.*

O'Callaghan (1994) notes that a successful outcome in the management of a production process requires attention to a small number of key factors - the so-called 80/20 rule. This presupposes that 80% of the variation in outcome depends on only 20% of the factors, the most important ones which affect the outcome of the process. In biological systems, there are usually a large number of factors which have the potential to affect the outcome particularly where, as in a pathosystem, organisms interact. Many of these factors change with time and, as they might be quantified in order to aid the management process, would be better termed variables (France and Thornley, 1984). Some consensus is emerging about the key variables involved in determining the variation in severity of *S. tritici* epidemics, and, therefore, to some extent, the need for fungicidal intervention. These are considered, briefly, in this section.

Inoculum

Although the role of ascospores in the spring epidemic is now being recognized (see Hunter *et al.*, Chapter 7 this volume) during the period when the upper canopy of winter wheat is developing, pycnidia, residing in diseased leaves lower down the canopy are still thought to provide the main source of *S. tritici* inoculum (Royle *et al.*, 1986). As the production of pycnidia is associated with the expression of visual symptoms, observations of disease lower in the canopy might be of value in predicting future disease severity in the upper canopy. This notion underlies treatment thresholds (Schofl and Morris, 1994). Practical experience and survey information (Smith and Webster, 1986) suggest that crop managers consider the level of disease in the crop to be the most influential variable in the decision to apply fungicide treatment.

Despite various limitations (Campbell and Madden, 1990), Van der Plank's simple theoretical relationships between initial inoculum and epidemic development may be sufficient to help test whether the confidence in the predictive value of disease observations, described above, seems justified in the *S. tritici* case.

In Fig. 16.2 (based on a diagram by Van der Plank, 1984), curves A and B represent the progress of disease in two crops. The growth rates are assumed to be constant for simplicity. The first crop is exposed to an arbitrary number of spores arriving from an adjacent source, whereas the second is exposed to only 10% of that number of spores, an initial inoculum ratio of ten. The delay (delta-*t*) in epidemic B represents the time taken for the epidemic in the second crop to multiply tenfold. Thereafter, the curves are the same. Each successive tenfold reduction in initial inoculum delays the epidemic by the same time period, until the epidemic is delayed beyond the time of natural canopy senescence (arbitrarily represented in Fig. 16.3 by the end of the time axis). In the *S. tritici* case, the upper canopy can be considered as analogous to the crop and the lower canopy, the external source of inoculum.

Experimental data to support this model of the relationship between inoculum and subsequent disease severity are sparse for *S. tritici* (and for most other pathosystems, Berger, 1988). However, the generally accepted notion that *S. tritici* inoculum is seldom limiting fits with the theoretical outcome that, at high relative growth rates, a very wide range of inoculum levels should allow substantial epidemics to occur within the life-time of the upper canopy.

Further experimentation in this area would be of value because, if the model is substantially correct, a number of important practical consequences follow. First, the relationship between inoculum and future disease severity is exponential with an upper asymptote. Crop managers behave as though the relationship is linear. Second, differences between very small amounts of observed disease lower in the canopy would determine whether or not a substantial epidemic could occur within the lifetime of the upper canopy. Third, the model deals with the amount of inoculum **arriving** in the upper canopy. It has become clear that the position of the inoculum source in relation to the upper canopy can have a profound influence on the efficiency of transfer of spores to newly-emerged leaves (Lovell *et al.*, 1997). Fourth, the effect of inoculum would be expected to vary according to the relative growth rate of the epidemic. Figure 16.4 shows the effect of relative growth rate on delta-*t* for an inoculum ratio of ten. If other variables cause the rate to vary, particularly in the range 0-0.2, where the effect on delta-*t* is most profound, these would be

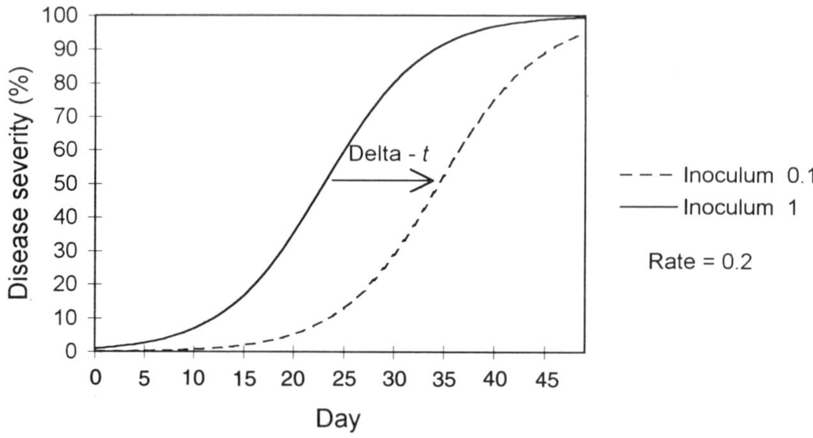

Fig. 16.2. Delay in epidemic development (delta-*t*) caused by a one order of magnitude reduction in initial inoculum (after Van der Plank, 1984).

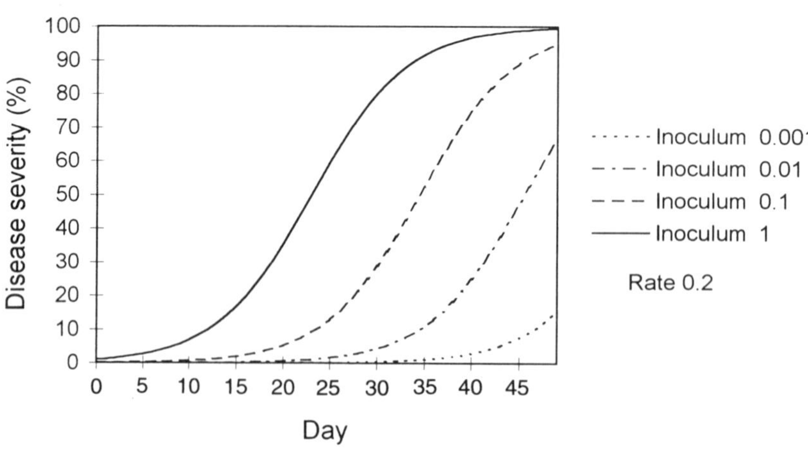

Fig. 16.3. Delay in epidemic development caused by successive one order of magnitude reductions in initial inoculum.

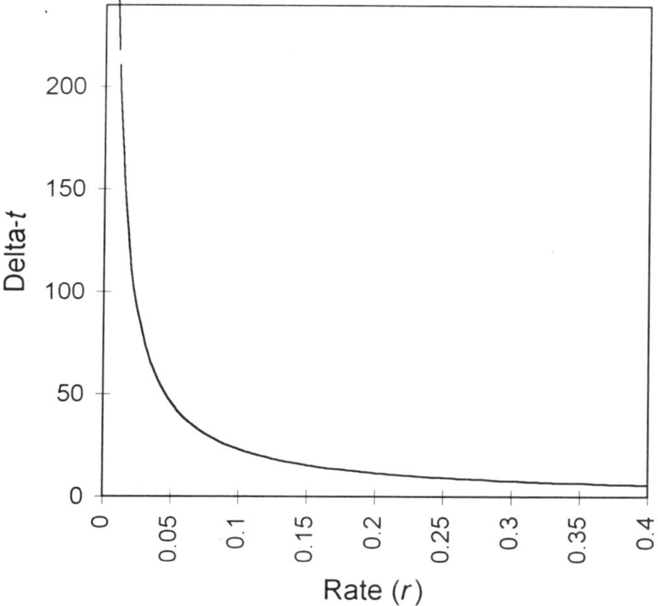

Fig. 16.4. Effect of epidemic rate (r) on delay (delta-t) caused by a one order of magnitude reduction in inoculum (after Vanderplank, 1984).

expected to interact with variation in inoculum in determining future epidemic progress. The main sources of variation in rate are host resistance and weather.

Resistance

Figure 16.5 shows the effect of four contrasting host genotypes on epidemic progress in the upper canopy at one experimental site. Logistic functions fitted to *S. tritici* disease progress curves on 20 wheat cultivars, selected for variation in resistance (N.D. Paveley, unpublished results), gave estimates of relative growth rate varying from 0.06-0.2 and differences in date of disease onset (arbitrarily defined as the date on which disease severity exceeded 0.5%) of up to 25 days. Differences in the upper asymptote or 'carrying capacity' were also noted (Gilligan, 1990). Clearly, host genotype is an important risk variable. However, there are problems associated with the use of genotype information to aid disease prediction.

If disease progress curves are plotted for a range of genotypes across a range of experimental sites and seasons, the most and least resistant types tend to behave consistently, but there is considerable variation in the rank order of the intermediate types (see Parker *et al.*, Chapter 6, this volume). Predictive precision could be improved by understanding and accounting for the causes of these changes in rank order of resistance. Two candidate causes await further investigation. First, possible interactions between host genotype and site/seasonal variation in pathogen genotype (see Mundt *et al.*, Chapter 8 this volume), and second, genotypic variation in disease escape, the expression of which is dependent on environment (Tavella, 1978; Parker *et al.*, Chapter 6 this volume). Until such mechanisms are elucidated, cultivar disease resistance ratings, such as those produced in the UK by the National Institute of Agricultural Botany (Anon., 1997), provide the most robust measure of relative disease risk. These ratings are derived from the mean response of each genotype across a wide range of sites and seasons. It should be noted, however, that the relationship between resistance ratings and the disease severity data from which they are derived, is not linear (R.A. Bayles, NIAB, 1996, personal communications).

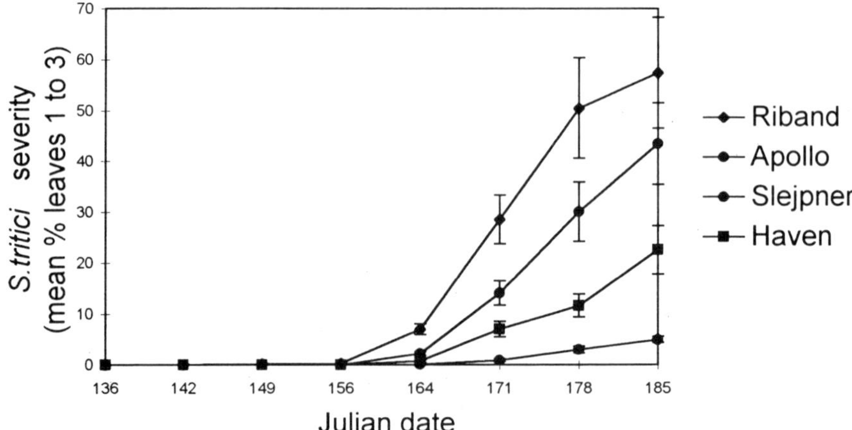

Fig. 16.5. Example disease progress curves for four winter wheat cultivars of contrasting resistance to *Septoria tritici*. Error bars are standard errors.

Relative growth rate measurements on more resistant types suggest that inoculum is more likely to be limiting, due to the increased impact on delta-*t*, than on the highly susceptible types on which most epidemiological studies have been undertaken.

Weather

Campbell and Madden (1990) point out that, because almost all abiotic and biotic factors can influence the relative growth rate of an epidemic, researchers should be cautious in attributing causes for differences in rate between epidemics. In general, a valid experimental design should be used to attribute properly one or more causes for the magnitude of rate. In field experimentation, it is difficult to arrange for variation in weather, in order to elucidate associations between weather variables and epidemic progress, whilst keeping all else constant. The effects of variation in weather across sites and seasons will tend to be confounded by variation in inoculum amount and position and, possibly, pathogen genotype.

Figure 16.6 shows data describing epidemics of *S. tritici* measured in the upper canopy of the susceptible cultivar Riband, at two experimental sites in each of three seasons. The variation in disease progress due to site and seasonal factors is of similar magnitude to the effect of host genotype at a single high disease site.

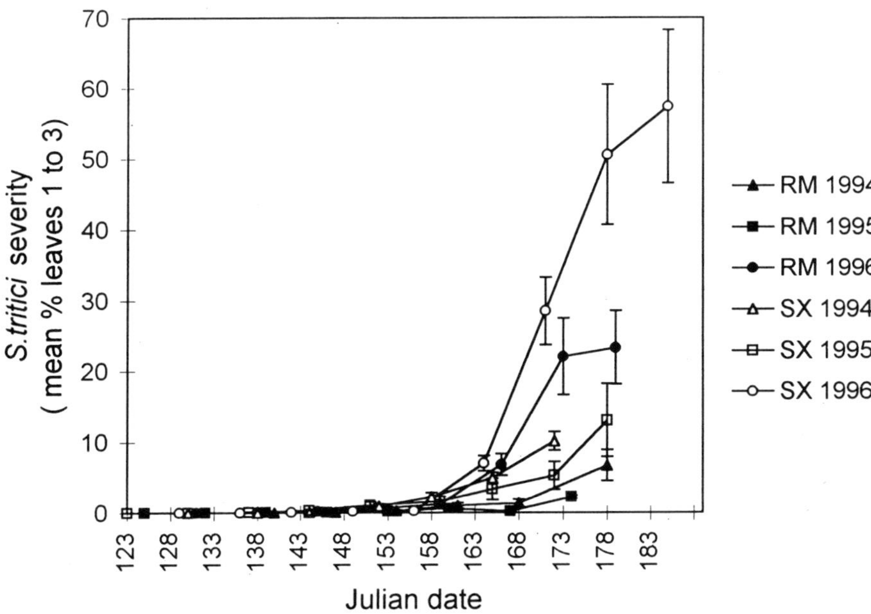

Fig. 16.6. *Septoria tritici* progress curves at two UK sites in each of three seasons. RM = ADAS Rosemaund, Herefordshire; SX = Starcross, Devon. Error bars are standard errors.

Despite the experimental limitations described above, good progress has been made by several workers in attributing such variation to weather variables; predominantly rainfall during the period of upper canopy expansion (Shaner and Finney, 1976; Eyal, 1981; Coakley *et al.*, 1985; Shaw, 1987; Thomas *et al.*, 1989; Hansen *et al.*, 1994). More recently, conditions during the autumn and winter have been found to associate with disease severity in the following season (see Parker *et al.*, Chapter 6 this volume), opening the possibility of a strategic estimate of disease risk. Such correlations await mechanistic interpretation.

Although it is difficult to ameliorate the effect of confounding variables in field experiment design, it may prove possible to reduce their impact on unexplained variation in the analysis; for example, by using measurements of inoculum as covariates.

Disease Progress and Variation in the Dose-Response Curve

Epidemiologists often seek to demonstrate the applied value of their work, by showing how their findings can effect an improvement in crop management decisions. This is a worthy objective, but success is often prejudiced by failure to understand, or recognize the importance of, the steps between predicted severity of an epidemic and yield response to fungicide treatment. These steps can be subdivided into those that link disease progress curves and dose-response curves (described here) and those mechanisms which link disease to yield loss (summarized in the next section).

If an extra dimension, fungicide dose, is added to the disease progress curve, a pattern like the one shown in Fig. 16.7 typically results. The data show a time segment of disease progress curves fitted to *S. tritici* data from an experiment on cv. Riband where a range of doses of propiconazole (as the commercial product Tilt 250 EC, Novartis) was applied at a single timing. Depending on the time of fungicide application, the effect on the epidemic is to delay disease onset (as in the case illustrated) and/or reduce rate.

Fungicides take time to reach the active site in the pathogen and impede growth, and having done so are not able to convert leaf tissue that already exhibits symptoms back into healthy tissue. Also, translocation to leaves which emerge after treatment is limited. As a result, fungicidal activity on a given leaf is optimized by application after the leaf has emerged, but before the onset of expressed disease (sometime before day 156 in the example given in Fig. 16.7).

If a section is taken across the dose dimension of Fig. 16.7, at any point in time (most conveniently at Julian day 170), it can be seen that the effect of dose on the disease progress curves results in a dose-response curve of disease severity. Study of such curves from a wide range of circumstances (Paveley, 1998) suggests that they take the general form shown in Fig. 16.8, whereby the first increment of increasing dose has the greatest effect and each subsequent increment shows a diminishing return.

As the cost of fungicide increases linearly with dose, the operating point for maximum profit, where the marginal return from an additional unit of input is equal to the cost of that input, lies somewhere on the dose-response curve above the point of maximum disease suppression. Clearly, fungicide cost and grain value will influence the optimum or 'appropriate' dose, but the optimum is more sensitive to variation in the shape of the dose-response curve. A number of factors control the shape, including the sensitivity of the pathogen population to the active ingredient, and the proportion of the applied dose which reaches the site of activity in the pathogen. However, two important sources of variation fall within the remit of this chapter: the severity of untreated disease and the relationship between disease and yield loss.

Taking two extreme cases of untreated disease progress: if, because of host resistance, adverse weather conditions or the absence of initial inoculum, disease failed to develop, the dose-response curve would be flat and the optimum dose zero. Conversely, if the combination of susceptible host, conducive weather and abundant inoculum conspired to produce a severe epidemic, the dose-response curve would be steep and the optimum close to the full dose. Intermediate circumstances would produce intermediate dose optima. Where the objective is simply to decide whether or not to treat, rather than to optimize the dose, only the two ends of the dose-response curve need to be considered, but the principles remain the same.

Hence variation in the untreated disease progress curve relates to treatment decisions via the dose-response curve and is a key determinant of both the need to treat and the dose optima. However, as the dose-response curve is shallow in the region of the optimum, it follows that the position of the optimum will be sensitive to any variation in the relationship between disease and yield loss.

Variation in the Disease:Yield Loss Relationship

The notion of a relatively stable relationship between *S. tritici* severity and yield loss was encouraged by the publication, in the late 1980s, of average disease:yield loss equations (Shaw and Royle, 1989b; Thomas *et al.*, 1989). Perhaps because of resource limitations, such studies did not meet the requirement of repeatability across a wide range of sites, seasons, levels of disease intensity and host genotype set out by Madden (1983). Subsequent studies on *S. tritici* and other pathogens have shown that average disease:yield loss relationships can hide a substantial degree of site, season and host genotype variation (Paveley *et al.*, 1997; Bryson *et al.*, 1998).

Waggoner and Berger (1987) sought to introduce the concept that disease should be related to yield via its impact on the processes by which yield accumulates. There is little evidence until recently (Gaunt, 1995; Madden and Nutter, 1995; Madeira and Clark, 1995; Bryson *et al.*, 1998) that this concept has been adopted, so perhaps it bears précis here.

The growth of a crop is largely the integral of net photosynthesis in its leaves. Although total growth does not relate directly to yield, Watson (1947) found that yield was related to leaf area duration (LAD); the integral over time of the size of the crop canopy (leaf area index - dimensionless unit area of leaf per unit area of ground). The realization that the amount of photosynthesis related more directly to the amount of photosynthetically active solar radiation (PAR) absorbed by leaves, rather than their area, provided a quantification of the relationship between leaf area and yield. An analogy with Beer's Law

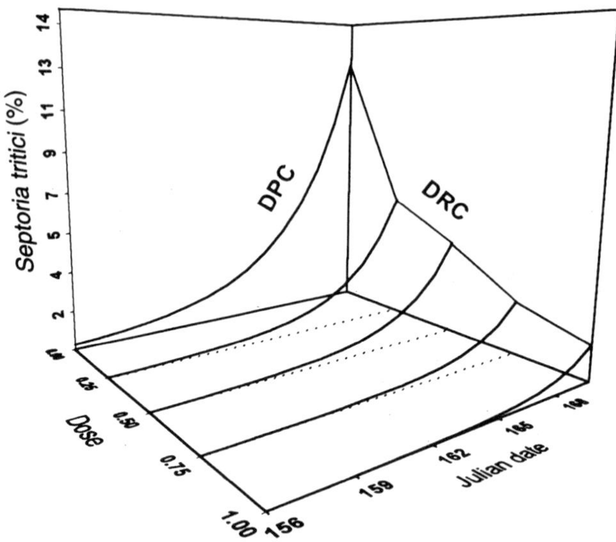

Fig. 16.7. Dose-response curve (DRC) arising from the effect of a range of fungicide doses on the onset and rate of *Septoria tritici* disease progress curves (DPC).

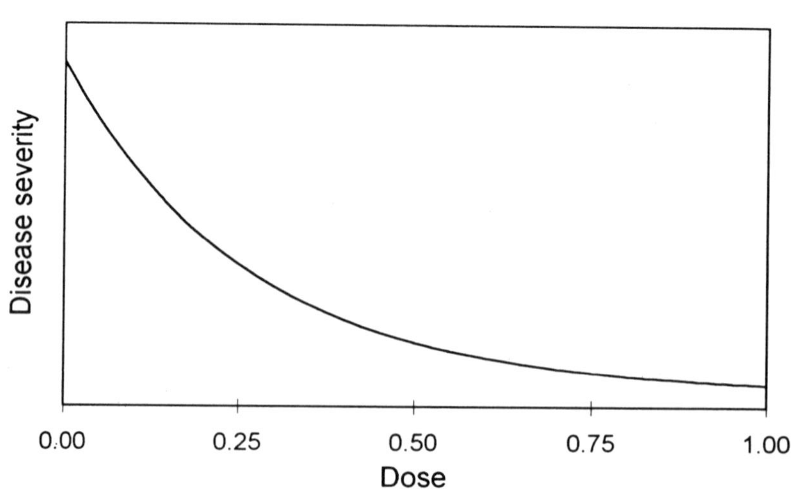

Fig. 16.8. General form for *Septoria tritici* dose-response curve.

(Monteith and Unsworth, 1990) was used to account for the relationship between leaf area and fractional interception of PAR.

Because pathologists are naturally most interested in disease, they have concentrated on the increase in the fraction of leaf area affected by disease with time. Progress curves of disease severity have been summarized by their relative rate of change or by the area under the disease progress curve (AUDPC). Unfortunately, single point, multiple point, rate of change or integral measures of disease do not tell of the size of the foliar factory or how much PAR it absorbs. It will, therefore, be necessary to re-express the measures of disease if they are to be more useful in predicting damage.

The next section will use the principles outlined above to illustrate how the position of the epidemic, both in space and time, might influence the relationship between disease severity and yield loss.

Disease Progress in Relation to Crop Growth and Development

Waggoner and Berger (1987) suggested that, if pathologists want to relate disease progress curves to yield, it would be logical to subtract the area of disease from the LAD by integrating the size of the healthy canopy during the season (healthy area duration - HAD, days). A logical extension then permits the calculation of the amount of PAR absorbed by healthy canopy during that period (healthy area absorption - HAA, MJ m^{-2}). Such an approach has two substantial limitations when applied to wheat; first, treating the canopy as a homogeneous unit could hide the effect of the position of the epidemic within the canopy; and second, integrals of HAA over time might only be expected to relate consistently to yield if a period over which dry matter accumulation was partitioned to reproductive growth (and hence the period over which HAA should be integrated) could be defined reliably.

Effect of Epidemic Position in Space, in Relation to the Crop Canopy

The majority of the canopy of a winter wheat crop consists of a succession of leaf layers which emerge at one phyllochron [approximately 110 day degrees (°C)] intervals between GSZ 31 and GSZ 39 (Kirby, 1984; Tottman, 1987). In a typical UK crop, leaf 1 (the flag leaf) emerges at GSZ 39 and leaves 2, 3, and 4 (counting down the shoot from the flag leaf) at around GSZ 33, 32 and 31 respectively. As leaves live for approximately 600 to 1000 day degrees, it follows that leaves 2, 3 and 4 are shaded by the leaves above for the majority of their lives. Figure 16.9 shows calculated values for the amount of PAR available at the level of the ear and at each of the upper five leaves integrated over the main period of grain filling (flowering to crop senescence) for a typical UK winter wheat crop. Note that the total available PAR exceeds the incoming radiation (amount available at the ear level), because a proportion of the available PAR is not intercepted at each layer and is, therefore, passed to the next. As the contribution of a leaf layer to crop growth is dependent on its capacity to intercept PAR, it is clear, from the shortage of resource at the lower levels of the canopy, that only the upper two or three leaves are going to make a substantial contribution. This outcome agrees with the findings of earlier studies (Shaw and Royle, 1989b; Seck *et al.*, 1991). If the effect of *S. tritici* on yield is expressed predominantly via a reduction in green leaf area, it is clear that a unit of disease on leaf 1 will have a much greater impact than a unit of disease on leaf 4. In summary, less light is wasted by hitting disease symptoms on lower leaves, as it has already been intercepted by green area on the leaves above. The disease:yield loss relationship is sensitive to the position of disease in the crop canopy.

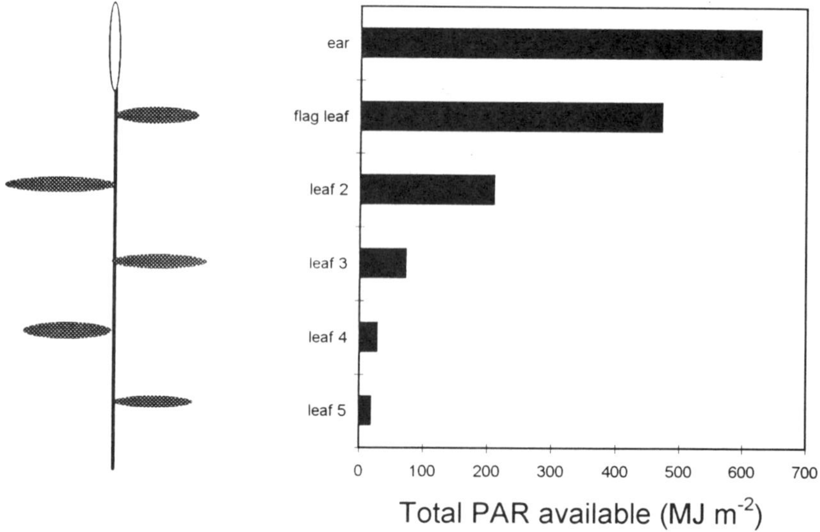

Fig. 16.9. Calculated values for the amount of PAR available at the level of the ear and each of the upper five leaves, integrated from flowering to senescence, for a crop of winter wheat.

Effect of Epidemic Position in Time, in Relation to Crop Growth and Partitioning

Figure 16.10 illustrates the dry matter growth of an undiseased winter wheat crop from GSZ 31 to senescence. During that period, total dry matter increases from approximately 3 t ha^{-1} to over 18 t ha^{-1}. Total dry matter growth can be broken down into three overlapping phases controlled by crop development: vegetative growth (production of leaves and stems), the accumulation of water-soluble carbohydrates (WSC), and reproductive growth. The last should be the focus of most attention, as by flowering (around 12 June in the example shown) the structure of the ear is complete and all subsequent growth is of grain, the only component with a substantial economic value. However, it is also important to note that WSC dry matter becomes zero before harvest and there is evidence that this dry matter can be readily translocated from the stem (where it accumulates) to the grain, and so can act as a reserve against adversity during grain filling (Gaunt, 1995). The potential size of this reserve is heritable (Foulkes, 1998), but the extent to which it is expressed is affected by environment (Beed, 1997).

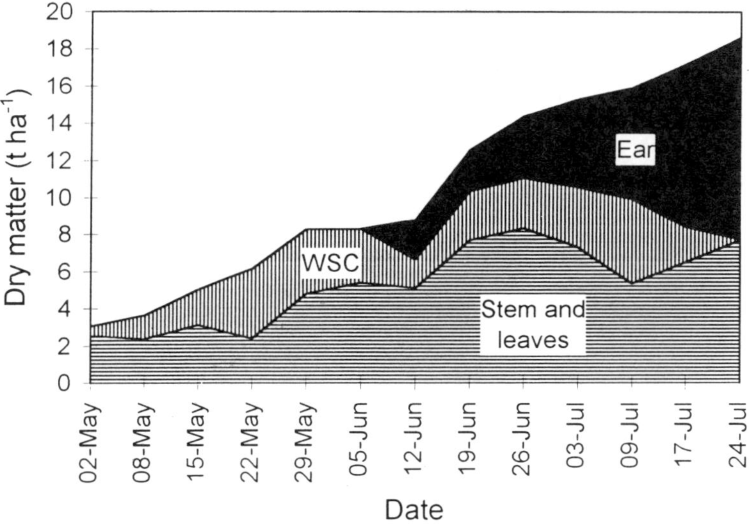

Fig. 16.10. Dry matter growth and partitioning for undiseased winter wheat, cv. Riband. WSC = water-soluble carbohydrate.

As dry matter accumulation is largely the product of photosynthesis, and most photosynthesis occurs in the upper leaf canopy, it is instructive to look at what is happening to green leaf area, in the presence and absence of disease, during the main period of crop growth.

In the absence of disease, the upper leaf canopy of a winter wheat crop expands by leaf emergence, survives for a period and dies by natural senescence. The effect of superimposing disease progress on this process is shown in Fig. 16.11. The undiseased data for Fig. 16.11 were collected from the same experimental plots as those for Fig. 16.10 and are shown on the same time axis. At typical temperatures during the period of upper canopy expansion, the pathochron (number of phyllochrons per latent period; Lovell *et al.*, 1997) for *S. tritici* is in the range two to three. As a result, maximum leaf canopy size is reached at GSZ 39 before substantial symptom expression can occur on the upper three leaves. After GSZ 39, disease-induced senescence precedes, and later supplements, natural senescence, leading to an earlier loss of green leaf area.

It is apparent, from considering the green leaf area and dry matter growth curves together, that most vegetative growth and the accumulation of WSC should be complete before *S. tritici* can begin to reduce the intercepting capacity of the upper leaves. Structural growth of the crop, canopy expansion and the accumulation of WSC are, therefore, not dependent on intervention with fungicides to prevent disease damaging the photosynthetic apparatus; protection of canopy expansion is provided **by** canopy expansion.

Unfortunately, the period of maximum disease-induced green area loss, later in the season, coincides with the period when the majority of the products of photosynthesis are being partitioned to grain growth. Figure 16.12 shows dry matter accumulation and partitioning data from the same experimental plots from which the diseased green leaf area index (GLAI) data in Fig. 16.11 were collected. The diseased and undiseased total dry matter curves separate shortly after disease impacts on GLAI. Although the effect appears small in relation to total dry matter, all of the accumulated loss (approximately 2 t ha^{-1}) is of grain.

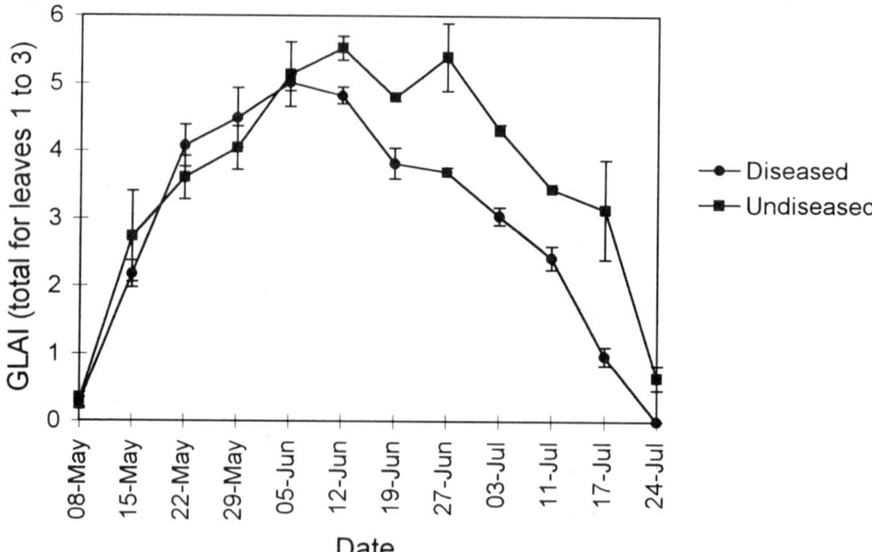

Fig. 16.11. The effect of *Septoria tritici* on the green leaf area index (GLAI) of the upper three leaves of a crop of winter wheat cv. Riband. Error bars are standard errors.

In the absence of adequate host resistance, prevention of disease-induced GLAI loss post-flowering requires intervention with fungicide during the period of leaf emergence (for the reasons outlined earlier).

In summary, the disease:yield loss relationship should be sensitive to the position of disease-induced green leaf area loss in time, in relation to dry matter growth and partitioning. But in practice the temporal position of the epidemic in relation to crop development is largely controlled by crop development, restricting disease-induced green area loss in the upper canopy to the critical period of grain growth.

Conclusions

The applied aim of improved predictive precision can best be furthered by identifying those key risk variables that control most of the variation in outcome, but about which understanding is least complete. Concentrating on these structural weaknesses will be of greater practical benefit than further refinement in understanding risk variables about which much is already known. Identification of these variables should be aided by the recognition that the risk of interest is not simply that of epidemic development, but of disease-induced yield loss. The latter is not consistently related to the former, so the study of variables that might influence the impact of a unit disease on yield should be a *bona fide* area of interest for pathologists and crop physiologists.

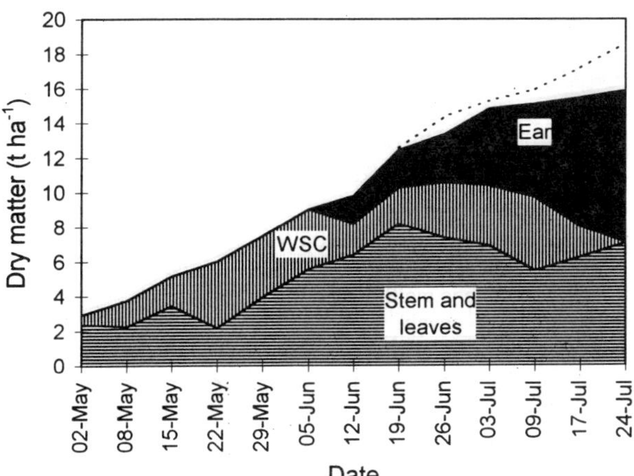

Fig. 16.12. Dry matter growth and partitioning for winter wheat cv. Riband affected by *Septoria tritici.* Dotted line represents total dry matter in undiseased plots for comparison. WSC = water-soluble carbohydrate.

Substantial progress has been made in associating variation in disease severity with weather factors. Less attention has been given to the effects of host resistance on epidemic development and, in particular, to unexplained variation in cultivar resistance when observed across sites and seasons. The role of variation in inoculum has also been neglected, possibly because it is seldom limiting in the highly susceptible cultivars commonly used in epidemiological research, or perhaps because the effect of inoculum position in the canopy was not recognized, leading to unexplained variation masking the effect of variation in amount. Confounding by density dependence of infection (Rouse *et al*., 1981; Shaw, Chapter 5 this volume) and dependence of rate on initial inoculum (Berger, 1988), may also have inhibited progress. Improved ability to predict epidemics might be achieved by integrating inoculum, host resistance and weather variables, and accounting for their potential to interact.

It has long been recognized that diseases reduce yield predominantly by causing premature loss of green leaf area, and that fungicides provide yield benefits by preventing a proportion of that loss (Jenkins and Morgan, 1969). It is unfortunate that pathologists have traditionally chosen to measure disease as a proportion of leaf area expressing symptoms. Where measures of the impact of disease on green area have been taken, these have also been expressed as proportions. Difficulties arise because leaf area is highly variable, between sites and seasons, between cultivars, and between leaf layers. The use of percentage disease or green leaf area is so well ingrained that we seldom ask: x% of what? Where this problem has been recognized the solution has sometimes been to make arbitrary adjustments (Shaw and Royle, 1989b) rather than to take more appropriate measurements. Converting percent disease and green leaf area measurements to absolute values, and noting the spatial and temporal location of those absolute areas in the canopy, provides a firm step towards being able to estimate the impact of disease on light interception and dry matter accumulation (Bryson *et al*., 1998).

Each unit of dry matter is born equal, but some become more equal than others - some become vegetative dry matter with little economic value whereas others form grain. The process of partitioning dry matter is controlled by crop development. Although the periods of vegetative and reproductive growth are distinct and the transition between the two (flowering) can be readily identified, the formation of WSC (which ultimately contributes to reproductive growth) during the vegetative phase complicates analysis of the relationship between disease severity and yield. It is too simplistic to believe that some measure of disease, green area or light interception integrated over a fixed developmental period will consistently relate to yield across sites, seasons and cultivars, without accounting for variation in the 'capacitance' built into the system in the form of WSC.

Returning to the beginning, rules of thumb used in practical disease management represent the shared experience of a large population of crop managers, accumulated over many years. If new understanding, hard earned through rigorous research, is not valued and taken up in farm practice, poor communication may be to blame. But we should be open to the possibility that our ability to integrate knowledge from epidemiology, crop protection and crop physiology, in order to predict economic outcomes, is still too poor to replace rules of thumb.

Fortunately for pathologists, the process of rule formation is compromised by the limitations of the observations on which it is based (Paveley *et al*., 1997). Experimentation removes many of these limitations. So research should be able to make a useful contribution, by identifying rules of thumb that are flawed. Where science agrees with rules of thumb, there may be comfort in understanding **why** existing management actions are appropriate. Where it disagrees, practitioners may be better persuaded to change their behaviour if the mechanisms underlying the need for change are understood and can be explained.

Acknowledgements

Funding by the Ministry of Agriculture, Fisheries and Food, and the Home-Grown Cereals Authority is gratefully acknowledged. Thanks are due to many colleagues for the collection of data, statistical analysis, figure preparation and constructive comments on the manuscript.

References

Anon. (1997) *NIAB Recommended List of Winter Wheat Varieties.* National Institute of Agricultural Botany, Cambridge, UK.

Bayles, R.A., Parry, D.W. and Priestly, R.H. (1985) Resistance of winter wheat varieties to *Septoria tritici.* *Journal of the National Institute of Agricultural Botany* 17, 21-26.

Beed, F. (1997) Forecasting the effect of disease as influenced by the host. Abstract in: *Proceedings of 1997 BSPP Presidential Conference.* British Society for Plant Pathology, UK, p. 16.

Berger, R.D. (1988) The analysis of effects of control measures on the development of epidemics. In: Kranz, J. and Rohem, J. (eds) *Experimental Techniques in Plant Disease Epidemiology.* Springer-Verlag, Heidelberg, pp. 137-151.

Bryson, R.J., Paveley, N.D., Clark, W.S., Sylvester-Bradley, R. and Scott, R.K. (1998) Use of in-field measurements of green leaf area and incident radiation to estimate the effects of yellow rust epidemics on the yield of winter wheat. *European Journal of Agronomy* 7, 53-62.

Campbell, C.L. and Madden, L.V. (1990) *Introduction of Plant Disease Epidemiology.* John Wiley & Sons, New York, 532pp.

Coakley, S.M., McDaniel, L.R. and Shaner, G. (1985) Model for predicting severity of *Septoria tritici* blotch on winter wheat. *Phytopathology* 75, 1245-1251.

Eyal, Z. (1981) Integrated control of Septoria diseases on wheat. *Plant Disease* 65, 763-768.

Forrer, H.R. (1992) Experiences with the cereal disease forecast system EPIPRE in Switzerland and prospects for the use of diagnostics to monitor the disease state. *Proceedings of the 1992 Brighton Crop Protection Conference - Pests and Diseases* 2, 711-720.

Foulkes, J. (1998) Responses to drought and nitrogen. In: *Proceedings of the 1998 HGCA R&D Conference.* Home-Grown Cereals Authority, London, pp. 3.1-3.17.

France, J. and Thornley, J.H.M. (1984) *Mathematical Models in Agriculture.* Butterworths, London, 15pp.

Gaunt, R.E. (1995) The relationship between plant disease severity and yield. *Annual Review of Phytopathology* 33, 119-144.

Gilligan, C.A. (1990) Comparison of disease progress curves. *New Phytologist* 115, 223-242.

Hansen, J.G., Secher, B.J.M., Jorgensen, L.N. and Welling, B. (1994) Thresholds for control of *Septoria* spp. in winter wheat based on precipitation and growth stage. *Plant Pathology* 43, 183-189.

Hearn, A.B. and Brook, K.D. (1994) A case study of the application of a knowledge-based system to cotton pest management: A tale of two technologies. *AI Application in Natural Resource Management* 3, 60-64.

Jeger, M.J. (1987) Modelling the dynamics of pathogen populations. In: Wolfe, M.S. and Caten, C.A. (eds) *Populations of Plant Pathogens: Their Dynamics and Genetics*. Blackwell, Oxford, pp. 91-107.

Jenkins, J.E.E. and Morgan, W. (1969) The effect of *Septoria* diseases on the yield of winter wheat. *Plant Pathology* 18, 152-156.

Jordan, V.W.L., Hunter, T. and Fielding, E.C. (1986) Biological properties of fungicides for control of *Septoria tritici*. *Proceedings of the 1986 British Crop Protection Conference - Pests and Diseases* 3, 1063-1069.

Kirby, E.J. (1994) Identification and prediction of stages of wheat development for management decisions. *Project Report No. 90*. Home-Grown Cereals Authority, London.

Lovell, D.J., Parker, S.R., Hunter, T., Royle, D.J. and Coker, R.R. (1997) Influence of crop growth and structure on the risk of epidemics by *Mycosphaerella graminicola* (*Septoria tritici*) in winter wheat. *Plant Pathology* 46, 126-138.

Madden, L.V. (1983) Measuring and modelling crop losses at the field level. *Phytopathology* 73, 1591-1595.

Madden, L.V. and Nutter, F.W. (1995) Modelling crop losses at the field scale. *Canadian Journal of Plant Pathology* 17, 124-137.

Madeira, A.C. and Clark, J.A. (1995) The principles of resource capture in relation to necrotrophic infection. *Aspects of Applied Biology 42, Physiological Responses of Plants to Pathogens*, 19-31.

Monteith, J.L. and Unsworth, M.H. (1990) *Principles of Environmental Physics*, 2nd edn. Edward Arnold, London, 291pp.

O'Callaghan, J.R. (1994) Simulation models of crop growth and technology transfer. *Computers and Electronics in Agriculture* 11, 291-292.

Passioura, J.B. (1996) Simulation models: science, snake oil, education, or engineering? *Agronomy Journal* 88, 690-694.

Paveley, N.D. (1998) Appropriate fungicide doses for winter wheat. *Project Report No. 166*. Home-Grown Cereals Authority, London.

Paveley, N.D., Lockley, D., Sylvester-Bradley, R. and Thomas, J. (1997) Determinants of fungicide spray decisions for wheat. *Pesticide Science* 49, 379-388.

Polley, R.W. and Thomas, M.R. (1991) Surveys of diseases of winter wheat in England and Wales, 1976-1988. *Annals of Applied Biology* 119, 1-20.

Reynolds, J.F. and Acock, B. (1985) Predicting the response of plants to increasing carbon dioxide: a critique of plant growth models. *Ecological Modelling* 29, 107-129.

Rouse, D.I., MacKenzie, D.R. and Nelson, R.R. (1981) A relationship between initial inoculum and apparent infection rate in a set of disease progress data from powdery mildew on wheat. *Phytopathologische Zeitschrift* 100, 143-149.

Royle, D.R. (1994) Understanding and predicting epidemics: a commentary based on selected pathosystems. *Plant Pathology* 43, 777-789.

Royle, D.J., Shaw, M.W. and Cook, R.J. (1986) Patterns of development of *Septoria nodorum* and *S. tritici* in some winter wheat crops in Western Europe, 1981-1983. *Plant Pathology* 35, 466-476.

Schofl, U.A. and Morris, D.B. (1994) The development of an integrated decision model based on disease threshold to control *Septoria tritici* on winter wheat. *Proceedings of the 1994 Brighton Crop Protection Conference - Pests and Diseases* 2, 671-678.

Scott, P.R., Sanderson, F.R. and Benedikz, P.W. (1988) Occurrence of *Mycosphaerella graminicola*, teleomorph of *Septoria tritici*, on wheat debris in the UK. *Plant Pathology* 37, 285-290.

Seck, M., Roelfs, A.P. and Teng, P.S. (1991) Influence of leaf position on yield loss caused by wheat leaf rust in single tillers. *Crop Protection* 10, 222-232.

Shaner, G. and Finney, R.E. (1976) Weather and epidemics of Septoria leaf blotch of wheat. *Phytopathology* 66, 781-785.

Shaw, M.W. (1987) Assessment of upward movement of rain splash using fluorescent tracer method and its application to the epidemiology of cereal pathogens. *Plant Pathology* 36, 201-213.

Shaw, M.W. and Royle, D.J. (1989a) Airborne inoculum as a major source of *Septoria tritici* (*Mycosphaerella graminicola*) infections of winter wheat in the UK. *Plant Pathology* 38, 35-43.

Shaw, M.W. and Royle, D.J. (1989b) Estimation and validation of a function describing the rate at which *Mycosphaerella graminicola* causes yield loss in winter wheat. *Annals of Applied Biology* 115, 425-442.

Smith, P.J. and Webster, J.P.G. (1986) Farmers' perceptions and the design of computerised advisory packages for disease control. *Proceedings of the 1986 British Crop Protection Conference - Pests and Diseases* 3, 1159-1167.

Sylvester-Bradley, R. (1991) Modelling and mechanisms for the development of agriculture. *Aspects of Applied Biology 26, The Art and Craft of Modelling in Applied Biology*, 55-67.

Tavella, C.M. (1978) Date of heading and plant height of wheat varieties as related to septoria leaf blotch damage. *Euphytica* 27, 577-580.

Teng, P.S. (1985) A comparison of simulation approaches to epidemic modelling. *Annual Review of Phytopathology* 23, 351-379.

Thomas, M.R., Cook, R.J. and King, J.E. (1989) Factors affecting development of *Septoria tritici* in winter wheat and its effect on yield. *Plant Pathology* 38, 246-257.

Tottman, D.R. (1987) The decimal code for the growth stages of cereals, with illustrations. *Annals of Applied Biology* 110, 441-454.

Van der Plank, J.E. (1963) *Plant Disease: Epidemics and Control.* Academic Press, London, 349pp.

Van der Plank, J.E. (1984) *Disease Resistance in Plants*, 2nd edn. Academic Press, London, 194pp.

Waggoner, P.E. (1986) Progress curves of foliar diseases: their interpretation and use. In: Leonard, K.J. and Fry, W.E. (eds) *Plant Disease Epidemiology: Population Dynamics and Management.* Macmillan, London, pp. 3-37.

Waggoner, P.E. and Berger, R.D. (1987) Defoliation, disease and growth. *Phytopathology* 77, 393-397.

Watson, D.J. (1947) Comparative physiological studies on the growth of field crops. I. Variation in net assimilation rate and leaf area between species and varieties, and within and between years. *Annals of Botany* NS 11, 41-76.

Chapter seventeen:

Decision Support Systems Featuring Septoria Management

L.N. Jørgensen[1], B.J.M. Secher[2] and H. Hossy[1]
[1]*Danish Institute of Agricultural Sciences, Flakkebjerg, DK-4200 Slagelse, Denmark.* [2]*Hardi International, Helgehøj Alle 38, DK 2630 Taastrup, Denmark*

Introduction

Diseases caused by *Septoria* and *Stagonospora* are very common in winter wheat in Denmark. Visual symptoms of *Septoria tritici* infections are present on the lower leaves in early spring (GSZ 25-30) in most Danish winter wheat fields. Symptoms of *Stagonospora nodorum* are less frequently recognized early in the season, but ELISA methods have confirmed that inoculum is present in many, but not all, fields. Inoculum of the Septoria pathogens is considered to be present at levels sufficient to create an epidemic if optimal conditions for sporulation and infection exist between stem elongation and the start of grain filling (GSZ 32-71). Over the past 15 years Septoria tritici leaf blotch has been the dominant and most economically important of the two Septoria diseases in Denmark.

The fact that there is no need to control Septoria every year makes it very important to have decision rules for guidance when assessing disease risk (Jørgensen *et al.*, 1996). Successful control has been shown to be possible in order to optimize timing and dose within the first half of the latent period (Jørgensen, 1991).

Based on historical trial data, a simple recommendation model was developed in 1991. When data from fungicide trials were correlated with precipitation data, it was shown that in Denmark an economically important attack requires a relatively large number of days with precipitation in the period from GSZ 32 and 30 days thereafter (Hansen *et al.*, 1994). In the analysis, the computer program 'Disease and Meteorology Correlation' based on a similar program, WINDOW (Coakley *et al.*, 1988), was used. Based on these data, spraying against Septoria is recommended in the most susceptible cultivars, if more than 7 days with rain (≥1 mm rain) have occurred in this period. In moderately resistant cultivars, more than 8 days with rain should be registered. When spraying decisions are made, the estimated number of days with precipitation in the forecasting period (5 days) is included. This model has been incorporated into the decision support system (DSS) PC-Plant Protection (Secher, 1991). Similar risk assessments for Septoria diseases in other countries are also based on precipitation data (Cook, 1977; Tyldesley and Thomsen, 1980).

Ten out of the past 17 years were assessed as Septoria years with a need for control of an economically important attack. It was found that the years with 8 days precipitation correlated well with the years in which Septoria was more severe (Hansen *et al.*, 1994). In 1994 and 1995, however, the prediction did not give the expected yield returns, because a period with drought started just before symptoms were expected to develop. Until weather prediction is improved, such a problem cannot be solved.

Field validation of PC-Plant Protection for control of pests and diseases in cereals has been previously presented (Secher *et al.*, 1995). The program has been able to adjust fungicide use to large yearly variations and to provide recommendations for the control of diseases at a satisfactory level, without affecting farmers' gross margins.

This chapter aims to validate the Septoria model as a single, simple model as well as discussing its advantages and disadvantages and indicating the need for future adjustments.

Dissemination of the Septoria Model

Since 1991, the Septoria model has been integrated in PC-Plant Protection which gives recommendations on all crop protection problems in cereals. PC-Plant Protection is sold to approximately 2000 advisors and farmers and used directly or indirectly by 28% of farmers (Anon., 1997). From 1998, the original DOS-version will be replaced with a Windows version. A simplification of the computerized model has been published on a single sheet using a turning disk to show thresholds related to growth stages in the crop (Secher and Jørgensen, 1995).

The simple forecasting model used in PC-Plant Protection for predicting the risk of Septoria attack is also widely used by farmers and advisors on the DSS. The model is being communicated in newsletters from The Danish Agricultural Advisory Centre (DAAC) to advisors as well as from local advisors to farmers. Twenty thousand farmers (i.e. more than 50% of all full-time farmers) receive the plant protection newsletter.

Due to the simplicity of the risk model it has been easy to disseminate and to persuade the farmers to adopt. In the newsletters, maps show the actual Septoria risk, but as rain events vary considerably even within small areas, farmers are recommended to record rain periods on individual farms.

Planteinfo is a new information system, which can be reached on the Internet (Jensen *et al.*, 1996). This brings up-to-date information for plant protection. The system presents information textually as well as graphically (www.planteinfo.dk/). Farmers using Planteinfo have the option of combining weather data from their own region with the risk models. The system has included the Septoria model since 1996. A printout from the program is shown in Fig.17.1.

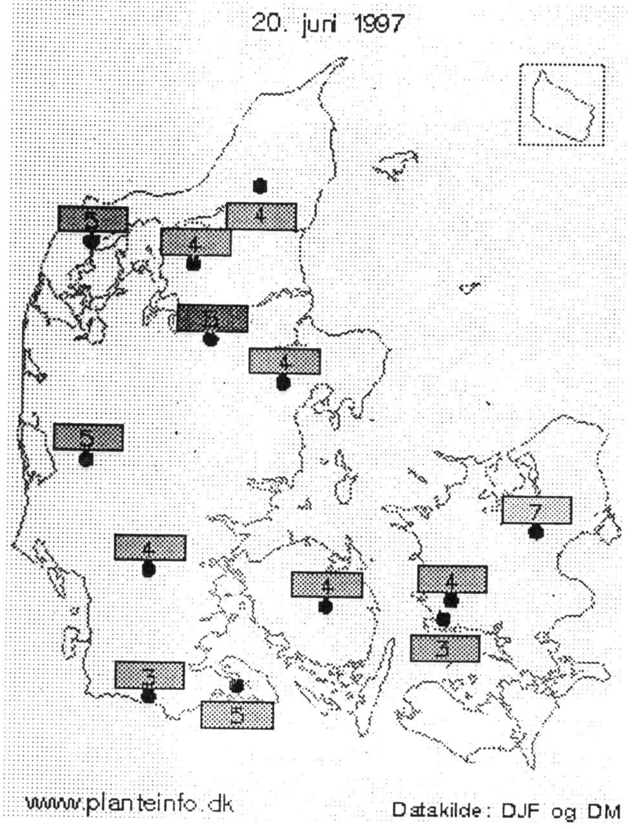

Fig. 17.1 Graphical presentation of the Septoria model as it can be seen in Planteinfo on the www.

Along with data from the model, the national monitoring system run by DAAC helps to give a picture of disease development in the whole of Denmark as well as in different regions (Fig. 17.2). West Jutland generally has most precipitation and the most severe attacks of the Septoria diseases (Jørgensen *et al.*, 1996). From the monitoring system it can be seen that, not all years are Septoria years.

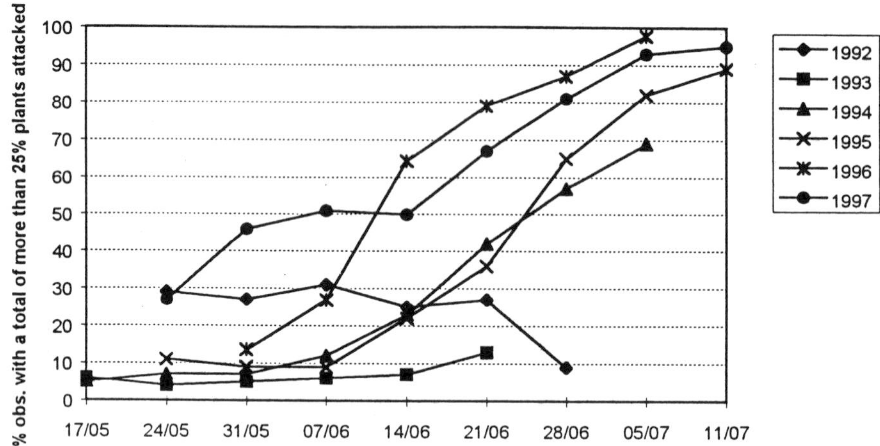

Fig. 17.2. Development of Septoria during the growth season in winter wheat 1992-97. Average of all observations in the national disease survey with incidences higher than 25%.

Methods

Element 1

Trials were carried out at The Danish Institute of Agricultural Sciences (DIAS) and DAS to validate the PC-Plant Protection model in winter wheat. Forty-one trials in the period 1991-1997 have had recommendations for the control of Septoria based on days of precipitation. Yield and disease data were collected from the trials. In some trials, control of mildew was also carried out, often at an earlier timing. For comparison, a standard treatment of two to three reduced rate applications of Tilt Top (propiconazole + fenpropinorph) was applied. When calculating the net yields, 335 Danish krone (dkr) for a full dose of fungicide, 60 dkr for applications per hectare and 900 dkr for a tonne of grain were used.

Element 2

Disease data were collected in 1990-1994 to investigate the spread and epidemic development of Septoria from three to nine localities per year. Forty plants were sampled per date. The number of sampling dates varied between four and eight per year. Overall, 277 disease assessments from nine localities each consisting of ten plants and four replicates were carried out. As no yield data were available from these localities, area under disease progress curve (AUDPC) on individual leaves, as well as on the upper two leaves, was used to validate the Septoria recommendation model. AUDPC values were correlated to different sets of precipitation data, with the Septoria model using number of days with ≥1 mm rain in the period from GSZ 32 and 35 days forward as one of the analyses. Disease attack using AUDPC has been transformed to percentage days with a threshold of 150 being equivalent to 5-10% disease at GSZ 75.

Element 3

Trials carried out at DIAS have been investigating the need for adjustment of the Septoria model using new strobilurin fungicides. Two trials were carried out in 1996. In these trials the fungicide treatments were applied after 2, 4 or 8 days with precipitation. The trials were assessed for disease and yields were measured. The residual effect of azoxystrobin (full dose = 250 g a.i. ha^{-1}) was tested in a specific semi-field trial. Spring wheat was grown in 8 l pots and all plants were sprayed with azoxystrobin using 100%, 50%, 25% and 12.5% of normal rate at GSZ 39-45. Plots were artificially inoculated with *S. nodorum* 4, 7, 12, 17 and 22 days after spraying. The latent period varied between 6-10 days. Disease assessment was done approximately 15-20 days after inoculation.

Results

Element 1

Forty-one trials which were treated with fungicides following the Septoria model gave, on average, 0.45 t ha^{-1} yield response (Table 17.1). This was slightly less than the standard treatment. In 70% of the 41 trials, a positive net return was found following PC-Plant Protection. Only 51% of standard treated trials gave a positive net return. In particular, in 1994 and 1995, due to drought late in the season, the Septoria model did not give the expected disease levels and, therefore, provided a lower response to Septoria control.

Generally, good control of Septoria was obtained in the trials from both PC-Plant Protection and the standard treatment. On average, more than 86% of the trials ended up with more than 5% disease assessed in untreated crops at GSZ 75.

Element 2

Disease data on Septoria, apart from one locality in one year, were dominated by *S. tritici*. An analysis of variance showed that there were significant differences in the level of disease between years and localities, and interactions between year and locality. No significant difference was found between cultivars (Table 17.2).

Table 17.1. Average yield and disease of 41 trials of winter wheat which were treated against Septoria in the period 1991-1997.

Treatment	Yield and yield increase (t ha^{-1})	Net yield (t ha^{-1})	% trials with positive net yield	% Septoria at GSZ 75
Untreated	7.50	-	-	23.5
PC-Plant Protection	0.45	0.15	70	4.6
Standard	0.62	0.06	50	4.8

Table 17.2. Results from analysis of variance on the effects of cultivar, year and locality on the disease variables expressed as AUDPC.

	Effect on:				
	Cultivar	Year	Locality	Year x locality	R2
Leaf F	ns	***	***	***	0.96
Leaf F-1	ns	***	***	***	0.80
Leaf F-2	ns	***	ns	***	0.61
Leaf F-3	ns	***	***	***	0.51
Leaf F-4	***	***	***	***	0.66

Due to the significant differences between years, and possibly correlated with precipitation, a further analysis of the model was made. The best correlation was found using AUDPC of the two upper leaves. The lower correlation coefficients and regression coefficients found for the lower leaves indicate that weather, in the period from GSZ 32 and 35 days forward, has less influence on infection on these leaves (Table 17.3).

Disease data were used to validate whether the model in PC-Plant Protection had been giving correct recommendations. The relationship between disease (AUDPC) on the two upper leaves and days with precipitation in a 35 day period starting at GSZ 32 and finishing at GSZ 65-70 is shown in Fig. 17.3. AUDPC data were transformed to threshold values, based on experience from fungicide trials which indicate that control can be justified if disease is higher than 150% days. The lines in Fig. 17.3 give the thresholds. Correct answers are located in the upper right or the lower left areas. Recommendations on treatment were correct in 75% of the 30 events studied.

Table 17.3. Parameters from regression analysis between disease variables and the precipitation model: disease = a x precipitation + b; n=174.

Disease variable AUDPC	Regression coefficient a	Intercept b	R2
Leaf F	24.8***	- 42	0.54
Leaf F-1	29.4***	- 9	0.39
Leaf F-2	22.0***	- 24	0.52
Leaf F-3	15.0***	66	0.22
Leaf F-4	2.3 ns	111	0.01
Leaf F+ F-1	47.0***	- 66	0.67

Element 3

From two trials in 1996, it was determined that the optimal timing using strobilurins was to apply them 4 days after precipitation. This was despite the fact that application 2 days after precipitation gave overall the best effect on the upper three leaves. However, 4 days gave both the best disease control on the flag leaf and ears and the best yield response. When using PC-Plant Protection for recommendation, rates between 0.36-0.45 l ha^{-1} were used depending on the growth stage and the need for residual effect (Table 17.4).

The semi-field trials with azoxystrobin compared with ergosterol biosynthesis inhibitor (EBI) compounds showed a much longer preventive or residual effect on *S. nodorum* (Fig. 17.4). Even 3 weeks after application, control of disease following artificial inoculation with *S. nodoum* was greater than 10% using full, half or quarter rates of fungicide. This indicates that if one treatment with azoxystrobin has been made after flag leaf emergence, it is not likely that another treatment will be needed.

Discussion

PC-Plant Protection gives advice on all plant protection problems in cereals. Approximately 2000 copies of the model have been sold to farmers and advisors. The simple Septoria rule has also been widely distributed in newsletters from advisors and is now well known to most farmers. It is expected that, in the near future, PC-Plant Protection will change from being a PC-based program only to a system which can be reached on the Internet.

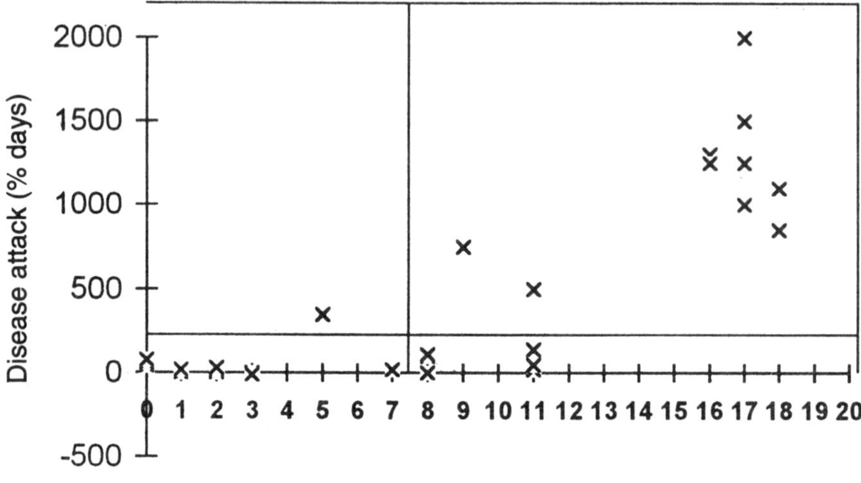

Fig. 17.3. The relationship between number of days with precipitation $\geq$1 mm and the development of disease on the upper two leaves. The current threshold value is indicated by the vertical bar.

Although a very simple model, validation has shown that it is relatively robust, giving correct recommendations in approximately 70-75% of the tested cases. There is, however, a clear wish to improve the model in order to use weather data in a more sophisticated manner, as well as to include specific recommendations depending on the fungicide used. A similar approach is adopted in the German DSS Pro-Plant (Frahm and Volk, 1993).

Collaboration between national meteorological institutes and DSS-developers has, or will, lead to the distribution of data from national automatic weather stations via modem. This, together with the commercial availability of cheap field-based weather stations which automatically transmit data to the local PC (Høstgaard, 1993), will stimulate the development of more sophisticated models based on detailed weather data, and also take into account the physiological development of the crop to compensate for plant contact and more detailed modelling of disease development and dispersal. It may also be necessary to distinguish between the two diseases, Septoria leaf blotch and Stagonospora leaf blotch.

Adjustments to the models for strobilurins will need to take into consideration the longer residual effect of these products with the potential use at lower rates (see Godwin *et al.*, Chapter 21 this volume). Even 3 weeks after application, more than 90% control was obtained with one quarter of the recommended rate. Propiconazole has shown a much shorter residual effect,

Table 17.4. Amount of Septoria and yield response in two trials using different Septoria control strategies based on days with precipitation.

Treatment: Product, dose l ha^{-1}, time of application	% Septoria[a] GSZ 75	% on flag leaf	% on second leaf	% ears diseased	Yield (t ha^{-1})
Untreated	19.6 a	20.3	66.5	12.6	6.99
Tilt top, 2 x 1.0 l, GSZ 31 & 57	3.7d	4.7	20.6	2.5	1.02
Amistar, 2 x 1.0 l, GSZ 31 & 57	1.2e	3.6	13.8	0.8	1.47
PC-P (Tilt top) 8 days, 0.36 l, GSZ 65	13.0b	6.1	36.8	4.4	0.62
PC-P (Amistar) 8 days, 0.36 l, GSZ 65	9.8b	5.3	42.6	3.7	0.61
PC-P (Amistar) 4 days, 0.45 l, GSZ 57	6.1c	6.2	34.4	1.4	1.06
PC-P (Amistar) 2 days, 0.42 l, GSZ 39	3.9d	9.9	26.7	7.1	0.71
LSD 95		3.0	15.6	4.5	0.33

[a]Attack on three upper leaves.

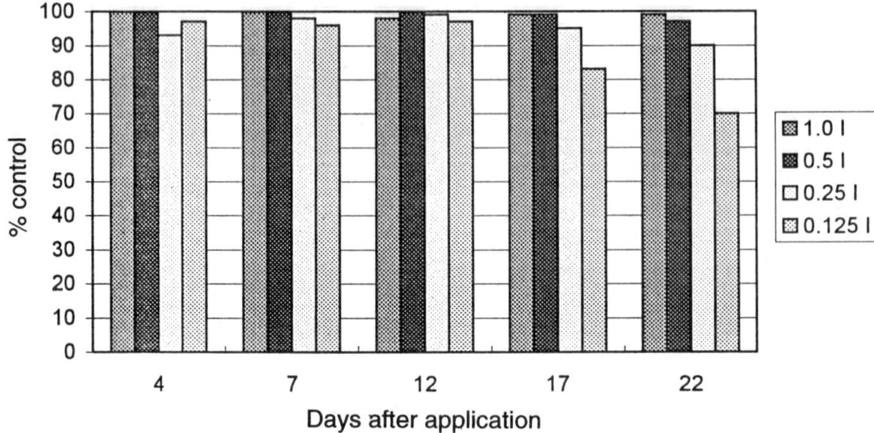

Fig. 17.4. Control of *Stagonospora nodorum* after preventive application of azoxystrobin at GSZ 39-45 and artificial inoculation of *Stagonospora nodorum* 4, 7, 12, 17 and 22 days after application.

especially when used at reduced rates (Jørgensen, 1994). In semi-field trials using artificial inoculation with *S. tritici* after ear emergence, the curative effect of azoxystrobin has been shown to be similar to propiconazole and much better than chlorothalonil (Jørgensen and Secher, 1996). Experiences from several field trials have, however, shown that early application at GSZ 30-31 often gives a slightly lower effect of azoxystrobin compared to propiconazole, probably due to more latent disease being present. In Danish trials using later applications (GSZ 39-55) for the control of Septoria, azoxystrobin has given better protection than propiconazole, indicating that the good preventive effect of azoxystrobin plays a major role at later applications. As the yield potential of strobilurins is significantly higher (0.5-0.6 t ha^{-1}), this suggests a lowering of the threshold. Preliminary results from 1996 confirm this and suggest that the control threshold might be reduced from 8 to 4 days with no need to consider further application.

ELISA diagnostic kits have been evaluated for the two Septoria diseases (Ellerton and Allen, 1994; Obst, 1994) with the aim of enabling a more accurate timing of fungicides. These techniques have shown the possibility of establishing new thresholds, but will have to prove their advantage over the conventional advisory methods using weather data. The time spent on collecting and processing samples imposes severe restrictions on the utility of such diagnostic kits (see Hollomon *et al.*, Chapter 19 this volume).

Should *S. nodorum* become more widespread, the model will need adjustment as it has been developed in a period dominated by *S. tritici*. In both 1996 and 1997, *S. nodorum* has been more widely seen in Denmark compared with previous years. ELISA tests in 1997 have confirmed the widespread occurrence of *S. nodorum*. The shorter latent period of *S. nodorum* also suggests a lowering of the threshold, as more life cycles of the disease are possible during a season.

Another way of optimizing the Septoria model will be to differentiate the dose in relation to the amount of inoculum. It is well known that several agronomic parameters influence severity, including previous cropping, time of sowing (Shaw and Royle, 1993), nitrogen strategies and plant density (Jørgensen *et al.*, 1997). The recent development of technologies for site-specific applications has made it possible to adjust applications according to the specific demand on the land (Secher, 1997) although, so far, this has not been tried for the control of Septoria.

References

Anon. (1997) Status report on the Danish Pesticide Action Plan. *Danish Ministry of Environment* (in press).

Coakley, S.M., Line, R.F. and McDaniel, L.R. (1988) Predicting stripe rust severity on winter wheat using an improved method for analysing meteorological and rust data. *Phytopathology* 78, 543-550.

Cook, R.J. (1977) Effect of timed fungicide sprays on yield of winter wheat in relation to Septoria infection periods. *Plant Pathology* 26, 30-34.

Ellerton, D.R. and Allen, H.J. (1994) Commercial use of diagnostic kits to make decisions on the use of fungicides. In: Arseniuk, E., Goral, T. and Czembor, P. (eds) *Proceedings of the 4th International Workshop on Septoria of Cereals.* Plant Breeding and Acclimatization Institute, Radzicow, Poland, pp. 285-292.

Frahm, J. and Volk, T. (1993) PRO-PLANT - a computer-based decision support system for cereal disease control. *EPPO Bulletin* 23, 685-693.

Hansen, J.G., Secher, B.J., Jrrgensen, L.N. and Welling, B. (1994) Thresholds for control of *Septoria* spp. in winter wheat based on precipitation and growth stage. *Plant Pathology* 43, 183-189.

Høstgaard, M.B. (1993) A climatic station with wireless transmission. *EPPO Bulletin* 23, 681-683.

Jensen, A.L., Thysen, I. and Secher, B.J.M. (1996) Decision support in crop production via World Wide Web. In: Secher, B.J.M. and Frahm, J. (eds) *Proceedings of the Workshop on Decision Support Systems in Crop Protection, SP Repport* 15, 39-47.

Jørgensen, L.N. (1991) *Septoria* spp. - use of different dosages and timing for optimal control. *Danish Journal of Plant and Soil Science - Special Series Report* 85 (S-2161), 159-172.

Jørgensen, L.N. (1994) Duration of effect of EBI-fungicides when using reduced rates in cereals. *Proceedings of the 1994 Brighton Crop Protection Conference - Pests and Diseases* 2, 703-710.

Jørgensen, L.N. and Secher, B.J.M. (1996) The Danish Pesticide Action Plan-Ways of reducing input. In: *Proceedings of the Dundee Conference, Crop Protection in Northern Britain*, Volume 1, 63-70.

Jørgensen, L.N., Secher, B.J. and Nielsen, G.C. (1996) Monitoring diseases of winter wheat on both a field and a national level. *Plant Protection* 13, 383-390.

Jørgensen, L.N., Secher, B.J.M., Olesen, J.E. and Mortensen, J. (1997) Need for fungicide treatments when varying agricultural parameters. *Aspects of Applied Biology 50, Optimising Cereal Input: Its Scientific Basis, Part 2: Crop Protection and Systems*, 285-292.

Obst, A. (1994) Early detection of *Septoria nodorum* by ELISA technique in comparison to visual assessment. In: Arseniuk, E., Goral, T. and Czembor, P. (eds) *Proceedings of the 4th International Workshop on Septoria of Cereals.* Plant Breeding and Acclimatization Institute, Rądzicow, Poland, pp. 259-262.

Secher, B.J.M. (1991) The Danish plant protection recommendation models for cereals. *Danish Journal of Plant and Soil Science - Special Series Report* 85 (S-2161), 127-134.

Secher, B.J.M. (1997) Site specific control of diseases in winter wheat. *Aspects of Applied Biology 48. Optimising Pesticide Applications,* 57-64.

Secher, B.J.M. and Jørgensen, L.N. (1995) Current development in fungicide use - success or failure. In: Hewitt, H.G., Tyson, D., Hollomon, D.W., Smith, J.M., Davis, W. and Dixon, K.R. (eds) *A Vital Role for Fungicides in Cereal Production.* BIOS Scientific Publishers, Oxford, pp. 241-250.

Secher, B.J., Jørgensen, L.N., Murali, N.S. and Boll, P.S. (1995) Field validation of a decision support system for control of pests and diseases in cereals in Denmark. *Pesticide Science* 45, 195-199.

Shaw, M.W. and Royle, D.J. (1993) Factors determining the severity of epidemics of *Mycosphaerella graminicola* (*Septoria tritici*) on winter wheat in the UK. *Plant Pathology* 42, 882-899.

Tyldesley, J.B. and Thomsen, N. (1980) Forecasting *Septoria nodorum* on winter wheat in England and Wales. *Plant Pathology* 19, 9-20.

Chapter eighteen:

Disease Management in Less-intensive, Integrated Wheat Systems

V.W.L. Jordan and J.A. Hutcheon
IACR-Long Ashton Research Station, Department of Agricultural Sciences, University of Bristol, Long Ashton, Bristol BS41 9AF, UK

Introduction

Current agro-economic pressures are encouraging the farming industry to review fungicide inputs, but other socio-economic factors argue against change from the somewhat routine, growth stage-orientated managed disease control strategies. Disease forecasting techniques, thresholds, decision support systems and the potential for diagnostic immunoassays provide practical options for integration and timing of fungicides in relation to economy and effectiveness of use. In addition, farmers are increasingly prepared to adopt more rational approaches by exploitation of alternative measures and changing land use patterns, to minimize disease risk.

There can be little doubt that the improved disease control achieved by the use of fungicides in recent years has contributed greatly to the increased yields of arable crops in Europe. Accordingly, their continued use will be pivotal to both efficient disease control and a productive agriculture for the foreseeable future. However, whilst agrochemicals are a necessity to maintain the required level of crop productivity, their intensive use has been deemed, by some, to be socially and environmentally unacceptable. In order to address this situation, the farming industry continues to seek ways to implement lower agrochemical input options and to protect the environment.

Recent attempts to rationalize and/or lower the use of fungicide inputs in arable crop production have aimed at decreasing the number of treatments or amounts of active ingredient applied, either by modifying treatments against specific disease targets or by using threshold values based on integrated understanding of pathogen population dynamics, crop agronomy and environmental parameters (Verreet and Hoffmann, 1990; Royle *et al.*, 1995). This approach has been complemented by utilizing the inherent biological activities of fungicides that underlie their action to improve timing and control response

(Jordan *et al.*, 1986; 1988), or by using reduced dose fungicide programmes (Wale, 1992).

This chapter presents information on an alternative approach to control *Septoria* in wheat through integration and manipulation of crop management and husbandry inputs within less-intensive production systems, in order to reduce disease progress and permit more optimal fungicide use.

Crop Management Strategies

Within IACR-Long Ashton Research Station's Less Intensive Farming and Environment (LIFE) project, conventional production systems, i.e. those managed by a technically competent farm manager reflecting current practice to achieve optimal yields and maximize profits, are compared, on a farm scale, with less-intensive, advanced integrated production systems designed to be less reliant upon agrochemical inputs and to be more environmentally benign, while maintaining profitability.

Influence of Cultivars and Sowing Date

As resistance to disease, in both breadmaking and feed quality wheat cultivars, is currently insufficient to prevent epidemic development in some years, the manipulation and integration of crop management and husbandry practices in less-intensive wheat production systems are targeted to minimize disease constraints at early stages in crop development and to focus on the need to protect the topmost leaves from flag leaf emergence until the end of grain filling. *Septoria tritici* is the main target disease for control in winter wheat, as it occurs regularly and severely under the mild wet climate of south-west England.

In conventional wheat production systems, cultivars such as Hereward, selected for high yield potential, are sown in mid September following traditional ploughing and associated cultivations. In such winter wheat crops, *Septoria* and *Stagonospora* frequently prevail in early autumn, with all plants being infected by mid December in most years (Table 18.1), providing sufficient inoculum potential during early spring (Table 18.2) for infection of newly emerging leaves. On conventionally-grown wheat crops with much nitrogen, subsequent disease progress, especially for *S. tritici*, is sufficient in spring and summer to justify annual managed disease control intervention.

By comparison, in wheat crops grown under less-intensive, integrated systems of production, cultivars, selected primarily for disease resistance, are sequenced as 'first wheats' within a multifunctional crop rotation and sown later (early October) using a one-pass non-inversion tillage system. In such crops, primary disease spread occurs later (Table 18.1), and infection frequency is lower during the autumn period, but by GSZ 31 (April) all plants are usually infected. Although inoculum potential in integrated wheat crops is initially lower for both *S. tritici* and *S. nodorum* (Table 18.2) than in conventionally-grown wheat crops, the

Table 18.1. Percentage of wheat plants infected by *Septoria tritici* in conventional and integrated wheat crops within the LIFE systems comparisons (mean 1990-1997).

Cultivar	November	December	March	April
Conventional				
Hereward	69	100	100	100
Mercia	88	100	100	100
Haven	28	100	100	100
Integrated				
Hereward	11	16	88	100
Estica	5	12	100	100
Spark	10	96	100	100
Genesis	19	24	100	100
Hunter	16	48	100	100

Table 18.2. Inoculum potential of *Septoria tritici (St) and Septoria nodorum (Sn)* (spores/tiller x 1000) in wheat systems, 1994-1996.

	January		February		March		April	
Conventional	***St***	***Sn***	***St***	***Sn***	***St***	***Sn***	***St***	***Sn***
1994	75	16	917	18	592	79	1194	99
1995	250	158	1110	330	1270	110	1053	937
1996	322	44	733	46	1100	16	nr	Nr
Integrated								
1994	13	<1	86	<1	354	55	542	66
1995	230	67	608	91	975	138	1340	538
1996	121	0	590	5	808	16	nr	Nr

nr = not recorded.

size of the inoculum pool at stem extension (GSZ 30) is considered sufficient for crops to be 'at risk' from the dispersal processes that drive seasonal disease development (Shaw and Royle, 1986).

In both wheat production systems, spring inoculum is not a limiting factor for disease development in most years. Subsequent disease progress must, therefore, depend upon the interactions between cultivar resistance, the influence of nutritional status on canopy structure and physiology, and the frequency and intensity of rainfall.

Influence of Crop Nutrition

In conventional wheat production systems within the LIFE project, an economic optimal supply of nitrogen is usually obtained by applying an early top dressing in late February (40 kgN ha^{-1}) with the balance (120-160 kgN ha^{-1}) at stem extension (GSZ 30). However, this strategy promotes early vegetative growth and vigour, making crops more liable to lodge with the consequent need for plant growth regulator applications. It also increases the severity of *Septoria* attack, in part through increased tissue susceptibility but also because modifications in crop canopy structure (large leaves, dense canopy) create changes in microclimate which favour disease becoming severe.

By comparison, nitrogen supply in less-intensive systems is based upon meeting the crop demands by utilization of residual soil available nitrogen reserves supplemented by appropriate fertilizer amounts and modified timing of spring applications, with amounts applied seldom exceeding 140 kgN ha^{-1} in total, in order to achieve targeted yields (Jordan and Hutcheon, 1994). Depending upon other interacting components, such as weed management strategies, two different options are implemented. Nitrogen may be applied at the same timing as in conventional wheat, but at lower amounts - an early top dressing in late February (40 kgN ha^{-1}) with the balance (80-100 kgN ha^{-1}) applied at stem extension (GSZ 30). Alternatively, if mechanical intervention (spring harrowing) is used for weed control, then sufficient nitrogen is mineralized from soil reserves to sustain early growth. This allows the early top dressing to be omitted and the allocated amount applied later, at GSZ 37, for yield and quality improvement. Whilst these nutritional strategies have provided integrated wheat crops with smaller leaves, fewer ears per square metre and, hence, a more open crop canopy than in conventional wheat systems, they markedly reduced disease progress and development and, thus, the need for fungicide intervention.

In the 1995 cropping year, disease incidence in crops managed using the mechanical weed control strategy were included for comparison. When assessed in mid May 1995, the topmost three leaves of conventional wheat crops had no symptoms of *S. tritici* infection whereas 1% and 9% of the areas of leaves 4 and 5 respectively, were affected. By comparison, the topmost four leaves of integrated wheat crops remained symptomless with only 3% diseased leaf area (DLA) present on leaf 5. By 27 June, although conventional wheat crops had received two fungicide sprays; tebuconazole + triadimenol (250:125 g a.i. ha^{-1}) on 15 May, and fenpropimorph + chlorothalonil (500:500 g a.i. ha^{-1}) on 13 June, *S. tritici* was present on all leaves, with <1% DLA on the flag leaf and leaf 2 and 12% DLA on leaves 3 and 4 (Fig. 18.1). By comparison, the topmost two leaves were symptomless in both integrated wheat crops, but had 2% DLA on leaf 3 (the fungicide intervention threshold) (Fig. 18.1). As a consequence, integrated wheat crops received a single, appropriate dose, fungicide spray (either propiconazole + tridemorph [63:175 g a.i. ha^{-1}] or tebuconazole + triadimenol [188:94 g a.i. ha^{-1}]) on 9 July. All treatments were sufficient to control disease until the end of grain filling.

Fungicide Performance

Previous research on the activity of certain triazole fungicides against phases of disease cycles and disease development of *Septoria* and *Stagonospora* (Jordan *et al.*, 1986) provided opportunities for modifying disease progress and for exploitation in more efficient decision-making control processes. For example, under optimum conditions for disease development these triazole fungicides prevented symptom expression and pycnidia development when applied 14 days after infection, thus providing curative and 'kick-back' activity within a fairly long spray timing 'window'.

More recently, this research was extended to include the chemical activity of the 'new-generation' triazoles, cyproconazole, epoxiconazole and tebuconazole (Jordan and Hutcheon, 1994). These compounds provided greater longevity of control such that, when applied at manufacturers' recommended rate, the protectant activity lasted for about 21 days after application and when applied either 7 days or 14 days after infection, symptom expression was delayed for 56 days. It also provided information on the most appropriate (reduced) dose requirement for different situations.

Disease Control Strategy in Less-intensive Wheat

Yield losses associated with *S. tritici* are largely attributable to infection of the topmost two leaves (Shaw and Royle, 1989). Fungicide intervention before flag leaf emergence (GSZ 37) appears, therefore, to be inappropriate and was omitted in less-intensive wheat production to minimize and lower fungicide requirement. Accordingly, the following disease control strategy has been evaluated and adopted to provide cost-effective disease control.

In order to minimize disease-induced yield loss, it is necessary to protect the topmost two leaves (flag leaf and leaf 2) until the end of grain filling. Data generated from research into the phenology of wheat across the UK (Porter *et al.*, 1987) have shown that the period from anthesis to the end of grain filling requires 650 day degrees which, averaged over the past 10 years at IACR-Long Ashton Research Station, equates to 45 days. Similarly, the accumulated day degree data

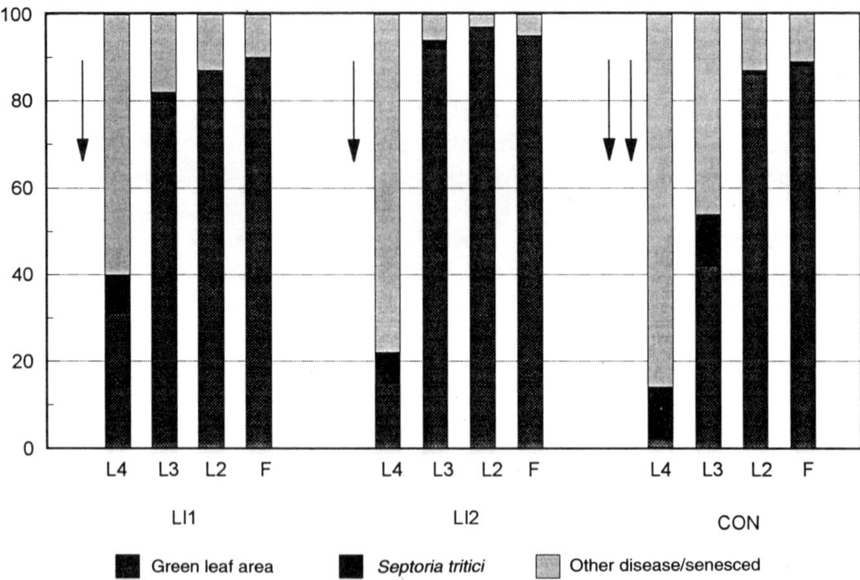

Fig. 18.1. Green (GLA) and diseased (DLA) leaf areas on 27 June on the topmost four leaves (F = flag-L4 = leaf 4) on conventional wheat crops (CON) and less intensive wheats with reduced nitrogen (LI1) and with spring harrowing (LI2). Arrows indicate fungicide applications.

from flag leaf fully emerged (GSZ 39) to anthesis equates to 16 days. Therefore, a 60-day period of crop protection is required from flag leaf fully emerged (GSZ 39) to the end of grain filling to avoid disease-induced yield loss. As the optimum regeneration time for *S. tritici* in field crops is about 21 days (Verreet and Hoffman, 1990; Royle *et al.*, 1995), protection of the topmost two leaves from infection is only required from emergence until 21 days before the end of grain filling. Thus, infection and/or cyclic regeneration of *S. tritici* needs to be prevented for a period of 40 days from GSZ 39 (Fig. 18.2). The recently introduced triazole fungicides have this capability (Jordan and Hutcheon, 1994). A single application at manufacturers' recommended rate at GSZ 39 is justified when crops are at risk of disease at this stage. When crops are not at risk, fungicide applications can be delayed, and dose rates reduced, provided the time of infection can be correctly identified (Fig. 18.2). Biologically-based disease threshold decisions have provided acceptable control in conventional wheat production systems (Verreet and Hoffman, 1990) and, using these data, a 'threshold-based' system has been developed and successfully implemented in less-intensive wheat production systems within IACR-Long Ashton Research Station's LIFE project and on commercial farms in south-west England (Hutcheon and Jordan, 1994). Adoption of different crop management strategies from crop establishment until flag leaf emergence resulted in different *Septoria* disease progress curves in the two contrasting wheat production systems (Jordan and Hutcheon, 1994). For

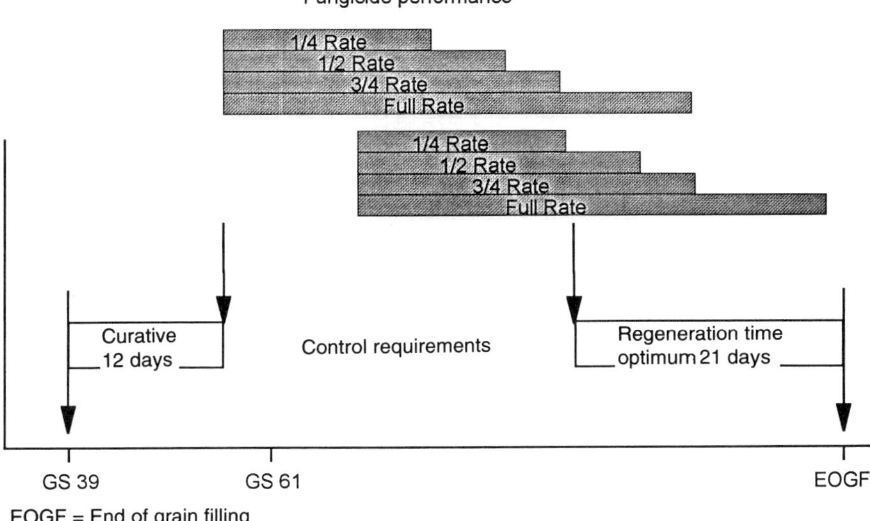

Fig. 18.2. Control requirement (*Septoria tritici*) in relation to crop phenology and fungicide performance. EOGF = end of grain filling.

conventional wheats, disease was sufficient in all years to justify managed disease programmes. In less-intensive wheats, disease development was such for substantial reductions to be made in application frequency and in the amount of fungicide applied.

In conventional wheats, three sprays were applied annually in the 1990, 1991 and 1992 cropping years (GSZ 31 + GSZ 39 + GSZ 59), but thereafter, as predominantly 'first wheats' were grown, only two sprays were applied (GSZ 37/39 + GSZ 59) (Table 18.3). In the less-intensive wheat systems, but with the exception of 1996 when fenpropimorph was applied to one wheat crop for late-season control of powdery mildew, a single spray application was sufficient for effective disease control. As cv. Pastiche was predominantly grown in the first three years (1990-1992), *Septoria* and *Staganospora* seldom reached damaging levels on this cultivar and fungicide application was, therefore, mainly targeted for powdery mildew control (with morpholines), occasionally supplemented by propiconazole or chlorothalonil. Thereafter, less-intensive wheat cultivars included Hereward, Hunter, Genesis, Spark, Encore and Estica, which were more susceptible to *Septoria* than cv. Pastiche and hence, cyproconazole, epoxiconazole or tebuconazole, often at reduced rates, predominated in the fungicides selected. Over all the 8 cropping years, 1990-1997, a 79% reduction in fungicide inputs (g a.i. ha^{-1} $crop^{-1}$) was achieved with the aforementioned disease control strategy in less-intensive wheat crops.

Table 18.3. Number of fungicide sprays and quantities of a.i. applied to conventionally managed and to less-intensively managed wheat in each cropping year from 1989/1990 to 1996/1997 in the LIFE project.

Year	Conventional wheat			Less-intensive wheat		
	Number of crops	Number of applications	g a.i. ha^{-1} $crop^{-1}$	Number of crops	Number of applications	g a.i ha^{-1} $crop^{-1}$
1989/1990	4	8	1763	5	5	315
1990/1991	5	15	2283	5	5	412
1991/1992	4	12	1176	4	2	145
1992/1993	4	8	677	4	5	229
1993/1994	6	12	871	6	6	84
1994/1995	3	6	1181	6	6	334
1995/1996	3	6	975	6	7	241
1996/1997	3	6	925	6	6	375
Total	32	73	1231	42	42	267

Discussion

Although *Septoria* and *Stagnospora* can infect wheat crops throughout the year, evidence to date indicates that they exert a major constraint to yield and quality at specific periods only during crop growth and development. There is little evidence to show that infection is a constraint to yield at early stages in crop development prior to tillering, but from late spring (GSZ 37) onwards, these pathogens can cause much loss of green leaf tissue on the topmost leaves, can affect ears (*S. nodorum*), and can impair both grain yield and quality. Various strategies, based upon data generated from numerous fungicide trials in conventional crop management systems, have been adopted by farmers for disease control directed towards economically justified maximum yield response. Such strategies, in the early 1990s, were reliant upon somewhat routine prophylactic spray schedules and managed disease control programmes or were targeted to specific risks, with a wide range in frequency of applications. Indeed, at that time, 24% of wheat crops received between three and seven fungicide sprays annually (Polley, 1991), but these invariably involved applications for selective and/or collective control of stem, leaf and ear diseases at pre-determined stages of plant development. As disease epidemics differ from year to year, depending mainly on the interactions between pathogen biology and climatic parameters, these growth stage-orientated treatments were randomly timed with respect to disease development and, thus, were only partially successful.

Whilst optimal disease control strategies in conventional cereal production systems have been directed towards economically justified maximum yield responses, they currently involve either arbitrary dose reductions targeted to stages in crop development, or are based upon rational fungicide use in relation to prediction and forecasting schemes against specific diseases. Such strategies pay little attention to interactions, utilization and integration of potentially beneficial husbandry factors that decrease disease incidence and spread. Exploitation of some component interactions in less-intensive integrated wheat crops has a marked influence on crop canopy structure and leaf susceptibility, sufficient to reduce subsequent progress and severity of *S. tritici*, and allow exploitation of a minimum fungicide use strategy.

The aim should be to achieve the minimum use required to provide adequate control in order to limit disease-induced yield loss, and to strike the optimum balance between cultivar, disease risk, methods of chemical and biological control, and crop management practices. This has been achieved through an integrated farming systems approach to crop production within IACR-Long Ashton Research Station's LIFE project, where the effective use and manipulation of crop husbandry practices has decreased disease sufficiently to make routine fungicide treatments inappropriate for cost-effective control. The threshold decision models thus employed have given satisfactory and cost-effective disease control in wheat crops grown under less-intensive management, within the integrated farming systems LIFE project during the past 8 years. Significant reductions have been achieved both in the number of fungicide treatments (42%), and in the amounts of fungicide active ingredient applied per hectare (79%).

Acknowledgements

The authors thank Mr J.V.G. Donaldson and Mr D.R. Iles for component research studies and technical assistance. This research and development is supported by the Ministry of Agriculture, Fisheries and Food. IACR receives grant-aided support from the Biotechnology and Biological Sciences Research Council of the United Kingdom.

References

Hutcheon, J.A. and Jordan, V.W.L. (1994) Strategies for reduced-input disease control in integrated wheat production systems. *Aspects of Applied Biology 40, Arable Farming Under CAP Reform,* 293-296.

Jordan, V.W.L. and Hutcheon, J.A. (1994) Strategies for optimal fungicide use in less-intensive cereal growing systems. *Proceedings of the 1994 Brighton Crop Protection Conference - Pests and Diseases* 2, 687-694.

Jordan, V.W.L., Hunter, T. and Fielding, E.C. (1986) Biological properties of fungicides for control of *Septoria tritici*. *Proceedings of the 1986 British Crop Protection Conference - Pests and Diseases* 3, 1063-1069.

Jordan, V.W.L., Hunter, T. and Fielding, E.C. (1988) Interactions of fungicides with nitrogen and plant growth regulators for cost effective control of wheat diseases. *Proceedings of the Brighton 1988 Crop Protection Conference - Pests and Diseases* 2, 873-880.

Polley, R.W. (1991) *Winter Wheat Disease Survey 1990-1991.* Agricultural Development and Advisory Service, MAFF, Harpenden, UK.

Porter, J.R., Kirby, E.J.M., Day, W., Adam, J.S., Appleyard, M., Ayling, S., Baker, C.K., Beale, P., Belford, R.K., Biscoe, P.V., Chapman, A., Fuller, M.P., Hampson, J., Hay, R.K.M., Hough, M.M., Matthews, S., Thompson, W.J., Weir, A.H., Willington, V.B.A. and Wood, D.W. (1987) An analysis of morphological development stages in Avalon winter wheat crops with different sowing dates and at ten sites in England and Scotland. *Journal of Agricultural Science* 109, 107-121.

Royle, D.J., Parker, S.R., Lovell, D.J. and Hunter, T. (1995) Interpreting trends and risks for better control of *Septoria* in winter wheat. In: Hewitt, H.G., Tyson, D., Holloman, D.W., Smith, J.M., Davies, W.P. and Dixon, K.R. (eds) *A Vital Role for Fungicides in Cereal Production.* BIOS Scientific Publishers, Oxford, pp. 105-115.

Shaw, M.W. and Royle, D.J. (1986) Saving *Septoria* fungicide sprays: the use of disease forecasts. *Proceedings of the 1986 British Crop Protection Conference - Pests and Diseases* 3, 1193-1200.

Shaw, M.W. and Royle, D.J. (1989) Estimation and validation of a function describing the rate at which *Mycosphaerella graminicola* causes yield loss in winter wheat. *Annals of Applied Biology* 115, 425-442.

Verreet, J.-A. and Hoffmann, G.M. (1990) Threshold based control of wheat diseases using the BAYER cereal diagnosis system. *Proceedings of the 1990 Brighton Crop Protection Conference - Pests and Diseases* 2, 745-750.

Wale, S.J. (1992) An evaluation of the potential of reduced rate fungicide programmes in winter wheat. *Proceedings of the 1992 Brighton Crop Protection Conference - Pests and Diseases* 2, 603-608.

Chapter nineteen:

Detection and Diagnosis of Septoria Diseases: The Problem in Practice

D.W. Hollomon[1], B. Fraaije[1], E. Rohel[2], J. Butters[1] and S. Kendall[1]
[1]*IACR-Long Ashton Research Station, Department of Agricultural Sciences, University of Bristol, Long Ashton, Bristol BS41 9AF, UK.* [2]*INRA-Rennes, BP 29, 35650, Le Rheu, France*

Introduction

Disease identification is often a problem, and accurate, pre-symptomatic detection and quantification of Septoria diseases of wheat should improve their control through better choice, dose rate and timing of fungicide sprays. Physical damage caused by frost and hail can cause necrotic symptoms similar to those caused by the leaf blotch pathogen. Fallen pollen remaining on leaves provides a nutrient source for fungal saprophytes and causes disease symptoms (pollen scorch, which over recent years has become common on cv. Riband particularly) similar to the lesions of Septoria and Stagonospora species. In damp locations, and in wet seasons, *Didymella exitialis* can be found on damaged and senescing tissue where it causes brown oval, or irregular, shaped lesions with brown or black pseudothecia easily confused with the pycnidia of *Septoria tritici*. Furthermore, the symptoms of *Stagonospora nodorum* are often non-specific, and the coalescing lesions and pale pycnidia are difficult to distinguish from natural senescence. Therefore, when necrosis occurs it can be difficult to diagnose the exact cause and decide on the best course of action.

Although wheat cultivars carrying resistance to Septoria are available, control relies predominantly on the use of fungicides and, in this respect, the triazole group of fungicides have been particularly successful. However, for optimum control, the most appropriate fungicide must be applied at the correct dose rate and timing, and this will vary between individual crops depending on local weather conditions, cultivar and growth stage of the crop (Eyal, 1981).

Despite differences in their physiology and development, *S. tritici* and *S. nodorum* have similar life cycles. Immediately after infection and during the early stages of hyphal colonization and initiation of pycnidia within leaf tissue, there are no visible signs that infection has occurred. These latent periods differ; under optimum conditions for *S. nodorum* it can be 7-10 days, while for

S. tritici it can be 14-21 days. Triazole fungicides differ in efficacy against these two diseases. The duration of the eradicant activity of the triazoles varies, but for both diseases, the activity of even the most effective triazole is limited to the period of hyphal extension and colonization. As soon as pycnidia are initiated, none of the fungicides presently available are effective in controlling the generation of Septoria to which they were applied (i.e. the established lesions). To accurately target fungicide application to protect the grain filling capacity of the flag leaf and leaf 2, and subsequently optimize yields, it is necessary to detect, pre-symptomatically, which diseases are present, how much, and where they are within the crop canopy, taking into account local climatic conditions, cultivar morphology and susceptibility (Royle *et al.*, 1995). To achieve these aims, rapid, specific and precise techniques are needed.

Serological methods exist that specifically identify *Stagonospora nodorum* (=*Phaeosphaeria nodorum*; glume blotch) and *Septoria tritici* (=*Mycosphaerella graminicola*; leaf blotch; Petersen *et al.*, 1990; Joerger *et al.*, 1992), but their application to disease management has so far largely been limited to charting regional changes in Septoria levels ('Septoria watch', Smith *et al.*, 1994).

PCR technologies have created many new opportunities for diagnosis, and these may be better suited to pre-symptomatic detection of Septoria diseases. PCR primers based on polymorphic sequence regions of internal transcribed spacer (ITS) regions of ribosomal DNA were used by Beck and Ligon (1995) to specifically amplify 448 bp and 345 bp fragments from *S. nodorum* and *S. tritici* respectively. Subsequently, these authors integrated this PCR amplification with enzyme linked immunosorbent assay (ELISA) methodology in an effort to quantify the PCR products (Beck *et al.*, 1996). Other approaches to PCR-ELISA have employed biotinylated PCR primers, coupled with incorporation of digoxigenin-UTP into the PCR products captured in streptavidin-coated microtitre plate wells, followed by quantification of the product by ELISA using a ·digoxigenin-specific alkaline phosphatase conjugate. To compensate for sample to sample variation, internal standards were used in quantitative competitive PCR (Holmstrøm *et al.*, 1993).

The aim of our research is to explore the value of both serological and DNA-based diagnostic tools to improve fungicide efficacy, primarily through more accurate timing of applications. These diagnostic tools may help answer key questions concerning the epidemiology of Septoria diseases, and which should aid the forecasting of disease development. How much inoculum is present? Where is the inoculum? Where is it likely to go within the crop? We present results from several years' field work with ELISA showing that disease levels measured on leaf 3 can be useful, and report on the development of a PCR assay which can be quantified simply with the fluorescent cyanine dye, PicoGreen (Singer *et al.*, 1997).

Methods

Immunoassay

Detection and measurement of *S. tritici* and *S. nodorum* was carried out using kits (Diagnolab) supplied by Dupont Agrochemicals (Stevenage, UK). Depending on the cultivar, leaf samples were collected from field plots at weekly intervals prior to fungicide treatment (generally around GSZ 59 for moderately resistant cultivars such as Hereward but earlier (GSZ 39-41) for susceptible cultivars such as Riband), and fortnightly thereafter. Samples were separated into leaf layers, homogenized (5 ml buffer per leaf) in a Waring blender, and the extracts used directly in a double antibody sandwich (DAS)-ELISA protocol according to the manufacturer's instructions. Optical densities were measured at 405 nm in a microtitre plate reader, and quantification derived from the standard curve obtained from the antigen standards in the kit.

DNA Extraction

DNA was extracted from samples comprising a single leaf which was powdered in liquid nitrogen by using a pestle and a mortar. Each sample was mixed, in additional steps, with 35 μl 1% (v/v) β-mercaptoethanol, 350 μl TEN buffer (500 mM NaCl, 400 mM Tris-HCl, 50 mM EDTA, pH 8.0) and finally 350 μl 2% (w/v) SDS and then incubated for 30 minutes at 65°C. 350 μl ammonium acetate (7.5 M) was mixed with heat-treated samples and kept on ice for 20 minutes before centrifugation at 10,000 rpm for 10 minutes in a microfuge. An equal volume of cold (-20°C) isopropanol was added to the supernatant, and the extract shaken at room temperature for 15 minutes before centrifugation at 6000 rpm for 5 minutes. After washing with 70% (v/v) ethanol, final pellets were dissolved in 200 μl purified water (Millipore Q, Uxbridge, UK). A typical DNA yield was approximately 20 μg per leaf.

PCR Protocols

PCR reactions were carried out with 0.625 units of Red Hot DNA Polymerase (Advanced Biotechnologies, Leatherhead, Surrey, UK) using 50 mM KCl, 1.5 mM $MgCl_2$, 10 mM Tris-HCl, pH 8.3, containing 120 μM of each dTTP, dATP, dCTP and dGTP, 0.5 μM primers and 100-150 ng DNA template in a final volume of 50 μl. Two different primer pairs were used; primers JB446 and ITS1 designed to ITS regions of ribosomal DNA of *S. tritici* by Beck and Ligon (1995), and primers E1 (5'-CGGTATG GGAACACTTCTCATCAG-3') and E2 (5'-CCTTCATGGACACCTTT CCGCGGT-3') coding for a conserved region of the β-tubulin of *S. tritici* (the complete coding sequence of the genomic β-tubulin gene has been submitted to the EMBL database). The PCR conditions with a Perkin Elmer DNA Thermal Cycler 480 were 94°C for 2.5 minutes, followed by 40 cycles at 94°C for 30 seconds, 60°C for 45 seconds and 72°C for 1 minute. The PCR was terminated with a DNA extension at 72°C for 9 minutes. PCR products were analysed by electrophoresis of 10 μl samples on 1.3% (w/v) agarose gels containing ethidium bromide and were further quantified using the dye PicoGreen (Molecular Probes, Leiden, The Netherlands).

Quantification of PCR Products Using PicoGreen

PicoGreen is a cyanine dye which probably intercalates specifically with double-stranded DNA and can be used, therefore, to quantify the product of PCR reactions where only a single fragment is produced (Ahn *et al.*, 1996; Singer *et al.*, 1997). Its properties are summarized in Table 19.1. Up to 3 μl of a PCR were incubated at room temperature in a microtitre plate well with 150 μl diluted PicoGreen (1:500 with water) for 15 minutes. Fluorescence was measured at 523 ± 2.5 nm after excitation at 480 ± 2.5 nm using a Perkin Elmer LS50B fluorimeter with microtitre plate attachment (Perkin Elmer, Seer Green, UK).

Table 19.1. Properties of PicoGreen double-stranded DNA quantification agent (from Singer *et al.*, 1997).

- Asymmetric cyanine dye which fluoresces with double-stranded DNA
- Excitation maximum 500 nm, emission maximum 523 nm
- High extinction coefficient
- Negligible intrinsic fluorescence of unbound dye
- Large fluorescence enhancement on binding to DNA
- High affinity for double-stranded DNA
- Brief incubation period
- Concentration used not critical
- Suitable for high throughput systems using microtitre plates

Results

Immunodiagnostics

Preliminary Experiments

In growth room experiments, wheat (cv. Riband) was inoculated by placing 2 x 10 μl droplets of a *S. tritici* spore suspension (5 x 10^5 spores ml^{-1}) onto each leaf. Antigen was detected pre-symptomatically at 16°C, but only 10-12 days after inoculation when pycnidial initials were forming (Fig. 19.1). Application of 'Sanction' (a.i. flusilazole) 1 day after inoculation prevented the appearance of antigens. Transferring these experiments to field plots in 1995-1996, and sampling leaves 1 (flag) to 4 separately, identified the need to focus immunoassay measurements on leaf 3, in order to protect the upper two leaves from infection. This field work also suggested that any rise in antigen levels above background signalled the need to apply DMI fungicides. In these field trials, *S. nodorum* was sometimes detected well before the appearance of *S. tritici*, even though glume blotch did not subsequently become a serious disease within the crop. Consequently, decisions to treat the crop were based on the detection of *S. tritici* antigens and not *S. nodorum.*

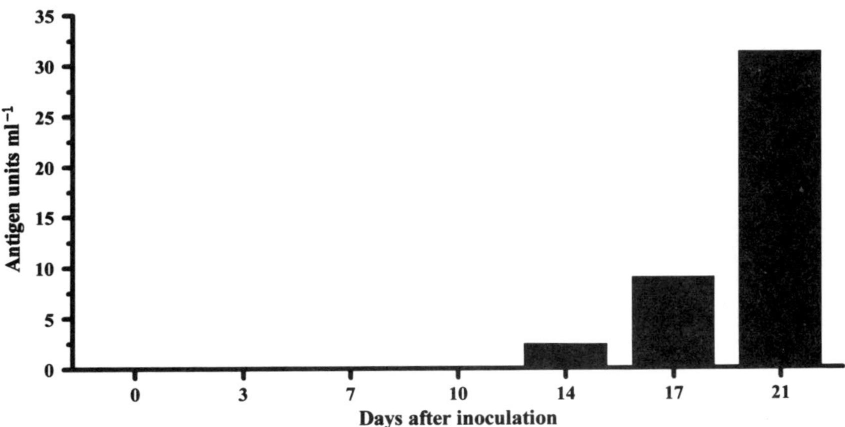

Fig. 19.1. Pre-symptomatic detection of *Septoria tritici* by immunoassay.

Fungicide Timing

In 1997, weekly sampling of a crop of cv. Riband from late April (GSZ 31) onwards identified an immunodiagnostic threshold for *S. tritici* at GSZ 41. The recommended field rate of 'Sanction' was applied at this stage, at 19 days prior to this (GSZ 33), at 9 days post-diagnostic (GSZ 55) and at 14 days post-diagnostic (GSZ 65). Assessments made during grain filling (GSZ 69) showed that, as expected, there was no control of *S. tritici* on leaf 3 from any of the spray timings, as the disease was already established before the sprays were applied. There was no control of Septoria on leaf 2 following the pre-diagnostic spray and both post-diagnostic treatments, but there was significant disease control on this leaf (P=0.01) in plots treated at the immunodiagnostic threshold (Fig. 19.2). Better disease control was achieved on the flag leaf following

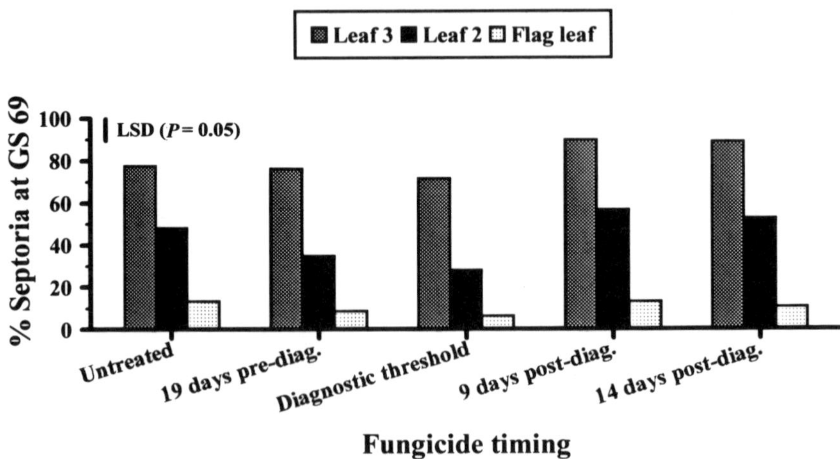

Fig. 19.2. Efficacy of Sanction (full rate) at different immunodiagnostic timings for the control of *Septoria tritici*.

sprays based on a diagnostic timing than at other timings, but differences were small, and not significant (P=0.05).

Fungicide Dose

Full, half and a quarter of the recommended field rate of ‘Sanction’ and ‘Alto’ (a.i. cyproconazole) were applied at the immunodiagnostic threshold at GSZ 41 in early June. ‘Sanction’ and ‘Alto’ applied at full-rate, and ‘Alto’ at half-rate, gave significant control (P=0.05) on leaf 2, whereas there was no control on this leaf with the other dose rates (Fig. 19.3).

DNA Diagnostics

Primer Specificity

Both primer pairs specifically amplified fragments of the expected size from *S. tritici* DNA, and not from templates obtained from other cereal pathogens. But whereas the ITS primers JB446 and ITS1 amplified only a single fragment of 345 bp from heavily-infected leaves, they also amplified a smaller fragment (approximately 280 bp) when DNA extracted from healthy plants was used as a template. At intermediate levels of infection competitive PCR occurred, and two fragments could be separated on agarose gel electrophoresis (Fig. 19.4A

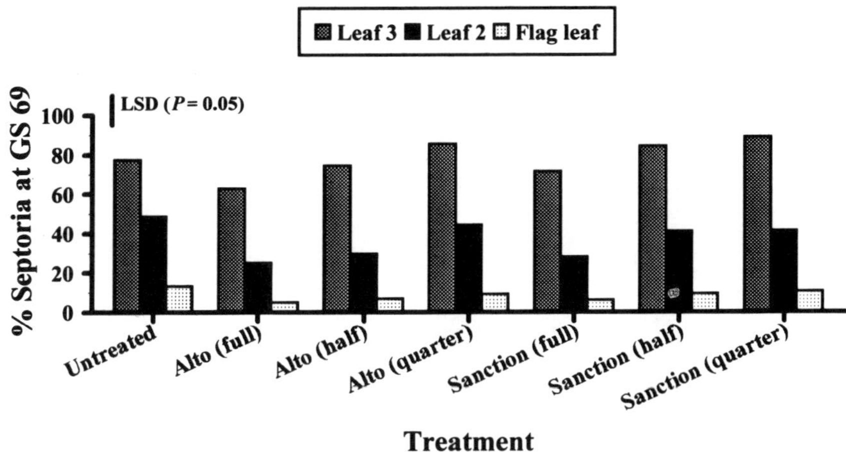

Fig. 19.3. Efficacy of dose rate of two triazole fungicides applied at immunodiagnostic threshold (GSZ 41) for the control of Septoria on wheat cv. Riband

and B). The 'β-tubulin' primer pairs E1 and E2 produced a single band of 539 bp with no significant background amplification.

Sensitivity of PicoGreen Quantification

PCR amplification of samples derived from leaves inoculated with increasing numbers of *S. tritici* spores (1 to 10^6 in tenfold increments) showed that 100-1000 spores could be detected (Fig. 19.5) using primer pairs E1-E2. This experiment also confirmed that, at least after 40 amplification cycles, the final fluorescence was correlated with the amount of *S. tritici* initially present. PicoGreen quantification was quick, and at least as sensitive as staining agarose gels with ethidium bromide. ELISA detection of *S. tritici* in these same samples was 10-100-fold less sensitive than PCR (Fig. 19.5).

Testing Field Samples

DNA was extracted not only from plants grown in the greenhouse, but also from leaves of samples collected from field crops. None of the leaves of the field samples showed symptoms caused by *S. tritici* for the β-tubulin primers, but in all except one (cv. Drake) the pathogen was detected, while with the ITS

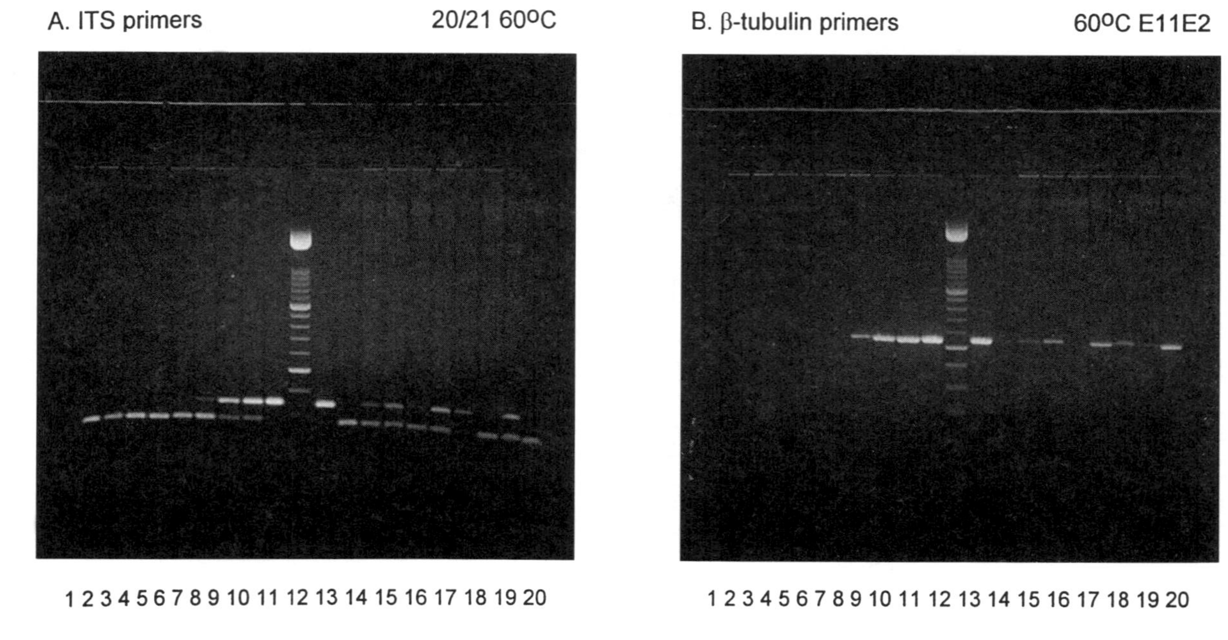

Fig. 19.4. Detection of *Septoria tritici* in wheat leaves by PCR, 'ITS' primers (A) compared with 'β-tubulin' primers (B). Cultivar (cv.); greenhouse (gh); field sample (fs). Lanes: 1 = cv. Riband uninoculated (gh), 2 = 0, 3 = 1, 4 = 10, 5 = 10^2, 6 = 10^3, 7 = 10^4, 8 = 10^5 and 9 = 10^6 *Septoria tritici* spores per leaf. Lane 11 = 100-bp DNA ladder. Lanes: 11 = cv. Cadenza with pycnidia (gh), 12 = cv. Cadenza uninoculated (gh), 13 = cv. Cadenza (fs), 14 = cv. Cantata (fs), 15 = cv. Drake (fs), 16 = cv. Madrigal (fs), 17 = cv. Soissons (fs), 18 = cv. Spark (fs), 19 = cv. Hereward (fs) and 20 = cv. Maris Otter (barley; gh).

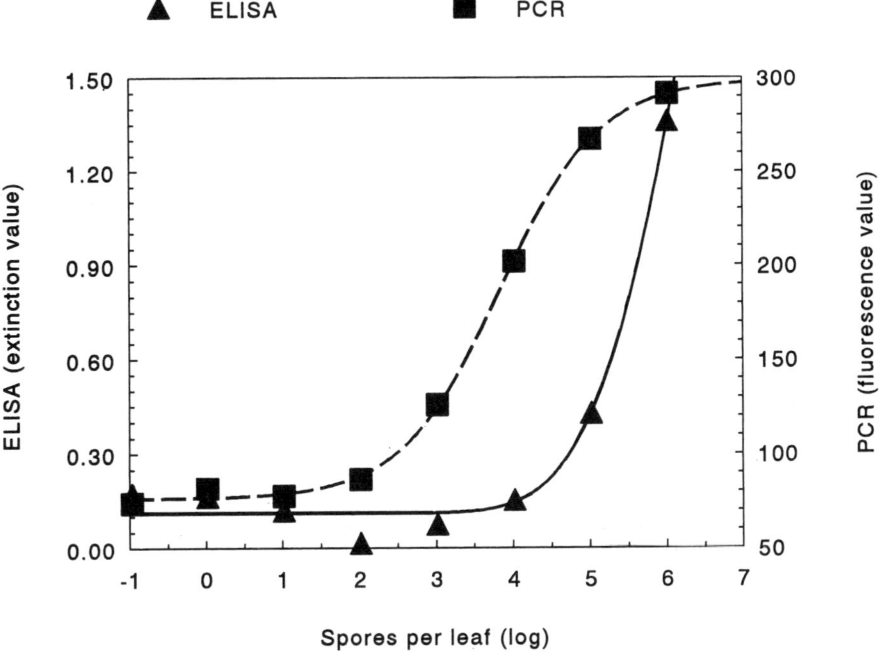

Fig. 19.5. Determination of detection thresholds of ELISA and PCR in inoculated wheat leaves of cv. Riband.

primers all samples were positive (Fig. 19.4A and B, lanes 13-19). In these samples, no specific background amplification products were observed suggesting that other micro-organisms on the leaf surface did not interfere with the result. The PicoGreen double-stranded DNA quantification assay corresponded well with the results obtained with agarose gel electrophoresis especially for the strongest positive samples 11, 14, 16 and 19 (Fig. 19.6). The slightly higher than expected fluorescence values of samples 15 and 18 might be caused by a higher initial total DNA concentration in comparison with the other samples.

Discussion

Pre-symptomatic detection and quantification of both Septoria diseases of cereals is feasible using either immunoassay or PCR-based technologies. Results obtained with these novel diagnostic methods may correlate with visual symptoms, but both technologies measure characteristics of the pathogen only, whereas disease results from the interaction between host and pathogen. Before new diagnostic methods can help define spray timing more precisely, they need to be evaluated under field conditions, over a number of seasons, and with different cultivars, to establish the relationship between antigen or DNA levels and the benefits gained through fungicide treatment. This relationship will depend not only on cultivar susceptibility, but also on the protectant and eradicant properties of the fungicides.

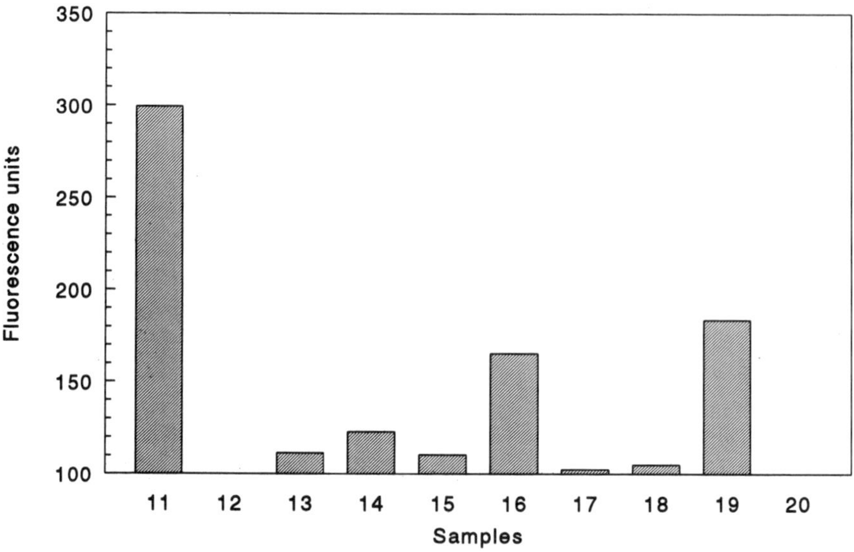

Fig. 19.6. PigoGreen double-stranded DNA quantification of PCR products amplied with 'β-tubulin' primers E1 and E2. Samples: 11 = cv. Cadenza with pycnidia (gh), 12 = cv. Cadenza uninoculated (gh), 13 = cv. Cadenza (fs), 14 = cv. Cantata (fs), 15 = cv. Drake (fs), 16 = cv. Madrigal (fs), 17 = cv. Soissons (fs), 18 = cv. Spark (fs), 19 = cv. Hereward (fs) and 20 = cv. Maris Otter (barley; gh).

Septoria immunodiagnostics have been available since 1991, and some experience of their use in field crops is available, although little of this information has been published. It is not a viable option to sample and assay crops weekly, and although 'Septoria Watch' has provided useful new information on the epidemiological progress of Septoria diseases, it does not help identify spray thresholds for individual crops. Our own work with winter wheat cultivars, such as Hereward and Riband, has indicated that sampling should focus on leaf 3, guided by rainfall events between GSZ 39 and GSZ 59 which favour splash dispersal and infection (Shaw, 1987). A rise in antigen units above background is sufficient to trigger a spray threshold. Good control was achieved following treatment at the diagnostic threshold, but spray timings on either side of this threshold were less successful, emphasizing the importance of spray timing in Septoria control. No clear picture has emerged, as yet, on how diagnostic results might influence dose rate, but our experience with several DMI fungicides has exposed their different eradicant properties, and suggests that knowledge of antigen levels may well influence dose rates. One difficult practical question centres around which Septoria disease should trigger treatment. *Stagonospora nodorum* generally precedes the appearance of *S. tritici* antigens. However, antigens of the two diseases are not equivalent, and the same value for both Septoria and Stagonospora does not indicate similar amounts of disease. So far, we have no information on what would constitute a diagnostic threshold for *S. nodorum*.

The technology surrounding PCR detection of Septoria diseases is still evolving rapidly but, although it provides a useful research tool, data are not yet available from field studies. Consequently, we have no information on the extent to which false positives may be a problem, because of fragments amplified from contaminating micro-organisms on the leaf surface. Quantifying PCR products with the fluorescent dye PicoGreen is quick, robust and greatly simplifies the protocol which would otherwise require competitive PCR (Nicholson *et al.*, 1997). However, it relies on a defined constant amount of DNA initially present in a sample and subsequently on specific amplification with no other PCR products produced. This emphasizes the need to have specific PCR primers, and to carefully establish the conditions required to generate a single PCR product. Whilst the ITS primers of Beck and Ligon (1995) yield a single product, when heavily infected leaves are used to extract template DNA these primers amplify a second fragment from wheat DNA when disease levels are low. It does not follow, of course, that all primers derived from the ITS region of ribosomal RNA genes will lack the necessary specificity. In our hands, however, the β-tubulin gene, which is present in the genome of all fungal cells, can provide suitably specific primers. PCR-based detection is very sensitive, indeed it may be too sensitive and potentially identify that all crops require treatment. As with immunodiagnostics, field work will be needed to establish the relationship between DNA levels and whether fungicide treatment is worthwhile. This work, involving cultivars differing in susceptibility to Septoria diseases, will be the subject of much of the research in this area over the next few years.

The Future

Quantification of PCR products may be simplified still further by incorporation of specific 'molecular beacons' in the PCR to provide a real time fluorescent measurement of amplification (Tyagi and Kramer, 1996). Providing that common annealing temperatures can be met, multi-plexing using different fluorescent probes should allow the detection of several diseases at the same time. However successful immuno- and DNA-diagnostics might be, results will not themselves trigger spray decisions, but the information will be incorporated into decision support systems, inputting better quality data especially on inoculum levels.

Much of the diagnostic work so far has been guided towards providing better timing of DMI fungicides in order to keep the upper two leaves free from Septoria infection. The introduction of systemic protectant fungicides for Septoria control, such as azoxystrobin (see Godwin *et al.*, Chapter 21 this volume) offers new dimensions for diagnostics in the management of Septoria diseases. Not only will diagnostics provide a rational basis for treating - or not - crops as early as GSZ 32, but they will identify if the protectant activity needs augmenting by GSZ 59, in order to preserve the grain filling properties of the upper part of the crop canopy.

Acknowledgements

The authors are grateful for financial support from H-GCA, MAFF and BBSRC, and encouragement as well as the supply of immunodiagnostic reagents by Dupont Agrochemicals.

References

Ahn, S.J., Costa, J. and Emanuel, J.R. (1996) PicoGreen quantification of DNA: Effective evaluation of pre- or post-PCR. *Nucleic Acids Research* 24, 2623-2625.

Beck, J.J. and Ligon, J.J. (1995) Polymerase chain reaction assays for the detection of *Stagonospora nodorum* and *Septoria tritici* in wheat. *Phytopathology* 85, 319-324.

Beck, J.J., Ligon, J.J., Etienne, L. and Binder, A. (1996) Detection of crop fungal pathogens by polymerase chain reaction technology. In: Marshall, G. (ed.) *Diagnostics in Crop Production.* BCPC Symposium No. 65, Farnham, UK, pp. 111-118.

Eyal, Z. (1981) Integrated control of Septoria diseases of wheat. *Plant Disease* 65, 763-768.

Holmstrøm, K., Rossen, L. and Rasmussen, O.F. (1993) A highly sensitive and fast non radioactive method for detection of polymerase chain reaction products. *Analytical Biochemistry* 209, 278-283.

Joerger, M.C., Hirata, L.T., Baxter, M.A., Genet, J.L. and May, L. (1992) Research and development of enzyme-linked immunosorbent assays for the detection of the wheat pathogens *Septoria nodorum* and *Septoria tritici. Proceedings of the 1992 Brighton Crop Protection Conference - Pests and Diseases* 2, 689-695.

Nicholson, P., Rezanoor, H.N., Simpson, D.R. and Joyce, D. (1997) Differentiation and quantification of the cereal eyespot fungi *Tapesia yallundae* and *Tapesia acuformis* using a PCR assay. *Plant Pathology* 46, 842-856.

Petersen, F.P., Rittenburg, J.H., Miller, S.A., Grothaus, G.D. and Dercks, W. (1990) Development of monoclonal antibody-based immunoassays for detection and differentiation of *Septoria nodorum* and *S. tritici* in wheat. *Proceedings of the 1990 Brighton Crop Protection Conference - Pests and Diseases* 2, 751-756.

Royle, D.J., Parker, S.R., Lovell, D.J. and Hunter, T. (1995) Interpreting trends and risks for better control of Septoria in winter wheat. In: Hewitt, H.G., Tyson, D., Hollomon, D.W., Smith, J.M., Davies, W.P. and Dixon, K.R. (eds) *A Vital Role for Fungicides in Cereal Production.* BIOS Scientific Publishers, Oxford, pp. 105-115.

Shaw, M.W. (1987) Assessment of upward movement of rain splash using a fluorescent tracer method and its application to the epidemiology of cereal pathogens. *Plant Pathology* 36, 201-213.

Singer, V.L., Jones, L.J., Yue, S.T. and Haugland, R.P. (1997) Characterization of PicoGreen reagent and development of a fluorescence-based solution assay for double-stranded DNA quantification. *Analytical Biochemistry* 249, 228-238.

Smith, J.A., Leadbeater, A.J., Scott, T.M. and Booth, G.M. (1994) Application of diagnostics as a forecasting tool in British agriculture. *Proceedings of the 1994 Brighton Crop Protection Conference - Pests and Diseases* 1, 277-282.

Tyagi, S. and Kramer, F.R. (1996) Molecular beacons: probes that fluoresce upon hybridization. *Biotechniques* 14, 303-308.

Chapter twenty:

Management by Chemicals

R.J. Cook
Morley Research Centre, Wymondham, Norfolk NR18 9DB, UK

Introduction

The Septoria diseases have been recognized as one of the most important groups of foliar pathogens on cereals for several years. In north-west Europe changes in cultural practices associated with an increase in the area of wheat since the 1970s has enhanced awareness of their significance. There is also an indication that many modern cultivars are more susceptible to *Mycosphaerella graminicola* (teleomorph of *Septoria tritici*) than traditional types (Bayles, 1991). The most important pathogens, *M. graminicola* and *Phaeosphaeria nodorum* (teleomorph of *Stagonospora nodorum*) are well adapted and able to cause substantial yield losses in diverse geographical regions of the world. In the UK, annual losses caused by *M. graminicola* are estimated at about 330 kt (Cook *et al.*, 1991), despite the use of fungicides on over 95% of the wheat crop.

Indications of the magnitude of yield losses which they caused in Britain began to accumulate during the early 1970s (National Agricultural Advisory Service, unpublished results). In these experiments, in which *S. nodorum* was the most cited cause of disease, protectant fungicides such as thiram, maneb and mancozeb were applied as repeated applications during the spring and early summer. Subsequent introduction of the locally systemic carbendazim (MBC) generating fungicides provided an opportunity for more comprehensive experiments. These led to recommendations for their use in mixture with maneb or mancozeb on commercial wheat crops in north-west Europe during the early 1970s (e.g. Lescar *et al.*, 1973). Studies by ADAS (UK extension service, unpublished results) associated significant epidemics of *Septoria* spp. with periods of rain during May or June (Tyldesley and Thompson, 1980). Application of fungicides immediately following the occurrence of similar periods of wet weather was associated with disease control and consequent yield increase (Cook, 1977) when fungicide was applied at about GSZ 39 (flag leaf ligule just visible; Tottman,

1987). These wet periods were later linked to rain splash and enshrined in ADAS guidance for Septoria control (Anon., 1986).

The Septoria diseases are part of the complex of diseases for which fungicides are used on winter wheat. Both are primarily leaf pathogens in the UK, although *S. nodorum* can cause significant ear disease. Their control must be integrated with the disease control strategy for the crop as a whole and this invariably requires fungicide treatment at about GSZ 39. Data from MAFF surveys show that the proportion of crops treated with one- or two-spray programmes, with one of the sprays at about GSZ 39, has increased from 9% in 1985 to 28% in 1995 (N.V. Hardwick, CSL York, 1997, personal communication). There is evidence, however, that the treatment regimes used on individual cultivars do not take adequate account of resistance to disease, so that the dose and programme regimes on cultivars with widely differing resistance types are the same (Stevens *et al.*, 1997).

Advances in understanding the epidemiology of the Septoria diseases on wheat have provided a sound basis for the development of effective control strategies. This chapter reviews the evolution of these strategies and considers current options for their control. It considers how approaches to disease management have adapted as our understanding of epidemiology and knowledge of fungicide activity have improved.

Wheat Development

Crop development is an important component of disease control. Wheat produces more leaves the earlier it is sown, leaves emerging at intervals of about 110 day degrees (Kirby, 1995). For crops sown in September or October, the final leaf 3 (leaf 1 = flag leaf) emerges at GSZ 32, usually between the last week of April and the first week of May, so that the final leaf 2 is often emerging during the second week of May. For winter wheat crops sown during September or early to mid October, the leaves exposed at GSZ 30-31 will remain in the lower parts of the crop canopy, as the final leaves 4, 5 or 6, and are unlikely to contribute to yield. By contrast, crops sown from late October onwards produce fewer leaves so that the final leaf 3 will be emerging in late April or early May. The crop is, however, more likely to be at GSZ 30 or 31 rather than at GSZ 32, as in crops sown earlier.

Evidence from MAFF surveys (Cook *et al.*, 1991) showed that where two-spray fungicide programmes were applied to wheat at GSZ 31 and GSZ 39, Septoria infection, and thus subsequent yield loss, was substantially lower when the first spray was applied in early May as opposed to mid April. Wheat phenology suggests that sprays applied in May are more likely to coincide with emergence of the final leaves 2 and 3, hence the subsequent reduction of infection on leaf 2.

Period of Maximum Disease Risk

Studies on the epidemiology of *Septoria* and *Stagonospora* spp. and yield losses caused by the diseases have shown consistently that adverse effects on yield are greatest when severe infections occur on leaf 1 and leaf 2. There are also suggestions that infection on leaf 3 may have an effect (Thomas, *et al.*, 1989). Results from 1995 and 1996, when low levels of disease established in eastern England, suggest that losses owing to disease on this leaf can be significant (Fig. 20.1). This has implications for disease control in open canopied crops, such as the susceptible cultivar Riband.

Flag leaf emergence occurs during mid to late May in most of the wheat-growing areas of Britain. The leaf remains green and viable until grain maturation in mid July. It is, therefore, essential that fungicide programmes protect this leaf from infection throughout that period. Given the need to provide effective protection from the emergence of leaf 3, fungicide programmes should incorporate overlapping eradicant and protectant activity between early May and late June. *Mycosphaerella graminicola* and *S. nodorum* have respectively latent periods of 21 and 10-14 days at that time of the year. In some conditions, therefore, when a single lesion develops at the tip of an emerging leaf, noticeable symptoms may not become visible on that individual leaf until two generations later, which may be up to 6 weeks after the initial infection has occurred. This can create a false sense of security, especially when the infection occurs on the top two leaves. In addition, during stem extension, one or two additional leaf layers, which may remain uninfected, could have appeared, so that the infection may not be of long term significance, unless there is time for a subsequent infection cycle. The period during which there is a risk that new infections will lead to damaging disease will last from the start of emergence of leaf 2 until about one fungus generation before ripening. In practice, there is little evidence of economic yield benefits to fungicide treatment later than watery ripe (Cook and Yarham, 1985), so that the last stage at which fungicide application is likely to provide a yield benefit will be the end of flowering (GSZ 69) at the latest. The period of risk of yield loss is thus a maximum of 6 weeks.

Inoculum is rarely limiting owing to the long period during which ascospores spread into the crop during the winter. Thus, any conditions which facilitate disease spread from leaf to leaf, either by movement in surface water layers or by upwardly driven rain splash (e.g. Royle *et al.*, 1986; Lovell *et al.*, 1997) during or immediately after stem extension, are likely to initiate yield-affecting disease on upper leaves. The confirmation (see Hunter *et al.*, Chapter 7 this volume) that *M. graminicola* ascospores are released on infected leaves throughout the spring and summer provides an additional dimension to the period of disease risk. It is possible that they may play an important part in secondary infection cycles on the upper leaves. Their role in the disease cycle may also explain those cases where infection is detected in crops unaffected by rain splash.

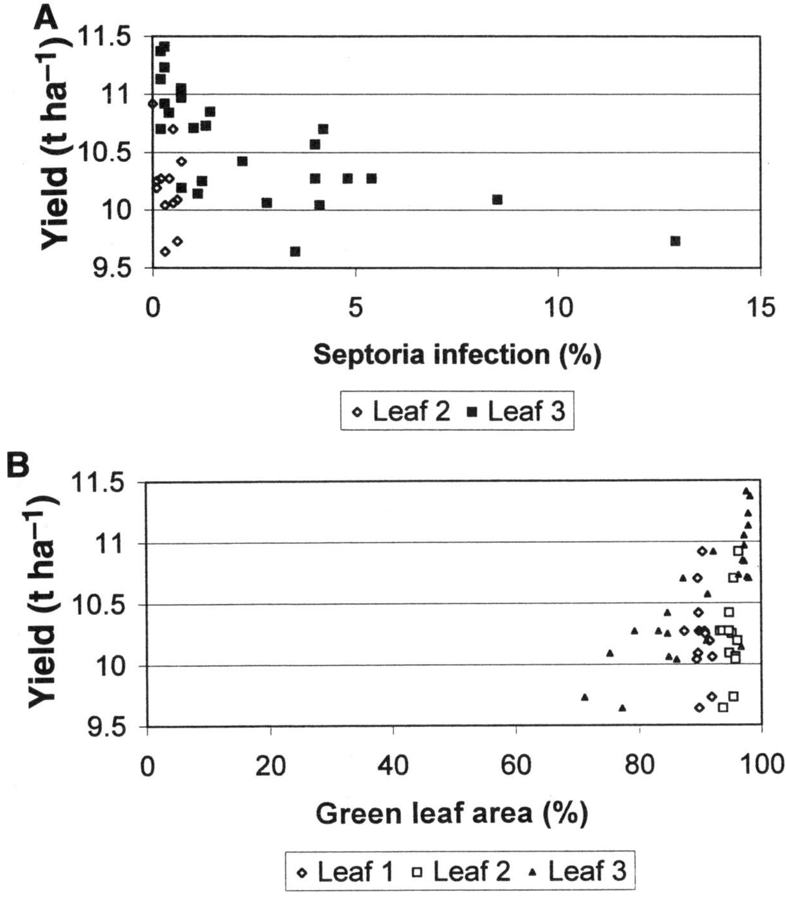

Fig. 20.1. Effect of fungicides applied at GSZ 32 on *Mycosphaerella graminicola* (A) and associated green leaf area (B) on leaves 2 and 3 at GSZ 75 and at yield.

Current Practice

Seed Treatment

A range of chemicals has been used to control *S. nodorum* since the introduction of organomercury seed treatments in the 1940s. These reduced primary inoculum but provided no protection against infection by ascospores or secondary spread by conidia on seedlings or overwintering crops by either *S. nodorum* or *M. graminicola*. The more recently introduced conazole fungicides are systemic when used as seed treatments. They control seed-borne infection and can protect the young crop from airborne spores for several weeks. They delay, but do not prevent, field infection from late ascospores so cannot remove the risk of severe disease later in the growing season. It is possible that at least one of the new fungicides currently under development may provide improved protection from a seed treatment.

Winter and Spring Disease

Control of eyespot (*Pseudocercosporella herpotrichoides;* teleomorph *Tapesia yallundae*) has been an important component of disease management strategies on wheat since the late 1970s. It is likely that widespread and routine use of MBC for eyespot control contributed to the resistance of *M. graminicola* to this group of fungicides. It is also possible that resistance of *M. graminicola* to MBC contributed, at least in part, to the upsurge of this disease during the early 1980s. There are few substantive data to illustrate the effect of MBC fungicides applied for eyespot control, at or about stem erect or first node detectable (GSZ 30-31), on *M. graminicola* incidence prior to the development of resistance in these pathogens. There are, however, experimental indications that these fungicides reduced Septoria on the lowest green leaves up to ear emergence.

Following the recognition of resistance of the eyespot pathogen to MBC, the alternative fungicides prochloraz and flusilazole were used. These also control *M. graminicola.* In the absence of significant eyespot risk their use on winter wheat may be justified on the grounds that they decrease Septoria during the early part of stem extension. Treatments with suitable fungicides at about GSZ 30-31 will reduce Septoria on the adult plant, on those leaves exposed at the time. For crops sown before November, that effect is unlikely to be associated with any effect on yield.

Attempts have been made to reduce inoculum and prevent over winter infection by the application of fungicide sprays during the winter or very early spring. The evidence from experiments in the UK suggests that although some reduction of inoculum may occur, this reduction is not translated into yield benefits (Thomas *et al.*, 1989). More recently, Hindley (1995) reported successful suppression of inoculum with chlorothalonil but failed to materially affect final disease levels or yield, compared with the untreated control. Similarly, Nuttall and Stevens (1997), using conazole fungicides alone and in mixture with chlorothalonil, have confirmed that treatment with fungicides before stem extension has little impact on final disease level or yield loss caused by *M. graminicola.* These results (Table 20.1) were obtained in an experiment over three years in which *M. graminicola* affected over 40% of leaf 2 by GSZ 75 in 1 year. These data support the contention that inoculum level is an inappropriate determinant of spray need for these diseases (Paveley *et al.*, 1997). They are consistent with the ability of the pathogens to spread quickly and effectively from low levels of inoculum in the base of the crop when suitable rain splash conditions occur (e.g. Royle *et al.*, 1986).

Table 20.1. Effect of early spring fungicides on wheat yield and control of *Mycosphaerella graminicola* in addition to standard treatment.

	Yield (t ha^{-1})			
	1993	1994	1995	Mean
Standard treatment[a]	10.33	8.89	9.97	9.73
Additional spray in March	+0.16	-0.05	+0.24	+0.11
Additional sprays in March and April	+0.17	-0.11	+0.24	+0.10
Untreated	7.17	7.99	8.15	-
LSD (*P*=0.05)	0.273	0.223	0.330	

[a]Fungicide at GSZ 32, 39 and 69.

Stem Extension

Consideration of wheat growth and leaf emergence suggests that it is prudent to prevent disease establishing on the mid and upper canopy leaves from the start of stem extension. Spray timing experiments have demonstrated that effective reduction of these infections by appropriately timed and suitable fungicides is an essential component of any strategy to control *M. graminicola.* Fungicide experiments with one- and two-spray programmes have shown that fungicide application at the start of stem extension (about GSZ 32) does allow the opportunity for greater flexibility in the timing of later sprays (Cook *et al.*, 1995). This may be of particular benefit when the flag leaf spray is delayed by adverse weather.

Fungicide application at about GSZ 39 remains the single most important time for treatment to eradicate any established infections on leaves already infected and prevent subsequent disease on the flag leaf.

Fungicide Activity

Chlorothalonil has strong protectant activity and provides comparable control of *M. graminicola* to the conazole fungicides when applied immediately before rain at flag leaf emergence (Hims and Cook, 1992). It can also protect individual treated leaves for several weeks (Table 20.2). It may also be redistributed, in either rain splash or leaf to leaf contact, to successively higher and later emerging leaf layers for several weeks even during prolonged heavy rainfall (V.W.L. Jordan, IACR-Long Ashton Research Station, Bristol, 1997, personal communication). Such redistribution probably accounts for the finding by Jarrett and Allard (1981) that chlorothalonil provided better control of ear infection with *S. nodorum* when applied at GSZ 37 than during ear emergence. For these reasons, substitution of chlorothalonil for a portion of a conazole fungicide is an option which may be considered for some treatments.

Table 20.2. Effect of fungicides applied to winter wheat at GSZ 32 on subsequent development of *Mycosphaerella graminicola* in 1995.

Fungicide	*Mycosphaerella graminicola* (% GSZ 75)		
	Leaf 4	Leaf 3	Yield (t ha^{-1})
Untreated	61.3	12.9	9.73
Chlorothalonil (1000 g)	38.1	4.8	10.27
Chlorothalonil (500 g)	42.4	8.4	10.09
Epoxiconazole (125 g)	2.4	0.3	10.92
Epoxiconazole (62.5 g)	5.6	0.7	10.19
Epoxiconazole + chlorothalonil (62.5 g + 500 g)	5.6	0.4	10.71
Flutriafol (125 g)	31.1	4.0	10.27
Flutriafol (62.5 g)	38.4	4.1	10.04
Flutriafol + chlorothalonil (62.5 g + 500 g)	33.6	3.8	10.31
LSD (P=0.05)	15.69	4.34	0.502

The conazole fungicides vary in activity and these differences can also be related to the time interval between infection date and fungicide application. Stevens and Nuttall (1997) compared a range of conazole fungicides applied either as a routine at GSZ 37 or about 20 days (180 to 200 day degrees) after rainfall believed to have caused infection of leaf 2. Yield response was related well to green leaf area on leaf 2 (r^2=0.765) but not on the flag leaf. The conazole fungicides tested performed better when applied at this later stage than as a routine at GSZ 37, before infection occurred.

Similar work to investigate the relationship between treatment date and fungicide dose suggests that dose required may be proportional to the number of days that have elapsed between the infection event or GSZ 39 spray and the date of fungicide application (Fig. 20.2). Advice is, therefore, that fungicide dose should be increased as the time interval between the infection event and application increases. The loss in yield averaged over the 3 years of this experiment represents 30 kg ha^{-1} for each day that treatment is delayed, a loss comparable to that found by Cook *et al.* (1998).

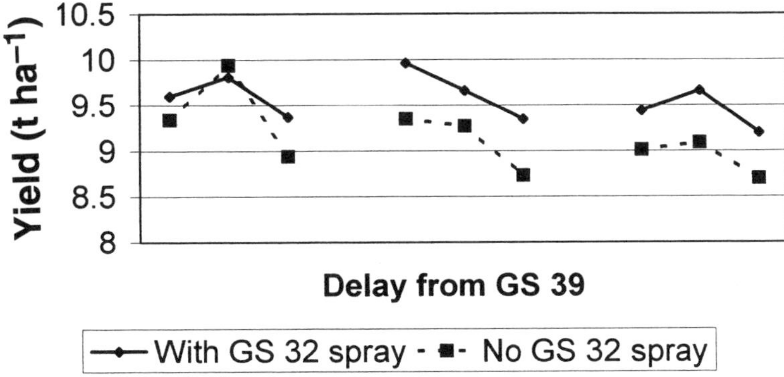

Fig. 20.2. Effect of delay in treatment on dose response of epoxioconazole to *Mycosphaerella graminicola*. Each successive point represents a reduction in dose from full to half and one quarter of recommended dose (125 g a.i. ha^{-1} epoxiconazole). The three small graphs represent sprays applied at GSZ 39, at GSZ 39 + 7 days and at GSZ 39 + 14 days, from left to right. Data are means of three years 1995-1997, when *M. graminicola* infection was 12.2% and 31.9% on leaf 2 in 1995 and 1997 respectively and 21.5% on leaf 3 in 1996.

Future Developments

Most work on disease control is undertaken using standard spray application techniques and uniform dose. There is some evidence, however, that the efficacy of different fungicides can be changed by manipulation of the spray volume and quality. These characteristics affect directly the distribution of fungicide within the crop canopy. This presents opportunities for spatial variation in fungicide dose so that a uniform amount of fungicide may be applied in relation to variations in crop structure.

The introduction, during 1997, of new protectant fungicides based on the strobilurin group offers additional opportunities to develop disease management strategies based on effective protectant fungicides. It is a reminder of the need to evaluate fungicides under a range of conditions to exploit their full potential as they are introduced into commerce.

Integrated Disease Management

A range of effective fungicides is now available for control of the Septoria diseases. Improved understanding of their properties and disease epidemiology provides opportunities to reduce fungicide inputs but maintain yield and grain quality.

For Septoria-susceptible cultivars, the first treatment should be applied as a mixture of chlorothalonil with an eradicant conazole when the third leaf has emerged either during late April or early May (at GSZ 31 or 32), irrespective of sowing date. This treatment provides for eradicant control of any disease present on emerged leaves where symptoms are not yet showing (i.e. on the final leaves 3 or 4) and may be delayed until GSZ 37 in the more *M. graminicola*-resistant cultivars.

The next phase of the programme is application of an appropriate dose of fungicide at GSZ 39; the aim being to protect the upper leaves from rain splash infection until at least mid June. Where this application can be applied promptly at GSZ 39 or is not more than 4 or 5 days after weather conditions are likely to constitute a Septoria risk, addition of chlorothalonil may be advantageous (Fig. 20.3) in case of subsequent severe infection conditions. There is, however, evidence (Morley Research Centre, unpublished results) that in situations where this treatment is delayed for more than 10-14 days after the infection event, chlorothalonil reduces the eradicant activity of the conazole fungicides. In high risk situations, a third application of reduced dose conazole may be applied in mid June (GSZ 69) to reduce the risk of late Septoria infection, glume blotch and brown rust. The strategy ensures that the upper leaves and ear receive appropriate fungicide treatments so that no part of the plant is unprotected during the critical period of development.

Host partial resistance plays an important part in selection of appropriate fungicide and dose. Current work is attempting to quantify disease resistance so that it may be equated to fungicide dose. Evidence to date suggests that the most effective way of exploiting cultivar resistance may be by maintaining fungicide timing with appropriate dose reduction, rather than by reducing the number of sprays applied.

Integrated crop management (ICM) requires that all factors involved in management of the crop should be considered in any risk assessment associated with treatment need. Attempts to manage Septoria by assessing the risk potential led to the splashmeter (Royle, 1990). The fact that *M. graminicola* inoculum is rarely limiting and that suitable rain often occurs during stem extension, may limit the value of such approaches. Factors such as rain splash, plant height and speed of stem extension (Eyal, 1981) as well as leaf morphology (Lovell *et al.*, 1997) all influence susceptibility to Septoria and the subsequent need for chemical intervention. ICM requires that fungicide treatments should be applied in response to identified risk based on these factors and at appropriate doses. The potential involvement of ascospores in the summer epidemic supports the need for effective fungicide risk management and the need to provide effective cover throughout the period the crop is at risk of yield loss.

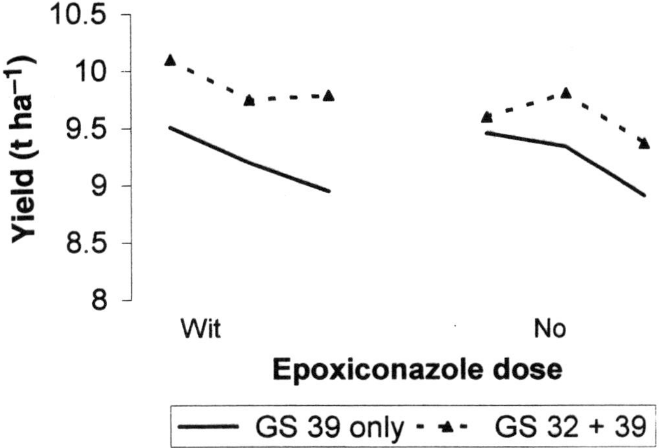

Fig. 20.3. Effect of addition of chlorothalonil to decreasing doses of epoxiconazole on yield of wheat affected by *Mycosphaerella graminicola*, mean for 1995 to 1997. Each successive point left to right represents a reduction in dose of epoxiconazole from full recommended (125 g a.i. ha^{-1}) to quarter dose; chl = chlorothalonil 500 g a.i. ha^{-1}, in mixture with epoxiconazole. Septoria infections for each year are as given for Fig. 20.1.

Recent Experience

The 1997 season illustrates the difficulties inherent in a simplistic adoption of risk assessment. Inoculum of *M. graminicola* was, as always, widespread in wheat crops throughout the prolonged late winter and early spring drought.

This perceived reduction in disease risk tempted some farmers, anxious to reduce costs and pesticide application, to believe that Septoria risk would be lower than average. Prudent approaches to disease management dictated that disease control programmes on susceptible cultivars should have started as usual, when the third final leaf was expanding. In practice, this proved the right course. Subsequent rain and unsettled weather throughout stem extension in the last week of April and most of May meant that those farmers who had not started their disease control programmes were unable to prevent effectively infection by rain splash, with consequent significant disease and subsequent yield losses. This experience confirms that ICM and the development of sustainable crop production systems depend upon a clear identification and understanding of disease risk. In this case, inoculum was not limiting and the occurrence of suitable weather coincided with the period of maximum disease risk during stem extension and gave the opportunity for disease to develop. The dilemma for ICM is to ensure that treatments are based on suitable balanced judgements of the elements of the host-pathogen-weather interaction. In this context, predictions of weather to be expected are of little value unless they can be accurate to periods substantially in excess of 5 days.

Conclusions

Improvements in understanding the biology of Septoria diseases and the properties of fungicides have led to substantial advances in control strategies in recent years. *Mycosphaerella graminicola* can be controlled and there should be little excuse for high levels of disease in commercial crops. Under northern European conditions, wheat is at risk from yield-damaging disease for a relatively short period of time. Control of inoculum before stem extension is of little value, but the ability to protect upper leaves during and immediately after this development stage is fundamental to effective disease control. It is the key component of effective disease management. In the more resistant cultivars, judgements leading to appropriate fungicide dose are more difficult, especially when disease risk is low. The next steps must be to develop ways of equating resistance levels to portions of fungicide dose and to continue work that provides clear understanding of the properties of new fungicides.

Acknowledgement

I am grateful to Mr D.B. Stevens for helpful discussions on an early draft of this chapter.

References

Anon. (1986) Use of fungicides and insecticides on cereals. *Booklet 2257*, MAFF Publications.

Bayles, R.A. (1991) Research note: varietal resistance as a factor contributing to the increased importance of *Septoria tritici* Rob. and Desm. in the UK wheat crop. *Plant Varieties and Seeds* 4, 177-185.

Cook, R.J. (1977) Effect of timed fungicide sprays on yield of winter wheat in relation to *Septoria* infection periods. *Plant Pathology* 26, 30-34.

Cook, R.J. and Yarham, D.J. (1985) Fungicide use in cereal disease control in England and Wales. *Fungicides for Crop Protection: 100 Years of Progress,* BCPC Monograph 31. BCPC Publications, Farnham.

Cook, R.J., Polley, R.W. and Thomas, M.R. (1991) Disease-induced losses in winter wheat in England and Wales 1985-1989. *Crop Protection* 10, 504-508.

Cook, R.J., Hims, M.J., Clark, W.S. and Stevens, D.B. (1995) Fungicides for controlling leaf diseases of winter wheat: evaluation, timing and importance of varietal resistance. *Project Report No. 103*, Home-Grown Cereals Authority, London.

Cook, R.J., Hims, M.J. and Vaughan, T.B. (1998) The effects of fungicide spray timing on winter wheat disease control. *Plant Pathology* (in press).

Eyal, Z. (1981) Integrated control of Septoria diseases of wheat. *Plant Disease* 65, 763-768.

Hims, M.J. and Cook, R.J. (1992) Disease epidemiology and fungicide activity in winter wheat. *Proceedings of the 1992 Brighton Crop Protection Conference - Pests and Diseases* 2, 615-620.

Hindley, R.I. (1995) Decision models for the integrated use of fungicides on winter wheat 1992-94. In: *87th Annual Report of Morley Research Centre,* 139-148.

Jarrett, G.A. and Allard, P.M. (1981) Control of *Septoria* spp. in winter wheat with chlorothalonil. *Proceedings of the 1981 British Crop Protection Council Conference - Pests and Diseases* 1, 259-266.

Kirby, E.J.M. (1995) Identification and prediction of stages of wheat development for management decisions. *Project Report No. 90*, Home-Grown Cereals Authority, London.

Lescar, L., Bouchet, F. and Faivre-Dupaigre, R. (1973) Fungicidal control of winter wheat foliage disease in France. *Proceedings of the 7th British Insecticide and Fungicide Conference* 1, 97-104.

Lovell, D.J., Barker, S.R., Hunter, T., Royle, D.J. and Coker, R.R. (1997) Influence of crop growth and structure on the risk of epidemics by *Mycosphaerella graminicola* (*Septoria tritici*) in winter wheat. *Plant Pathology* 46, 126-138.

Nuttall, M. and Stevens, D.B. (1997) Early control of septoria on winter wheat 1993-95. In: *89th Annual Report of Morley Research Centre,* 65-70.

Paveley, N.D., Lockley, K.D., Sylvester-Bradley, R. and Thomas, J. (1997) Determinants of fungicide spray decisions for wheat. *Pesticide Science* 49, 379-388.

Royle, D.J. (1990) Advances towards integrated control of cereal diseases. *Pesticide Outlook* 1, 20-25.

Royle, D.J., Shaw, M.W. and Cook, R.J. (1986) Patterns in development of *Septoria tritici* and *Septoria nodorum* in some winter wheat crops in Western Europe, 1981-1983. *Plant Pathology* 35, 466-476.

Stevens, D.B. and Nuttall, N. (1997) Evaluation of triazole fungicides in winter wheat 1993-95. In: *89th Annual Report of Morley Research Centre,* 55-64.

Stevens, D.B., Turner, J.A. and Paveley, N.D. (1997) Exploiting variety resistance to rationalise fungicide inputs - theory and practice. *Aspects of Applied Biology 50, Optimising Cereal Inputs; Its Scientific Basis, Part 2: Crop Protection and Systems,* 279-284.

Thomas, M.R., Cook, R.J. and King, J.E. (1989) Factors affecting development of *Septoria tritici* in winter wheat and its effect on yield. *Plant Pathology* 38, 246-257.

Tottman, D.R. (1987) The decimal code for the growth stages of cereals, with illustrations by Hilary Broad. *Annals of Applied Biology* 110, 441-454.

Tyldesley, J.B. and Thompson, N. (1980) Forecasting *Septoria nodorum* on winter wheat in England and Wales. *Plant Pathology* 29, 9-20.

Chapter twenty-one:

Azoxystrobin: Implications of Biochemical Mode of Action. Pharmacokinetics and Resistance Management for Spray Programmes Against Septoria Diseases of Wheat

J.R. Godwin, D.W. Bartlett and S.P. Heaney
Zeneca Agrochemicals, Jealott's Hill Research Station, Bracknell, Berkshire RG42 6ET, UK

Introduction

Azoxystrobin is a broad novel spectrum, systemic strobilurin fungicide being developed for use on a wide range of crops (Godwin *et al.*, 1992). Its biochemical mode of action (Baldwin *et al.*, 1996) and that of other strobilurins is inhibition of mitochondrial respiration by blocking electron transfer between cytochrome b and cytochrome c_1. The current chapter explains how our understanding of the biochemical mode of action of azoxystrobin, its uptake and redistribution properties and resistance management considerations have influenced Zeneca's recommendations for spray programmes against *Mycosphaerella graminicola* (anamorph *Septoria tritici*) and *Phaeosphaeria nodorum* (anamorph *Stagonospora nodorum*; syn. *Septoria nodorum*) on wheat.

Materials and Methods

Spore germination studies were carried out as previously described by Godwin *et al.* (1994).

Wheat seedlings (*Triticum aestivum L.*) for growth room efficacy tests (usually three replicates) were grown from seed in John Innes No.2 compost in 38 mm or 75 mm diameter pots. The seedlings were maintained in growth rooms prior to inoculation (16 hours under fluorescent lighting at 21°C, 60% R.H. followed by 8 hours dark at 17°C, 95% R.H. for plants to be inoculated with *S. tritici*; same conditions except 24°C day for plants to be inoculated with *S. nodorum*). Adult plants for glasshouse testing (eight replicates) were grown from seed in John Innes No.2 compost in 125 mm diameter pots and maintained under glasshouse conditions prior to inoculation with *S. nodorum* (16 hours under natural or fluorescent lighting at 21°C, 60% R.H. followed by 8 hours dark at 17°C, 80% R.H.). Fungicide treatments were applied in 0.05% w/v Tween 20 in water to maximum retention on seedlings using a hand-held pressurized sprayer. Adult plants were treated via a boom-mounted spray nozzle without the need for Tween 20 and subsequently misted daily until inoculation in order to simulate dew fall. Inoculation for all growth room and glasshouse tests was with a suspension of conidia from cultures raised on agar plates in 0.05% w/v Tween 20 in water (1.25×10^6 conidia ml^{-1} for *S. nodorum*; 10^6 conidia ml^{-1} inoculated twice within 1 hour for *S. tritici*) applied using a hand-held pressurized sprayer. Inoculated seedlings were placed in a mist chamber at room temperature for 48 hours and subsequently incubated (16 hours under fluorescent lighting at 21°C, 60% R.H. followed by 8 hours dark at 17°C, 95% R.H.) before assessment of percentage diseased leaf area 6 days later (*S. nodorum*) or percentage diseased leaf area with visible pycnidia 19-27 days later (*S. tritici*). Adult plants inoculated with *S. nodorum* were placed in a mist chamber at room temperature for 48 hours and subsequently incubated (16 hours under fluorescent lighting at 23-27°C, 80% R.H. followed by 8 hours dark at 17°C, 95% R.H.) before assessment 5 days later.

Foliar uptake of azoxystrobin was determined by applying ^{14}C-azoxystrobin via a microdroplet applicator to a zone 2 cm in length in the centre of the basal quarter of the youngest fully expanded leaf of wheat (GSZ 32). Unabsorbed residue on the foliar surface and in the epicuticular wax layer was removed by cellulose acetate stripping (Silcox and Holloway, 1986). The radiochemical content of the cellulose acetate removed was determined by liquid scintillation spectrometry and azoxystrobin uptake was quantified both by calculation of loss from the leaf surface and by recoveries from within treated leaves following analysis using validated Zeneca procedures. Systemic movement in wheat was quantified by microdroplet application of ^{14}C-radiolabelled fungicides to growth-room grown wheat seedlings followed by extraction of parent active ingredient and quantitative analysis.

The field trials were designed, analysed and reported in accordance with EPPO Guidelines Numbers 152 and 181. Each trial was of a randomized block design with four replicate plots (size range 10-84 m^2). Applications were made via hand-held small plot sprayers fitted with hydraulic nozzles producing a flat fan spray pattern of fine or medium quality, as defined by the British Crop Protection Council. Application volume was 200-300 l ha^{-1}. Crop growth stages were recorded using the Zadoks' decimal code (GSZ). Foliar disease control was determined by visual estimation of the percentage leaf area affected by individual diseases on 20-30 tillers per plot. Assessments were carried out in accordance with EPPO Bulletin 29 (*Septoria* in Wheat).

Where appropriate, data were subjected to analysis of variance and treatment separation was achieved via LSD. Treatment means with no letter in common are significantly different at the 5% level.

Factors Influencing Spray Programmes

Biochemical Mode of Action

The biochemical mode of action of azoxystrobin results in ATP synthesis being blocked, i.e. energy production within the fungus is halted. Therefore, highly energy-demanding stages of fungal development are particularly sensitive to the effects of azoxystrobin, for example spore germination of both *S. tritici* (Fig. 21.1) and *S. nodorum* (Godwin *et al.*, 1994). This is an interesting contrast with triazoles which inhibit the 14-demethylation step in the ergosterol biosynthesis pathway. Triazoles do not impair spore germination or the early stages of fungal colonization because the pathogen obtains a supply of ergosterol, or its precursors, from the spore (Hänssler and Kuck, 1987).

The potent inhibition of spore germination by azoxystrobin explains why it is most effective when applied in the early stages of disease development.

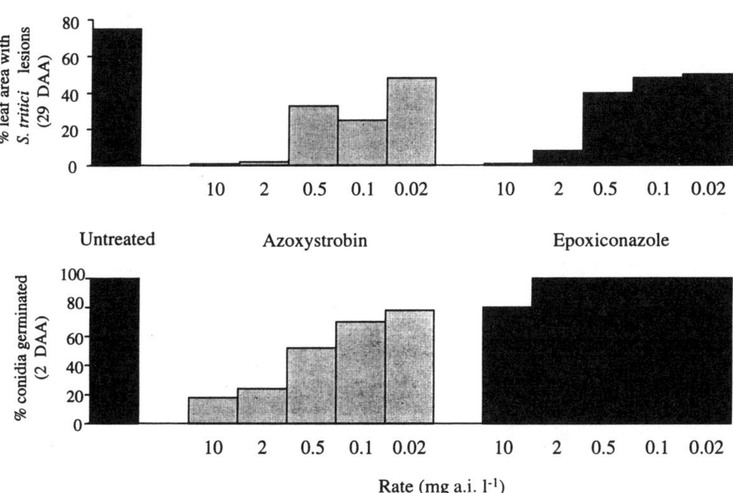

Fig. 21.1. *Septoria tritici* conidia germination and disease development on wheat seedlings inoculated 1 hour after treatment with azoxystrobin or the sterol biosynthesis inhibitor, epoxiconazole. Data show germination at 2 days after application (DAA) and corresponding disease levels 29 DAA.

However, unlike classic protectant fungicides which only inhibit spore germination, azoxystrobin also demonstrates significant curative activity against Septoria diseases of wheat. Field data against *S. nodorum* show azoxystrobin to give greater than 80% disease control when applied at up to 50% of the latent period (Fig. 21.2). Curative activity of azoxystrobin is less marked against *S. tritici*. Field observations suggest that azoxystrobin gives good control of *S. tritici* when applied at up to 20% of the latent period.

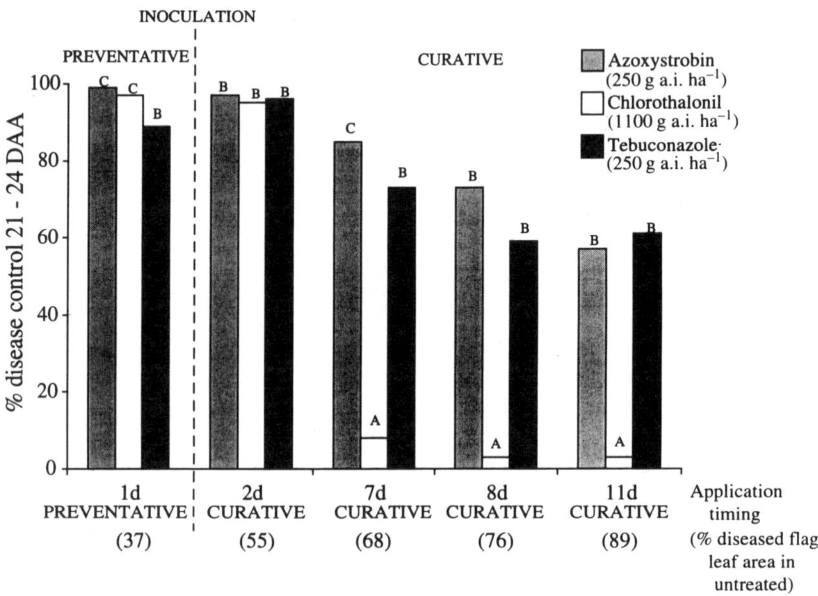

Fig. 21.2. Influence of application timing for control of *Stagonospora nodorum* on wheat. Disease control assessments were made on the flag leaf 21-24 days after fungicide application (DAA). Different letters within a timing indicate differences between treatments significant at the 5% level. *Stagonospora nodorum* had a 14 day latent period (cv. Bounty in the UK). Numbers in parentheses at the foot of the figure indicate per cent diseased flag leaf area in the untreated field plots at the time of assessment.

Triazoles inhibit ergosterol biosynthesis and are, therefore, well suited to curative or eradicant situations because at the time of application fungal mycelium is growing rapidly and ergosterol demands are high. Consequently, the most active sterol biosynthesis inhibitors demonstrate better late curative or eradicant activity against Septoria diseases than azoxystrobin.

Pharmacokinetics

Uptake of azoxystrobin is a gradual process, with approximately 20% of applied active ingredient (a.i.) absorbed into the leaves by 24 hours after foliar application to wheat under field conditions (Fig. 21.3). Uptake reached 45% of that applied after 8 days. The gradual uptake of azoxystrobin is in clear contrast to a number of triazoles for which published data clearly demonstrate rapid uptake, e.g. epoxiconazole (Akers *et al.*, 1990). Despite a relatively slow uptake into the leaf, azoxystrobin has been shown to give no significant loss in control of *S. tritici* when rainwashed 2 hours following application (Fig. 21.4).

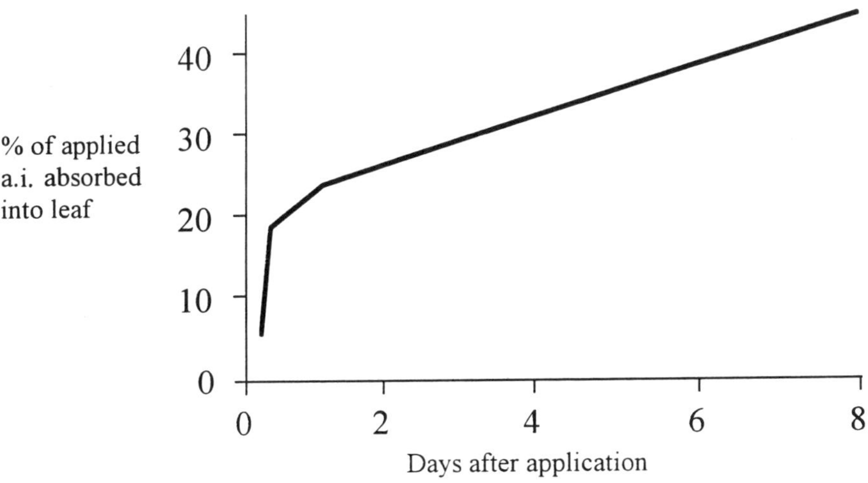

Fig. 21.3. Foliar uptake of azoxystrobin in field wheat.

Azoxystrobin is stable within wheat leaves, 77% of that taken up remaining as parent fungicide when quantified 21 days after application.

Once within the leaf, azoxystrobin moves systemically in the xylem to redistribute evenly above the point of uptake. Quantitative comparison with a range of triazole fungicides by direct chemical analysis has shown azoxystrobin to be more systemic than epoxiconazole and less systemic than tebuconazole or flutriafol (Fig. 21.5).

Treating leaves of wheat seedlings at their base with either azoxystrobin or chlorothalonil and inoculating the same leaves 24 hours later with a suspension of *S. nodorum* conidia has shown the systemic redistribution of azoxystrobin to be of clear significance for disease control (Fig. 21.6). Furthermore, examination by fluorescence microscopy revealed that the systemic disease control seen was at least partly due to azoxystrobin diffusing out to the leaf surface, because conidium germination and appressorium formation were markedly reduced in the (untreated) tip zone of the leaves (Fig. 21.7).

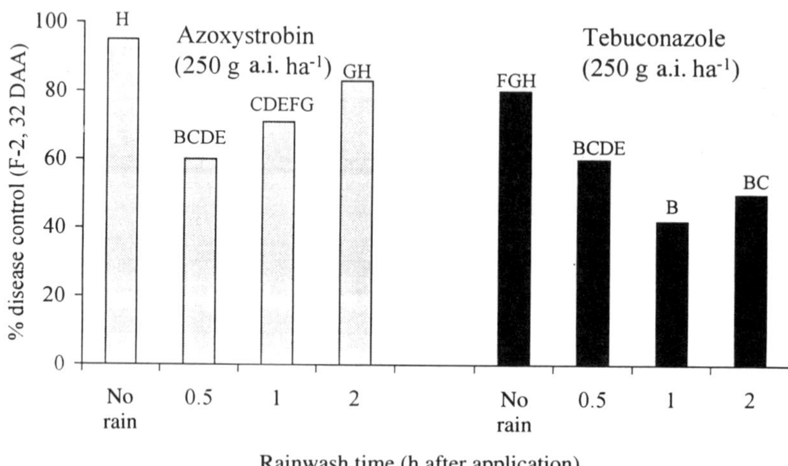

Fig. 21.4. Rainfastness of azoxystrobin and the sterol biosynthesis inhibitor, tebuconazole, on wheat. Artificial rain was applied at 7.5 mm h^{-1} for 2 hours at different timings following treatment at GSZ 43. Data show per cent control of *Septoria tritici* on leaf F-2, 32 days after application (DAA). Treatment means with no letter in common are significantly different at the 5% level. 88% diseased leaf area was recorded in the untreated plots of this UK field trial on cv. Beaver.

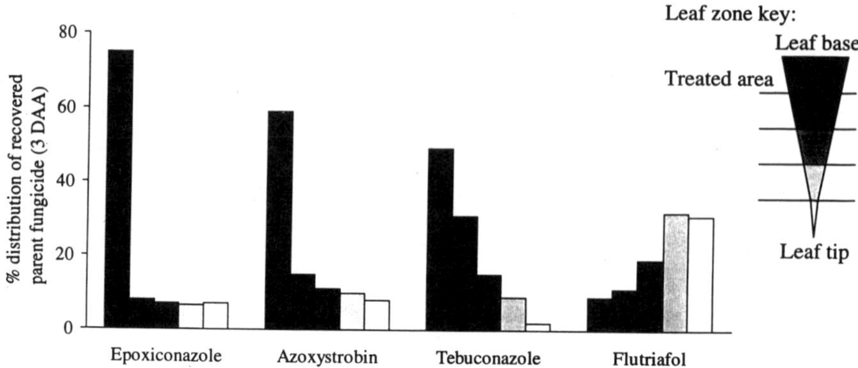

Fig. 21.5. Systemic movement of fungicides in wheat 3 days after application (DAA).

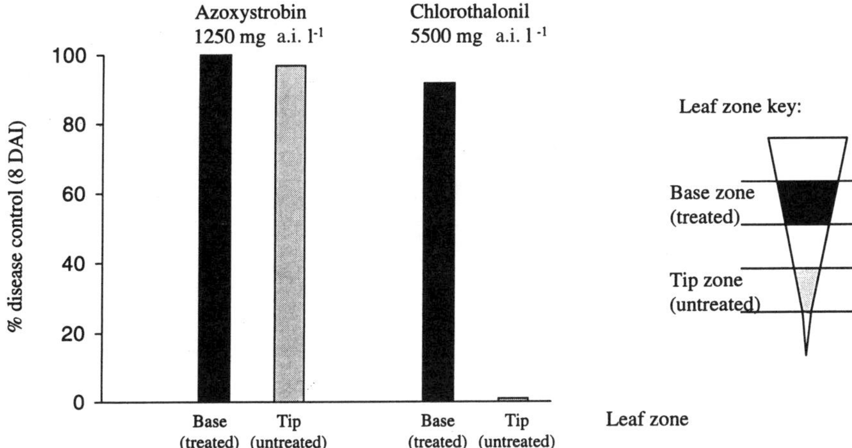

Fig. 21.6. Systemic control of *Stagonospora nodorum* on wheat by azoxystrobin and chlorothalonil. A zone at the base of individual leaves was treated 24 hours prior to inoculation of the complete leaf and disease control assessed 8 days after inoculation (DAI). 66% and 38% diseased leaf area was recorded in the base and tip zones respectively of untreated plants.

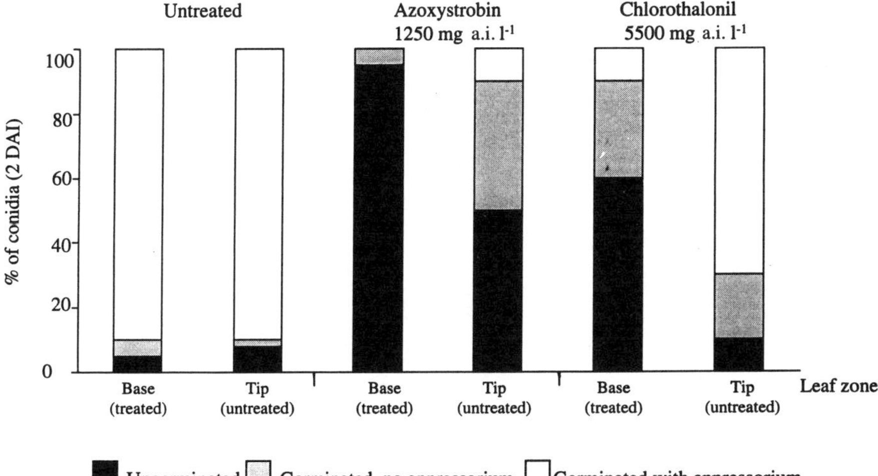

Fig. 21.7. Systemic effects of azoxystrobin and chlorothalonil on *Stagonospora nodorum* conidia development on wheat. A zone at the base of individual leaves was treated 24 hours prior to inoculation of the complete leaf and conidia development assessed in the treated base zone and untreated tip zone 2 days after inoculation (DAI).

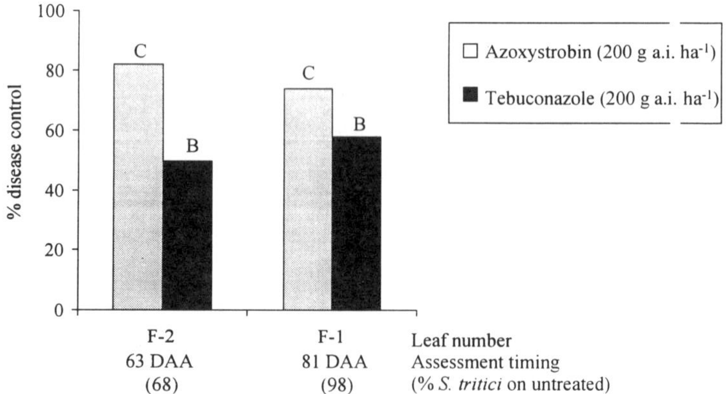

Fig. 21.8. Control of *Septoria tritici* by azoxystrobin and the sterol biosynthesis inhibitor, tebuconazole, on wheat leaves not unfurled at the time of application. Spray application was at GSZ 31 in this field trial on cv. Consort in the UK, with both treatments also receiving an overspray of azoxystrobin at 250 g a.i. ha^{-1} at GSZ 39, 36 days after application (DAA). Different letters within an assessment timing indicate differences between treatments significant at the 5% level.

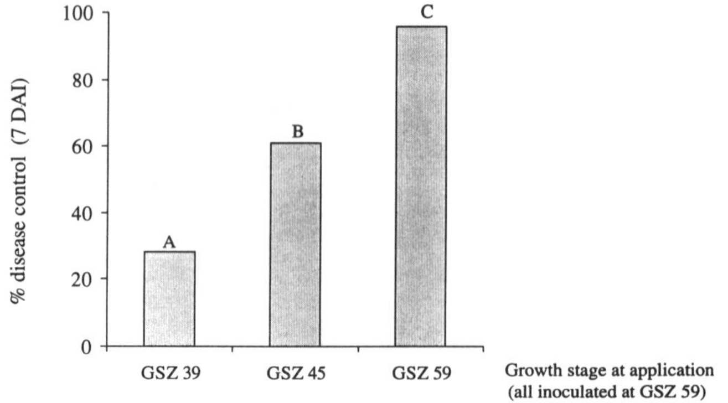

Fig. 21.9. Influence of application timing of azoxystrobin (250 g a.i. ha^{-1}) for the control of *Stagonospora nodorum* on ears of wheat. All treatments were inoculated at GSZ 59 and disease control was assessed 7 days later, at which time 43% ear area was infected with *S. nodorum* on untreated plants. Different letters indicate differences between treatments significant at the 5% level.

Radiochemical studies have confirmed that azoxystrobin moving in the xylem diffuses out to the leaf surface. Systemic inhibition of spore germination is a completely novel effect.

In addition to its movement within individual leaves, azoxystrobin has also demonstrated translaminar and systemic movement from stem or axil applications into new leaves which subsequently unfurl. Phosphor imaging of ^{14}C-labelled azoxystrobin applied to the sheath of the oldest leaf of a wheat seedling at GSZ 12 has demonstrated movement of a.i. through five leaf layers. Similar results were obtained following application directly into leaf axils. Field data strongly indicate that the movement of azoxystrobin into new leaves which subsequently unfurl is of consequence for control of *S. tritici*. Azoxystrobin applied at GSZ 31 and again at GSZ 39 gave significantly better control of *S. tritici* on F-2 and F-1, leaves not unfurled at the time of the first application, than did tebuconazole followed by azoxystrobin (Fig. 21.8).

Glasshouse studies have demonstrated that in order to achieve a high level of control of *S. nodorum* on the ear with azoxystrobin it is necessary to spray when the ear is exposed. Application of azoxystrobin at booting (GSZ 45) gave only moderate levels of control of *S. nodorum* on the ear when subsequently inoculated at GSZ 59 with a suspension of conidia, whereas application at GSZ 59 gave almost total control (Fig. 21.9). The moderate control achieved from the application at booting was probably due to translaminar movement of azoxystrobin through the foliage ensheathing the ear and perhaps physical redistribution on to the ear as it emerged from the sheath. Application at GSZ 39 gave poor control of *S. nodorum* on the ear, thereby indicating little movement of a.i. from foliage to ear via the xylem.

Resistance Risk

The biochemical mode of action of strobilurins is novel, and laboratory and glasshouse studies have demonstrated that there is no cross-resistance with other fungicides effective against Septoria diseases, notably sterol biosynthesis inhibitors and benzimidazoles.

Studies with yeast (Colson, 1993), *Rhodobacter* (Gennis *et al.*, 1993) and algae (Bennoun *et al.*, 1991) have shown that resistance to strobilurins is possible, but what level of risk is there with *S. tritici* and *S. nodorum*? In order to estimate this risk we must establish existing sensitivity levels within natural populations by baseline sensitivity monitoring, determine how readily resistant mutants can be generated and consider the results from other studies investigating resistance to this biochemical mode of action.

In vivo baseline monitoring of *S. tritici* populations in France for sensitivity to azoxystrobin prior to commercial use of any strobilurin has shown a normal distribution with a relatively narrow range in sensitivity (Fig. 21.10).

In mutation studies, it has proven relatively difficult to isolate stable mutants of *S. tritici* with reduced sensitivity to azoxystrobin. The mutants isolated have shown a relatively low level of resistance *in vitro* (typically x5 to x20 less sensitive to azoxystrobin than wild-type). Approximately 30% of putative mutants proved to be unstable in subculture, reverting to wild-type in the absence of azoxystrobin. Mutants were isolated at higher frequencies when lower selective doses of azoxystrobin were used (F. Corbisier-Colson, Louvain University, 1996, personal communication). It has not yet been confirmed that these mutant strains are also resistant *in vivo*.

The studies with *Saccharomyces cerevisiae, Chlamydomonas reinhardtii, Rhodobacter capsulatus* and *R. sphaeroides* referred to above have identified changes to the target site respiratory protein apocytochrome b, as the basis for resistance to strobilurins and other molecules with similar affinity for the target, e.g. myxothiazol. These mutants from several species represent a uniform collection in that they all exhibit a reduction in respiratory efficiency, typically being respiratory deficient. In simple terms, they are all significantly less fit than the wild-type. In *S. cerevisiae,* respiratory efficiency, and with it an improved ability to grow (fitness), can be partially restored by secondary and even tertiary changes at different codons in the apocytochrome b gene (Tron *et al.*, 1991). Interestingly, where such suppressor mutations restored respiratory efficiency to the level of the wild-type, resistance to myxothiazol was lost. Remarkably, the codon positions associated with amino acid substitutions that lead to strobilurin resistance are highly conserved between species, similar substitutions giving rise to strobilurin resistance in all of the above species. Similar patterns might, thus, be predicted for plant pathogens.

Although studies outside plant pathogenic fungi indicate that target site resistance is the most likely route for evolution of resistance to strobilurins, other mechanisms are possible, e.g. detoxification or efflux of active ingredient and alternative (but less efficient) mechanisms for respiration. Indeed, an alternative oxidase mutant of *S. tritici* with reduced sensitivity to strobilurins has been isolated from an *in vitro* system, but subsequent *in vivo* testing showed this isolate to be better controlled by azoxystrobin than the wild-type (Ziogas *et al.*, 1997). This suggests that the decreased ATP formation resulting from the alternative pathway of respiration was inadequate for normal fungal growth of this mutant on the host plant.

Considering all of the above information, it is predicted that the likelihood of *S. tritici* or *S. nodorum* populations developing practical resistance to strobilurins is moderate. Resistance development should involve several genetic changes making it more comparable to the gradual evolution of resistance in *S. tritici* to the ergosterol biosynthesis inhibitors (Hims, 1994) than to the rapid development of resistance seen in practice with *S. tritici* to the benzimidazoles (Locke, 1986). The latter probably reflected a single genetic change conferring high levels of resistance with little loss of fitness.

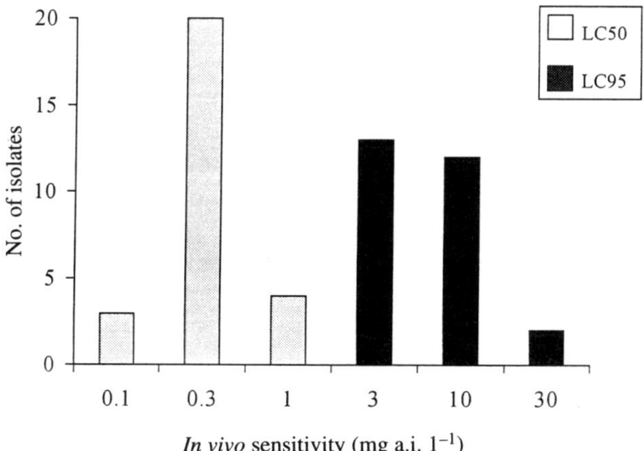

Fig. 21.10. Baseline monitoring of *Septoria tritici* field isolates from France for sensitivity of conidia to azoxystrobin, determined as lethal concentrations for 50% and 95% inhibition of disease development (LC 50 and LC 95 respectively) on wheat seedlings treated with azoxystrobin 24 hours prior to inoculation.

Discussion and Conclusions

Given the above understanding of its biochemical mode of action, pharmacokinetics and resistance management considerations, how should azoxystrobin best be used for the control of *S. tritici* and *S. nodorum*?

Based on these studies, the most important message for azoxystrobin on wheat is that it should be applied in the early stages of disease development for optimum efficacy. Application timing relative to disease development is likely to be an important issue for all fungicides which are potent inhibitors of spore germination. Azoxystrobin is no exception, although its uptake and stability in wheat ensure that sufficient a.i. is present within the leaf to give significant curative activity and, therefore, a degree of flexibility in application timing against Septoria diseases. Also, when considering application timing, it must always be remembered that disease development is a dynamic process, i.e. what starts as a curative application becomes more protectant as reinfection occurs. In this respect, the gradual (rather than rapid) uptake of azoxystrobin is important because it maintains a protective barrier on the leaf surface to prevent further incoming spores from germinating.

The biochemical mode of action of ergosterol biosynthesis inhibitors (EBIs) explains why the most active EBIs demonstrate better late curative or eradicant activity against Septoria diseases than azoxystrobin. Mixture of azoxystrobin with a strong curative product can result in improved control of *S. tritici* or *S. nodorum* in situations where the disease is well established in the crop prior to treatment. However, farmer practice will usually lead to a reduction in the rate of azoxystrobin applied in such mixtures, thereby reducing the disease control benefits that azoxystrobin brings, e.g. control of *S. tritici* and *S. nodorum* on leaves which emerge after application and persistence of effect on leaves already unfurled. Indeed, no significant differences in yield response were recorded between azoxystrobin applied alone at its recommended rate and when applied in mixture with hexaconazole (at reduced rates of both fungicides) in three *S. tritici* trials in France this year (Table 21.1). Although initially the mixture gave improved disease control in curative situations, later assessments demonstrated that the persistence of effect of azoxystrobin in combination with continued epidemic development can lead to little benefit with this mixture at the later assessments, e.g. N. France trial; see F-1 assessments at 33 DAA vs. 42 DAA. In all of these trials, a single application was made and no disease other than *S. tritici* developed to a significant extent.

Azoxystrobin applied during the stem extension phase of wheat will not only give disease control on leaves unfurled at the time of application but will also move into new leaves to give control of *S. tritici* and *S. nodorum* as they subsequently unfurl. However, dilution of the a.i. in these new leaves means that a gap of not more than 4 weeks is recommended between stem extension and 'flag leaf' sprays.

A good persistence of effect would be predicted from the gradual uptake, smooth redistribution and stability of azoxystrobin within the leaf. This is exactly what is observed in practice, with azoxystrobin applied at flag leaf emergence and beyond usually giving a greater persistence of effect against *S. tritici* and *S. nodorum* than triazole-based cereal fungicides (Table 21.2).

The persistence of effect of azoxystrobin and its potent inhibition of spore germination should be utilized by spraying early in disease development following flag leaf (F) emergence. Consequently, *S. tritici* and *S. nodorum* conidia will be prevented from germinating on the upper leaves that are important in determining yield (F, F-1, F-2: Thorne, 1974; Thomas *et al.*, 1989) and, therefore, these leaves will remain healthy and retain their green leaf area until late in the season, thereby maximizing grain yield and quality. This recommendation is supported by recent glasshouse studies (Schöfl and Zinkernagel, 1997) which have shown that treatment of wheat seedlings with a triazole alone prior to inoculation with conidia of *S. tritici* gave a good reduction in the number of pycnidia which subsequently developed but gave markedly higher levels of necrosis than when a fungicide which inhibits spore germination was also included in the treatment. They suggested that the difference in necrosis levels between the treatments was attributable to the triazole alone, allowing the host tissue to be penetrated by *S. tritici* spores early in their development before the inhibition of ergosterol biosynthesis became effective.

Table 21.1. Control of *Septoria tritici* on wheat by single applications of azoxystrobin alone vs. in mixture with hexaconazole at 3 field trial locations in France. Figures within the azoxystrobin and azoxystrobin + hexaconazole lines indicate per cent control of *S. tritici* or yield as a per cent of the untreated. Different letters within a column indicate differences between treatments significant at the 5% level.

Treatment	Rate (g a.i. ha^{-1})	Trial 1 (N. France) GSZ 46 at application				Trial 2 (E. France) GSZ 35 at application				Trial 3 (C. France) GSZ 33-34 at application			
		% disease control			% Yield	% disease control			% Yield	% disease control			% Yield
		F-1		F		F-1		F		F-2	F-1	F	
		33 DAA	42 DAA	42 DAA		46 DAA	55 DAA	55 DAA		45 DAA	58 DAA	66 DAA	
Untreated (% *S. tritici* or t ha^{-1})	-	(40)	(94)	(61)	(7.9)	(86)	(100)	(89)	(7.0)	(30)	(60)	(39)	(9.5)
Azoxystrobin	250	21A	65A	95A	126A	84A	41A	60A	125A	80A	74A	84A	112A
Azoxystrobin + hexaconazole	200 + 125	45B	65A	97A	124A	93B	65B	65B	129A	80A	72A	76A	112A
% *S. tritici* at application		0	0	0	-	0	0	-	-	0	0	-	-
Comments		Latest disease at application cv. Bourbon				2.5% *S. tritici* on F-3 at application cv. Paindor				Preventative situation cv. Scipion			

Table 21.2. Persistence of control of *Septoria tritici* or *Stagonospora nodorum* on the leaves or ears of wheat in field trials. Figures within the azoxystrobin, flusilazole/carbendazim and epoxiconazole lines indicate per cent control of *S. tritici* or *S. nodorum*. Different letters within a column indicate differences between treatments significant at the 5% level.

Treatment	Percentage disease control					
	Septoria tritici		*Stagonospora nodorum*: leaf		*Stagonospora nodorum*: ear	
	F 43 DAA2	F 51 DAA2	F-1 43 DAA2	F 63 DAA	33 DAA	42 DAA2
Untreated (% disease)	0A (79)	0A (94)	0A (54)	0A (18)	0A (52)	0A (30)
Azoxystrobin 250 g a.i. ha^{-1}	84C	74B	87C	78C	79C	79C
Flusilazole/ carbendazim 200/100 g a.i. ha^{-1}	65B	-	65B	-	56B	-
Epoxiconazole 125 g a.i. ha^{-1}	-	78B	-	55B	-	37B
Application timing	GSZ31 fb GSZ52	GSZ31 fb GSZ48	GSZ30-31 fb GSZ39	GSZ37	GSZ51	GSZ31[a] fb GSZ51[a]
Cultivar	Scipion	Scipion	Avalon	Hunter	Slejpner	Kanzler
Location	France	France	UK	UK	France	Germany

[a]Additional 375 g a.i. ha^{-1} fenpropimorph in epoxiconazole treatment.

It is clear from glasshouse studies that the ear must be exposed at application for azoxystrobin to give high levels of control of *S. nodorum* on the ear. Therefore, the field recommendation for optimum control of *S. nodorum* on the ear is application of azoxystrobin at GSZ 51 or beyond, ideally GSZ 55-59 when the majority of the ear is exposed. Again, the good persistence of effect of azoxystrobin and its potent activity against spore germination mean that applications to the ear should be made as early as possible after GSZ 51 if high levels of control of *S. nodorum* on the ear are desired. Application at booting (GSZ 45) may help reduce disease severity on the ear because azoxystrobin will move in a translaminar fashion through the foliage ensheathing the ear, perhaps be physically redistributed in moisture on to the ear as it emerges and also reduce the inoculum source on the upper leaves.

The likelihood of *S. tritici* or *S. nodorum* populations developing resistance problems to strobilurins in practice is considered to be moderate. This takes into account the inherent risk of the chemistry and the epidemiological characteristics of the pathogens. With a resistance risk considered to be similar to the sterol biosynthesis inhibitor fungicides still used effectively against Septoria diseases after over 20 years, it would seem reasonable to suggest that resistance management considerations should not have a major impact on recommendations for the use of azoxystrobin against *S. tritici* and *S. nodorum*. However, we believe that frequent applications of strobilurins at suboptimal doses are of concern. Setting aside the generic and contentious argument about the effect of dose rate on resistance development, we believe that there is an argument to support optimal doses which is specific to strobilurins. This argument is based on their mode of action and the most likely route for evolution of resistance to strobilurins, i.e. target site changes. The same argument might also apply to other inhibitors of proteins coded for by mitochondrial DNA (mtDNA).

The target site, cytochrome b, is contained within the mitochondrion where it is coded for by mtDNA. The frequency of mutations in mtDNA resulting in resistance to strobilurins can only increase if the cells that they originally arise in remain viable. In order to achieve this, the cell must contain functioning wild-type mitochondria, as it seems inconceivable that a single resistant mitochondrion could maintain the cell. The fact that all the available evidence indicates that the initial mutation will also impair respiratory efficiency further supports this argument. Thus, suboptimal doses are likely to increase the risk of resistance arising by providing more opportunities for the initial stages of resistance evolution to proceed, i.e. suboptimal doses provide more situations where the concentration of strobilurins at the cellular level is low enough to allow sufficient numbers of wild-type mitochondria to function. It is proposed that the application of azoxystrobin at its recommended rate and most effective timing, i.e. in the early stages of disease development, is the most effective approach to minimizing this risk by creating more situations where the initial resistance mutation is effectively lethal to the cell. Perhaps, for once, we are in the fortunate situation where good disease control and anti-resistance management go hand-in-hand.

Acknowledgements

The authors wish to thank all of their colleagues in Zeneca Agrochemicals who have helped with the preparation of this chapter or been involved in generating the data contained herein.

References

Akers, A., Köhle, H.H. and Gold, R.E. (1990) Uptake, transport and mode of action of BAS 480F, a new triazole fungicide. *Proceedings of the 1990 Brighton Crop Protection Conference - Pests and Diseases* 2, 837-845.

Baldwin, B.C., Clough, J.M., Godfrey, C.R.A., Godwin, J.R. and Wiggins, T.E. (1996) The discovery and mode of action of ICIA5504. In: Lyr, H., Russell, P.E. and Sisler, H.D. (eds) *Modern Fungicides and Antifungal Compounds.* Intercept, Andover, pp. 105-109.

Bennoun, P., Delosme, M. and Kuck, U. (1991) Mitochondrial genetics of *Chlamydomonas reinhardtii*: Resistance mutations marking the cytochrome b gene. *Genetics* 127, 335-343.

Colson, A.-M. (1993) Random mutant generation and its utility in uncovering structural and functional features of cytochrome b in *Saccharomyces cerevisiae. Journal of Bioenergetics and Biomembranes* 25, 211-220.

Gennis, R.B., Barquera, B., Hacker, B., van Doren, S.R., Arnaud, S., Crofts, A.R., Davidson, E., Gray, K.A. and Daldal, F. (1993) The bc_1 complexes of *Rhodobacter sphaeroides* and *Rhodobacter capsulatus. Journal of Bioenergetics and Biomembranes* 25, 195-210.

Godwin, J.R., Anthony, V.M., Clough, J.M. and Godfrey, C.R.A. (1992) ICIA5504: A novel broad spectrum systemic b-methoxyacrylate fungicide. *Proceedings of the 1992 Brighton Crop Protection Conference - Pests and Diseases* 1, 435-442.

Godwin, J.R., Young, J.E. and Hart, C.A. (1994) ICIA5504: Effects on development of cereal pathogens. *Proceedings of the 1994 Brighton Crop Protection Conference - Pests and Diseases* 1, 259-264.

Hänssler, G. and Kuck, K.H. (1987) Microscopic studies on the effect of ®Folicur on pathogenesis of brown rust of wheat (*Puccinia recondita* f.sp. *tritici). Pflanzenschutz-Nachrichten Bayer* 40, 153-180 (English Edition).

Hims, M.J. (1994) Monitoring sensitivity to the DMI group of fungicides in populations of the leaf blotch pathogen (*Septoria tritici*) and the glume blotch pathogen (*Septoria nodorum*) of winter wheat. *Project Report No. 91,* Home-Grown Cereals Authority, London.

Locke, T. (1986) Current incidence in the United Kingdom of fungicide resistance in pathogens of cereals. *Proceedings of the 1986 British Crop Protection Conference - Pests and Diseases* 2, 781-786.

Schöfl, U.A. and Zinkernagel, V. (1997) A test method based on microscopic assessments to determine curative and protectant fungicide properties against *Septoria tritici. Plant Pathology* 46, 545-556.

Silcox, D. and Holloway, P.J. (1986) A simple method for the removal and assessment of foliar deposits of agrochemicals using cellulose acetate film stripping. *Aspects of Applied Biology 11, Biochemical and Physiological Techniques in Herbicide Research,* 13-17.

Thomas, M.R., Cook, R.J. and King, J.E. (1989) Factors affecting development of *Septoria tritici* in winter wheat and its effect on yield. *Plant Pathology* 38, 246-257.

Thorne, G.N. (1974) Physiology of grain yield of wheat and barley. In: *Report for 1973 of Rothamsted Experimental Station* 2, 5-25.

Tron, T., Infossi, P., Coppee, J.-Y. and Colson, A.-M. (1991) Molecular analysis of revertants from a respiratory-deficient mutant affecting the center of domain of cytochrome b in *Saccharomyces cerevisiae. FEBS Letters* 278, 26-30.

Ziogas, B.N., Baldwin, B.C. and Young, J.E. (1997) Alternative respiration: a biochemical mechanism of resistance to azoxystrobin (ICIA5504) in *S. tritici. Pesticide Science* 50, 28-34.

Chapter twenty-two:

Septoria tritici and *Stagonospora nodorum* as Model Pathogens for Fungicide Discovery

J. Dancer, A. Daniels, N. Cooley and S. Foster
AgrEvo UK Ltd, Chesterford Park, Saffron Walden, Essex CB10 1XL, UK

Introduction

The current world wide market for fungicides has been estimated to be worth about £3 billion, reflecting the economic importance of crop losses through fungal diseases. The constant demand for new active ingredients with improved efficacy and environmental properties is being addressed through continued investment in fungicide discovery by the major agrochemical companies. The aim of this chapter is to describe how basic research on *Septoria tritici* and *Stagonospora nodorum* is contributing to this quest for novel fungicides.

Stagonospora nodorum as a Model System for Biochemistry and Molecular Biology

Whilst AgrEvo conducts research on all major fungal pathogens, *S. nodorum* has become the focus of work by our molecular biologists. The selection of *S. nodorum* was based on the following criteria.

Tractability

The ease with which a fungal pathogen can be grown and maintained, both *in vitro* and *in planta,* was a primary consideration in the choice of a model for biochemical and genetic studies. *Stagonospora nodorum* and *S. tritici* are both readily cultured *in vitro* on complex or defined media. However, the more rapid mycelial growth of *S. nodorum in vitro* was a decisive factor in making it the preferred species for biochemical and genetic studies. Semi-quantitative pathogenicity assays can be readily carried out either on detached leaves or on whole plants. Such assays are essential to the process of defining novel fungicide targets as discussed in the next section. In culture, spores may be readily contained so reducing the risk of cross-contamination or escape of genetically-modified strains.

The recognition of a distinct life-cycle is a further advantage, although it is noted that the lack of clearly defined infection structures may be a drawback. The life-cycles of *S. nodorum* and *S. tritici* share basic similarities and are discussed in the section on life cycle studies on p. 324. Such studies are essential to underpin biochemical and genetic research.

Stagonospora nodorum and *S. tritici* had already become the focus of more applied research such as monitoring for the development of resistance to fungicides. These species are particularly well suited to such applications as the diseases are easily recognized both in the field and in the glasshouse and it is relatively easy to isolate individual pycnidia. There are obvious opportunities for synergy between more applied and basic aspects of research when both concern the same species.

Genetics

Gene transfer systems are available for *S. nodorum* and, more recently, for *S. tritici,* although these are not yet efficient enough to allow cloning by complementation. Classical genetics through sexual crossing, although possible for *S. nodorum* and *S.tritici*, are not yet used routinely in most laboratories (see Halama, Chapter 4, and Kema *et al*., Chapter 10 this volume).

Industrial Relevance

The leaf blotches, which include *S. nodorum* and *S. tritici*, are major diseases in the important cereal segment of the market. The powdery mildews represent the largest market for cereal fungicides, however these are obligate biotrophs and are, therefore, not yet amenable to biochemical and genetic studies. *Magnaporthe grisea* (rice blast), another important cereal disease, is well suited to basic fungicide research and has been successfully adopted as a model both by workers at Du Pont in the USA and in academia (Valent, 1990; Talbot, 1995; Sweigard *et al*., Chapter 12 this volume). While an important international fungicide market, rice blast is not a major disease in western Europe, and it was anticipated that this could be a disadvantage when seeking support for UK-based research. The most economically important diseases of broad leaf crops are *Plasmopora viticola* (vine downy mildew) and *Phytophthora infestans* (potato blight). As Oomycetes, these are not representative of the majority of fungal phytopathogens since over 50% of the total fungicide market is directed towards the control of Ascomycetes, the subdivision to which *S. nodorum* belongs. Furthermore, *P. viticola*, the most important of these broad leaf diseases, is an obligate pathogen.

It has been estimated that the total cost of the research and development required to launch a novel agrochemical active ingredient is £20-60 million. There are few individual fungal diseases for which the potential market is sufficiently large to support this cost. Thus, although *S. nodorum* and *S. tritici* are important diseases, it is very unlikely that the commercialization of a fungicide restricted to these species alone would be profitable. It is, therefore, important that fundamental research on model species such as *S. nodorum* can lead to the discovery of broad spectrum fungicides.

Adopting a common system, such as *S. nodorum*, for studies on fungal pathogenesis has the obvious advantage that the results from diverse research strategies can be correlated, therefore, all things being equal, concentrating resources on a single model makes good sense. There may, however, be very good reasons for carrying out particular studies on an alternative pathosystem. For this reason, imposing a single model on the scientific community is unrealistic.

It should also be emphasized that the choice of model is partially subjective and that there are other fungi, most notably *M. grisea,* which satisfy the majority of the criteria outlined above. *Stagonospora nodorum* and *S. tritici* are proving to be useful models and, as is clear from the work described in this volume, the study of them is increasing our understanding of pathogenesis. In the remainder of this chapter we will demonstrate how research on these species is contributing to the discovery of novel fungicides. We will concentrate on two aspects of our research: the identification of new fungicide targets and a detailed description of the pathogen life-cycle.

Identification of Novel Fungicide Targets

Approaches to Fungicide Discovery

There are several different approaches by which novel fungicides can be identified. These may be largely chemistry driven, for example random, 'blue sky' synthesis, pharmacophore approaches, synthesis of analogues of successful compounds or those described in competitors' patents. Natural products can provide a valuable source of inspiration for new compounds, as has recently been demonstrated by the development of the strobilurin fungicides of Zeneca and BASF (Beautement *et al.*, 1991; Sauter *et al.*, 1995; Godwin *et al.*, Chapter 21 this volume). Two further approaches which are receiving significant attention are so called 'rational' or 'biochemical' design and high throughput biochemical screening. It is to these in particular that our research on *S. nodorum* is contributing, and which will now be discussed in more detail.

Target Validation

Fungicides act by interfering with essential biochemical processes, often through inhibition of a specific enzyme reaction. Through a knowledge of an enzyme mechanism or, in some cases, a crystal structure, it is possible to rationally design novel inhibitor types. There are plenty of examples in the literature of successful inhibitor design, although, as discussed below, few if any of these have translated into commercial successes. A pre-requisite for inhibitors with useful fungicide activity is that the cellular process targeted is necessary for pathogenicity. Selection of enzymes for these biochemical approaches to fungicide discovery is frustrated by the lack of information about even the most basic biochemical pathways in fungal phytopathogens. Furthermore, even if the pathways have been described, it is very difficult to predict which enzymes represent good targets based on purely theoretical grounds. Before investing time and money in a programme of inhibitor design or discovery, it is important that the selected enzyme is 'validated' as a good target. There are two methods by which enzyme targets can be validated: chemically or genetically.

Chemical Validation

Chemical validation of a target remains the most convincing approach at present. The biochemical targets for several commercial fungicides have been defined through mode of action studies and are obvious candidates for rational design approaches. A good example of successful inhibitor design against an established fungicide target is the work on sterol isomerase by Huxley-Tencer *et al.* (1992). Alternatively, a known inhibitor which is not a commercial fungicide can be used to demonstrate that a particular enzyme is essential for pathogenicity. An obvious consideration is that the chosen inhibitor is absolutely specific for its enzyme target. This approach is limited in that the supply of specific, potent enzyme inhibitors is limited. Furthermore, they may not reach their target in the intact fungus through poor uptake into the cell or rapid degradation once inside.

Genetic Validation

Genetic validation has the considerable advantage that, in theory, any protein can be targeted. The tractability of *S. nodorum,* as outlined in the section on Tractability above, has made it our choice as a model pathogen to develop a genetic approach for fungicide target validation. Here we review progress in the development of this approach and highlight further technical developments which are needed urgently.

Random Mutagenesis

There are two approaches to random mutagenesis as outlined below, which allow for the random identification of potential fungicide targets. An advantage of these random approaches is that they may reveal novel or unexpected targets.

Classical Mutagenesis

One basic approach for the identification of novel fungicide targets is the use of classical mutagenesis. We have used this route to isolate a number of mutant strains deficient in various biosynthetic pathways. The pathogenicity of these mutants has subsequently been shown to be reduced or abolished. Auxanography followed by biochemical analyses have been used to identify the specific metabolic step affected in selected mutants.

The identification of the site of the metabolic lesion can be a time-consuming and laborious process akin to a fungicide mode of action study. Definitive proof that the correct enzyme has been identified must then be provided by complementation of the mutant with the wild-type gene from any source, although a homologous gene is preferred. If classical mutagenesis is used to identify an enzyme, the gene must subsequently be cloned using conventional approaches which may again prove laborious or, worse, inapplicable. A clone providing a source of recombinant enzyme from the fungal pathogen can greatly facilitate the subsequent biochemical characterization of the enzyme and analysis of inhibitors, and provides the potential for high throughput screens and protein crystallization. This cloning step may, therefore, be critical to subsequent analyses.

Restriction Enzyme-mediated Integration (REMI) and Transposon Tagging

Tagging mutagenesis offers significant advantages over classical mutagenesis in that the lesion may be more readily identified and cloning the gene of interest is facilitated. Two such approaches, restriction enzyme-mediated integration (REMI) and transposon tagging, are, therefore, currently being investigated in *S. nodorum*. Examples of success in other pathogens are varied and we should be cautious about applying existing genetic methods to new systems. For example, the lack of facile sexual crossing in *S. nodorum* makes unequivocal interpretation of gene tagging difficult.

Targeted Mutagenesis

It is also useful to be able to investigate potential targets of known function. This can be achieved by the targeted mutagenesis brought about by gene disruption, a method which relies on homologous recombination to interrupt a specific target gene. This method relies on the availability of a clone of at least part of the target. It has been used successfully in a number of fungal pathogens (for review, see Oliver and Osbourn, 1995) and has now been demonstrated to work in *S. nodorum* both in C. Caten's laboratory (see Chapter 2 this volume) as well as ours at AgrEvo. The specificity achieved in gene disruption is of particular use in situations where the predicted mutant phenotype is not known. Molecular techniques can be used to confirm the occurrence of the disruption even if an unpredicted or indiscernible phenotype is produced.

Gene Silencing (Antisense and Quelling)

We use gene disruption in the first steps of target validation, not least because it is technically feasible. However, disruption will produce a null mutation and so is limited to non-lethal traits *ex planta* or those which can be compensated by supply of the appropriate nutrients. While it is useful for ruling out non-essential processes from rational design programmes, the production of null mutants is not, in itself, sufficient to validate a potential fungicide target. It is highly improbable, if not impossible, that a fungicide will achieve 100% inhibition of its target enzyme. A more representative genetic method would involve attenuating the activity of the chosen enzyme to determine how much reduction is necessary before pathogenicity is affected. This may be achieved using gene silencing technology (antisense and quelling). The potential utility of the antisense approach has already been demonstrated in the field of herbicide discovery. Researchers at AgrEvo have proved the validity of the approach using a known herbicide target: acetolactate synthase, the site of action for the highly successful sulfonylureas (Höfgen *et al.*, 1995). The activity of the enzyme was effectively titrated out using antisense technology and it was demonstrated that a relatively small reduction in activity was required to produce symptoms which were identical to those seen following treatment with a sulfonylurea herbicide. Thus, although disruption rules out non-essential processes and also provides a bench-mark for 100% inhibition, target validation in fungicide discovery would gain greatly from the gene silencing technology which is now widely used in the discovery of herbicides.

Selection of Targets

The success of these targeted approaches relies on careful selection of enzymes to target. A number of criteria can be used to define suitable candidate enzymes. Specificity to fungi should reduce the risk of mammalian, phyto- or eco-toxicity. From the brief discussion of markets above, it is obvious that a broad spectrum fungicide is most desirable, therefore, while being restricted to fungi, the enzyme should ideally be present in a wide range of different fungal pathogens. In practical terms, an enzyme which is readily assayed will make the subsequent analysis of potential inhibitors easier and, if high throughput biochemical screening is envisaged, an assay amenable to automation will be required. If the enzyme is to be the starting point for a biochemical design programme, the reaction mechanism and/or existing inhibitors must provide inspiration for further chemical synthesis. Where inhibitor design is to be based on the substrate(s) and product(s) of the reaction, it is worth remembering that highly hydrophilic compounds are unlikely to be taken up by the fungus, unless they are able to take advantage of a specific transport mechanism. Hydrophobic intermediates, therefore, are generally a better starting point for the design of inhibitors targeted at intracellular processes. Finally, while it is difficult to predict with any certainty which enzymes are essential for growth or pathogenicity, it is often possible to make an educated guess. The concept of a single 'rate limiting' enzyme in each pathway has been challenged by metabolic control theory (Kacser and Burns, 1973). While, according to metabolic control theory, regulation of flux may be shared between many enzymes, in reality, control is usually concentrated at a few. These enzymes with high control coefficients would be anticipated to represent attractive targets. It is worth emphasizing again that selection of enzyme targets on a theoretical basis is limited by a lack of basic knowledge about the metabolic pathways in the target fungi.

Analysis of Mutants

Whichever method of genetic validation is used, it is important that the resulting mutants are analysed under appropriate conditions. This has been demonstrated from studies on C14-sterol reductase, an enzyme in sterol biosynthesis which is one of the target sites for the morpholine fungicides. The viability of yeast mutants in which the gene encoding this activity had been insertionally inactivated was found to be dependent on the medium. While the mutant was viable on a defined, synthetic medium, surprisingly it did not survive following the addition of yeast extract and peptone (Parks *et al.*, 1995). As discussed above, one of the advantages of *S. nodorum* is that semi-quantitative pathogenicity assays are relatively easy.

A further consideration in genetic validation is that if several copies of the gene for an enzyme activity exist (isozymes) then reduction in the amount of a single gene product may have no effect. It is possible, however, that closely-related isozymes will all be sensitive to a single chemical inhibitor. In such cases, a genetic approach to validation could be misleading. While there are no examples yet from work on *S. nodorum*, this principle has been demonstrated by the disruption of the cutinase gene in *M. grisea*. In this case, the resulting mutants were as pathogenic as the wild-type. It was subsequently shown that a second enzyme activity was still present (Sweigard *et al.*, 1992). This emphasizes the need for careful biochemical analysis of mutants before conclusions can be drawn about the essentiality of a particular enzyme activity.

Supply of Target Protein

An important contribution of molecular biology to fungicide discovery is the provision of large amounts of recombinant enzyme through cloning and overexpression. Biochemical design approaches rely on a thorough characterization of the target enzyme and accurate enzyme inhibition data. The use of enzyme assays is not, however, limited to biochemical design. In all cases where the biochemical target has been identified, information on the activity of potential fungicides at the molecular level can prove invaluable in directing chemical synthesis. Inhibitors which fail to reach their target because of poor uptake or breakdown in the cell are overlooked in an *in vivo* screen. However, where the delivery problems can be addressed, these compounds can provide valuable leads. Furthermore, biochemical assays are generally quicker than *in vivo* screens so have the potential to provide very rapid feedback to synthesis programmes.

Extraction and purification of enzymes from their fungal sources can be time-consuming, but molecular biology can provide rapid access to large amounts of the target protein. This is illustrated using a current target enzyme which has been cloned from *S. nodorum* and expressed in *Escherichia coli* at AgrEvo. In this case, 5 ml of the expressing *E. coli* culture grown overnight provides an amount of enzyme equivalent to a 2 l culture of *S. nodorum* grown over 5 days. No attempt has been made to optimize the expression of this particular enzyme and it is almost certain that, if necessary, far greater yields could be achieved. Furthermore, the higher specific activity of the recombinant enzyme provides a better starting point for further purification. For crystallization studies large amounts of homogeneous protein are required. In such circumstances the use of modern protein expression/purification vectors facilitates production and purification.

A good supply of enzyme from a suitable source is also important for high throughput biochemical screening which represents a complimentary approach to rational design in fungicide discovery. High throughput biochemical screening involves testing large numbers of compounds, generally tens of thousands, against an enzyme target, or targets, to identify novel inhibitors. There are several recent reviews of this approach (see Ormrod and Hawkes, 1995; Plummer, 1996). High throughput biochemical screening has been made feasible by the availability of large compound libraries and advances in automated screening, as well as the ability to obtain large amounts of the target protein. As is the case for biochemical design, some confidence in the essentiality of the target is required before investing the resource necessary to conduct a large-scale screening campaign, and genetic validation of targets, therefore, has a role to play.

A more recent development is the introduction of so-called smart screens. These generally exploit cells which have been genetically modified in such a way that interference with the targeted process is readily detected, for example through use of reporter genes such as GUS and GFP (Timberlake, 1995). These approaches will broaden the potential sites of action beyond the primary target to include all functions associated with the correct expression of the target.

Success Record of Biochemical Design and High Throughput Biochemical Screening Approaches

While there are many examples of the successful rational design of enzyme inhibitors, how many of these have translated into fungicides? In terms of actual products, we are not aware of any rationally-designed fungicides. Owing, no doubt, to the greater tractability of plants and a more advanced basic understanding, more progress has been made with the rational design of herbicides. Inhibitors designed against glutamine synthetase (Wright *et al.*, 1991) and pyruvate dehydrogenase (Baillie *et al.*, 1988) proved to be promising herbicides. These failed to reach development because, in the former case, the compound had already been synthesized elsewhere and is now a commercially successful herbicide with the common name phosphinothricin. In the latter, field trials revealed unacceptable crop damage. For true success stories it is necessary to look to the pharmaceutical industry and HIV protease where a combination of mechanism- and structure-based design has yielded at least two commercial products to date (Erickson *et al.*, 1990).

A similar situation exists for high throughput screening where, again, we are not aware of any commercial successes in the agrochemical industry. The interest and investment by agrochemical companies in this technology does, however, demonstrate that there is confidence in the future success of this approach. The chances of commercial success for fungicide discovery using these biochemical approaches will become greater as our basic understanding of fungal pathogenicity continues to increase.

Life-cycle Studies

Underpinning all these aspects of fungicide discovery is a clear requirement to fully understand the basic mechanisms of cellular pathology in the susceptible host-pathogen interaction. The progress being made in the genetic manipulation of *S. nodorum* and other phytopathogenic fungi provides opportunities for understanding the biochemical basis of pathogenicity and it is important that these are fully exploited. If we are to understand the fundamental mechanisms behind key processes in pathogenicity, it is essential to define the sequence of infection structure differentiation and its likely role in pathogenesis.

As well as contributing to our basic understanding of fungal pathogenicity, life-cycle studies play an important role in the fungicide discovery process. One of the greatest challenges in fungicide discovery is optimizing delivery of the compound to its site of activity *in planta* within the constraints imposed by its physico-chemical properties, which include systemicity and vapour activity.

Often, the biological impact of a compound can be enhanced by the use of different formulation types and tank-mix adjuvants. This requires a knowledge of the stage(s) in the life-cycle which are more or less subject to chemical inhibition. If the target enzyme is located within the fungus, the compound must enter the fungus, possibly via the plant. *En route* it must avoid metabolism (unless of course it is a propesticide, in which case it must be effectively activated), photodegradation, evaporation and being washed off by rain. Identification of the sensitive stage of the life-cycle enables the compound to be formulated so that it is most effectively delivered.

Life-cycle of *Septoria tritici*

In recent years, *S. tritici* has become the predominant species in winter wheat in the UK and western Europe and was, therefore, selected as the model for our own microscopy-centred life-cycle studies. Despite its commercial importance, a complete and detailed description of the infection strategy of *S. tritici* has not yet been reported in the literature. Some information is available from Hilu and Bever (1957), Cohen and Eyal (1993) and Kema *et al.* (1996). At AgrEvo, novel microscopic approaches are being used to evaluate dynamic responses during interactions between individual spores of *S. tritici* and host cells in highly controlled and optimized pathosystems.

Historically, the life-cycle of *S. nodorum* has been better studied than that of *S. tritici.* Although *S. nodorum* and *S. tritici* have been confirmed as distinct species, there are still basic similarities in elements of their infection strategies - specifically in their ability to colonize mesophyll tissue without damaging host cells via the mechanism known as latency. (Compare studies on *S. nodorum* by Shearer and Zadoks (1972); Baber and Smith (1978); King *et al.* (1983) with those on *S. tritici* described in the following section.) It should be noted, however, that tissue specificity is much less distinct in *S. nodorum*, which infects seed, foliage and glumes, whilst *S. tritici* is primarily a foliar pathogen. In our own studies, we have observed that pycnidia of *S. nodorum* are formed within mesophyll tissue and not exclusively in the substomatal cavity as seen for *S. tritici.* Care should, therefore, be taken in extrapolating data derived from one species to another.

Another important point to make is that generalization with respect to host-pathogen interactions at the cellular level should be drawn from studies with several isolates, and not just one since this may be atypical in a variable pathosystem, or where host-cultivar adaptation is known to occur. We have been working with six recent UK field isolates of *S. tritici* adapted to growth on wheat cv. Brigadier.

Penetration

In contrast to the work of Kema *et al.* (1996), who used two isolates adapted to cv. Shafir, we have shown that penetration through stomata (via chance encounter) accounts for a relatively small proportion of successful infections. In the isolates surveyed in our work the majority of infections occur in anticlinal cell wall grooves, particularly those of the ridges bearing leaf trichomes. Superficial hyphae grow along these grooves and often elaborate simple swellings along their length from which infection hyphae are produced which grow down between cells (intramural growth).

Latency

Once beneath epidermal cell layers, thin infection hyphae switch to a rapidly growing thicker mycelial growth form similar to runner hyphae seen in other filamentous ascomycetes such as *Tapesia* spp. and *Venturia inaequalis*. These grow along the anticlinal cell wall grooves on the underside of epidermal cells and then develop lateral branches which grow around subtending mesophyll cells (Fig. 22.1A). When a junction between adjacent mesophyll cells is encountered, further branching is stimulated and so forth until the mesophyll tissue becomes extensively colonized from a single successful penetration site. The area of infection becomes extended in the long axis of the leaf due to the original longitudinal extension of subepidermal runner hyphae, this being ultimately expressed as the short 'stripe' lesions seen in field infections. This period of latency is a key feature of *S. tritici* infections and occurs in the absence of visual symptoms. Mesophyll cells are not penetrated or damaged in any way, and there are no haustorial bodies or similar feeding structures via which host cell nutrients might be made available. The supposition is that the pathogen survives on nutrients present in free apoplastic water. The duration of latency is clearly dependent on environmental conditions of which temperature and humidity are probably the most important, and in the field has been recorded to last up to 25 days.

Pycnidium Formation

Under optimum conditions (approximately 20°C and high relative humidity) there is a distinct 'switch' in the fungal lifestyle, possibly stimulated by attainment of a critical fungal biomass linked to a starvation response. This is characterized by mesophyll cell damage as the pathogen becomes necrotrophic. It should be noted at this point that susceptible interactions produce much less visible tissue damage than resistant interactions - the former primarily being characterized by high pycnidial density. Extensive early tissue browning appears to indicate hypersensitive cell death. Low-temperature scanning

A

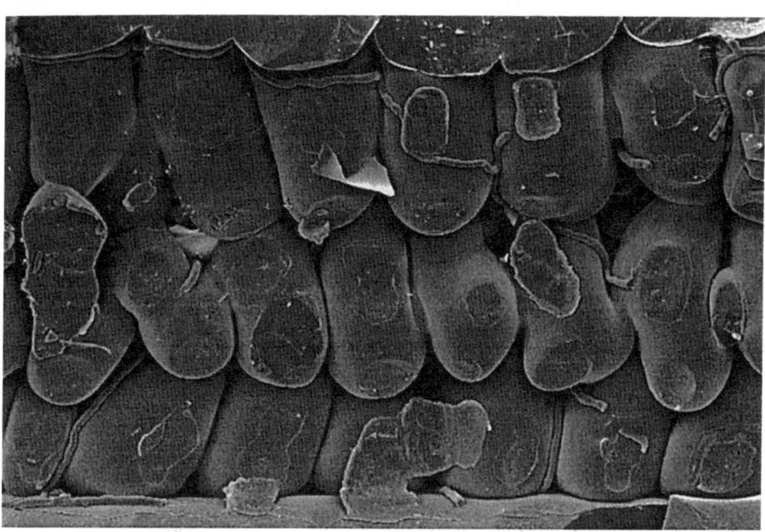

B

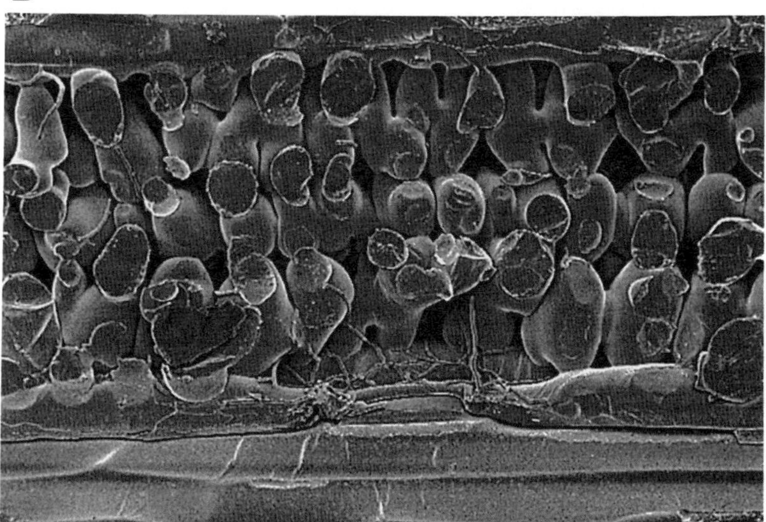

Fig. 22.1. Low-temperature scanning electron micrographs of fully frozen hydrated tissue. A: Cryofracture through a wheat leaf, 7 days after inoculation with *Septoria tritici*, showing random invasive mycelium in the absence of visible symptoms. B: Ten days after inoculation, showing mycelial proliferation in the substomatal cavity preceding pycnidium differentiation.

electron microscopy indicates extensive secretion of mucilaginous material during this phase. By this time, randomly invasive mycelium has coincidentally encountered most of the substomatal cavities (SSC) within an infected leaf area (as well as the contribution from directly penetrated stomata). It is not clear whether this newly acquired nutrient base provides a trigger for subsequent pycnidial differentiation in the SSC - but their exclusive location in this host tissue compartment indicates that it provides an optimum physical and/or chemical environment for their development. Pycnidial differentiation involves initial formation of a stomatal mass directly beneath guard cells (Fig. 22.1B) followed by a maturation phase in which vertically aligned parallel hyphae, similar to paraphyses, are formed which protrude through the ostiole located beneath the stomatal aperture. During this phase, large amounts of extracellular matrix (cirrhus) are produced which appear to stimulate conidiogenesis. Conidia appear to be formed by a process of microcycle conidiation and become aligned in parallel arrays using the 'paraphyses' as an initial scaffold. Spores are initially highly symmetrical and short (approximately 12 μm as seen in the yeast-like growth form *in vitro*), and elongate to become typically more curved and irregular once cirrhus extrusion has taken place. During the process of cirrus extrusion through the stomatal aperture, the cirrhus dries to a thin film on external surfaces indicating an antidessicant role. This is readily re-hydrated in water resulting in liberation of many thousands of spores from a single pycnidium.

Thus, it can be seen that key processes for potential targeted disruption are those involved in establishment and maintenance of latent phase - since this leads to leaf colonization, eventually resulting in host cell death and ultimately giving rise to pycnidium formation. An understanding of the mechanism of nutrient transfer during this phase may lead to the identification of novel targets. Furthermore, as noted by other workers, the nature of the necrosis-inducing compounds produced during the switch from latency to necrotrophy should also be elucidated by analysis of intracellular washing fluids.

Concluding Comments

The development of genetic approaches to investigate pathogenicity provides the potential to start addressing our lack of understanding about the basic biochemical processes involved. It is these processes which will form the basis for future fungicide design and discovery programmes. The development of gene silencing methods will allow new targets to be validated with the degree of confidence required to support inhibitor design and high throughput screening programmes. The use of strains carrying reporter genes under the control of promoters from target genes will allow us to assess when and where the target is expressed during pathogenicity. Such information may prove useful in the development and formulation of novel fungicides in order to ensure their efficient application as well as providing more basic information about the role of the target enzyme.

Stagonospora nodorum has been chosen as a model pathosystem for the identification of novel fungicide targets. In parallel, a detailed description of the life-cycle of the closely-related and commercially more significant *S. tritici* has been made. This has revealed a need for more detailed analyses of latency, which represents possibly the most important, but least understood, stage of pathogenesis. An exciting challenge for the future will be to discover the molecular mechanisms which determine such events in the life-cycle.

Our understanding of the mechanisms of cellular pathogenesis is advancing and many of the relevant biochemical and molecular biology tools for the design of fungicides are now in place. However, there are still significant gaps in our knowledge and a coordinated, interdisciplinary effort is now required to advance our understanding of pathogenicity and to continue to support fungicide discovery.

Acknowledgement

We thank Dr John Pillmoor for critically reading the manuscript.

References

Baber, E.A. and Smith, I.M. (1978) Development of resistant and susceptible reactions in wheat inoculation with *Septoria nodorum* (wheat glume blotch). *Transactions of the British Mycological Society* 71, 475-482.

Baillie, A.C., Wright, K., Wright, B.J. and Earnshaw, C.G. (1988) Inhibitors of pyruvate dehydrogenase as herbicides. *Pesticide Biochemistry and Physiology* 30, 103-112.

Beautement, K., Clough, J.M., Fraine, P.J. and Godfrey, C.R.A. (1991) Fungicidal β-methoxyacrylates: from natural products to novel synthetic agricultural fungicides. *Pesticide Science* 31, 499-519.

Cohen, l. and Eyal, Z. (1993) The histology of processes associated with the infection of resistant and susceptible wheat cultivars with *Septoria tritici*. *Plant Pathology* 42, 737-743.

Erickson, J., Neidhart, D.J., Van Drie, J., Kempf, D.J., Wang, X.C., Norbeck, D.W., Plattner, J.J., Rittenhouse, J.W., Turon, M., Widburg, N., Kohlbrenner, W.E., Simmer, R., Helfrich, R., Paul, D.A. and Knigge, M. (1990) Design activity and 2.8Å crystal structure of a C_2 symmetric inhibitor complexed to HIV-1 protease. *Science* 249, 527.

Hilu, H.M. and Bever, W.M. (1957) Inoculation, oversummering and suscept-pathogen relationship of *Septoria tritici* on *Triticum* species. *Phytopathology* 47, 474-480.

Höfgen, R., Laber, B., Schuttke, I., Klonus, A., Streber, W. and Pohlenz, H. (1995) Repression of acetolactate synthase activity through antisense inhibition. *Plant Physiology* 107, 469-477.

Huxley-Tencer, A., Francotte, E. and Bladocha-Moreau, M. (1992) 1(R)-(2,6-*cis*-dimethylmorpholino)-3(S)-(p-tert-butylphenyl)cyclopentane: a representative of a novel potent class of bio-rationally designed fungicides. *Pesticide Science* 34, 65-74.

Kacser, H. and Burns, J. A. (1973) The control of flux. In: Davies, D.D. (ed.) *Rate Control of Biological Processes, The Society for Experimental Biology Symposium No. 27.* Cambridge University Press, Cambridge, pp. 65-107.

Kema, G.H.J., Yu, D., Rijkenberg, H.J., Shaw, M.W. and Baayen, R.B. (1996) Histology of the pathogenesis of *Mycosphaerella graminicola* in wheat. *Phytopathology* 86, 777-786.

King, J.E., Cook, R.J. and Melville, S.C. (1983) A review of *Septoria* diseases of wheat and barley. *Annals of Applied Biology* 103, 345-373.

Oliver, R. and Osbourn, A. (1995) Molecular dissection of fungal phytopathogenicity. *Microbiology* 141, 1-9.

Ormrod, J.C. and Hawkes, T.R. (1995) Screening practices in the agrochemical industry. *Proceedings of the 1995 Brighton Crop Protection Conference - Weeds* 1, 97-103.

Parks, L.W., Crowley, J.H., Smith, S.J., Leak, F.W. and Palmero, L.M. (1995) The functions of sterols in antifungal control. In: Dixon, G.K., Copping, L.G. and Hollomon, D.W. (eds) *Antifungal Agents: Discovery and Mode of Action.* BIOS Scientific Publishers, Oxford, pp. 49-57.

Plummer, J. (1996) High throughput screening: an agricultural perspective. *Journal of Biomolecular Screening* 1, 39-41.

Sauter, H., Ammermann, E., Benoit, R., Brand, S., Gold, R.E., Grammenos, W., Köhle, H., Lorenz, G., Müller, B., Röhl, F., Schirmer, U., Speakman, J.B., Wenderoth, B. and Wingert, H. (1995) Mitochondrial respiration as a target for antifungals: lessons from research on strobilurins. In: Dixon, G.K., Copping, L.G. and Hollomon, D.W. (eds) *Antifungal Agents: Discovery and Mode of Action.* BIOS Scientific Publishers, Oxford, pp. 173-190.

Shearer, B.L. and Zadoks, J.C. (1972) The latent period of *Septoria nodorum* in wheat 1. The effect of temperature and moisture treatments under controlled conditions. *Netherlands Journal of Plant Pathology* 78, 231-241.

Sweigard, J.A., Chumley, F.G. and Valent, B. (1992) Disruption of a *Magnaporthe grisea* cutinase gene. *Molecular and General Genetics* 232, 183-190.

Talbot, N.J. (1995) Having a blast: exploring the pathogenicity of *Magnaporthe grisea. Trends in Microbiology* 3, 9-16.

Timberlake, W.E. (1995) Cellular reporters for antifungal drug discovery. In: Dixon, G.K., Copping, L.G. and Hollomon, D.W. (eds) *Antifungal Agents: Discovery and Mode of Action.* BIOS Scientific Publishers, Oxford, pp. 17-29 .

Valent, B. (1990) Rice blast as a model system for plant pathology. *Phytopathology* 80, 33-36.

Wright, K., Pillmoor, J.B. and Briggs, G.G. (1991) Rationality in herbicide design. In: Baker, N.R. and Percival, M.P. (eds) *Herbicides, Topics in Photosynthesis*, Volume 10, Elsevier, Amsterdam, pp. 337-369.

Chapter twenty-three:

Breeding for Resistance to *Septoria* and *Stagonospora* Diseases of Wheat

Z. Eyal
Department of Plant Sciences and the Institute for Cereal Crops Improvement, The George S. Wise Faculty of Life Sciences, Tel Aviv University, Tel Aviv 69978, Israel

Introduction

Host resistance is considered to be the 'main pillar of defense against disease' (Browning, 1979), yet detailed information on the types of resistance to *Septoria tritici* and *Stagonospora nodorum*, their inheritance, manipulation, accumulation, longevity, and on the mechanisms controlling resistance in important wheat accessions is scant. In any breeding programme resistance is being sought that is (a) stable over time and variable disease intensities; (b) easy to incorporate and identify in segregating populations; and (c) non-detrimental to yield potential under disease-free conditions (Nelson and Marshall, 1990).

Wheat cultivars can vary considerably in their response (necrosis and/or pycnidia) to each of the two pathogens. This variation has been used with variable success to improve disease resistance to the two pathogens. In certain cultivar x *S. tritici* isolate combinations seedling response can predict the adult plant response (Danon and Eyal, 1990; Kema and Van Silfhout, 1997). In most cases such correlations are low and appear not to be sufficient for use in breeding for resistance to Stagonospora nodorum blotch (Arseniuk *et al.*, 1991). The evaluation of germplasm under natural field conditions is affected by plant height and maturity level. In general, plant height and maturity level are negatively correlated with pycnidial coverage of *S. tritici* (Brokenshire, 1976; Tavella, 1978; Danon *et al.*, 1982; Van Beuningen and Kohli, 1990) and of *S. nodorum* (Scott *et al.*, 1982). Plant height is more influential where splashed pycnidiospores serve as the driving force in vertical dissemination. These factors may be less important where the epidemic is generated and perpetuated by incoming ascospores. Evaluation of germplasm under natural infection is done by exposure to a population of pathogen biotypes of unknown virulence spectra. In breeding programmes where screening for resistance is done routinely and in an orderly fashion, measures are taken to ensure selection pressure of a sort by artificially introduced inoculum (e.g. infected straw, inoculation with isolate mixtures). It is assumed that the screened germplasm is subjected to a population representing a segment or more of the natural pathogen population to which this genotype will be exposed upon gaining wider distribution. Discontinuous and low selection pressure due to poor establishment of the pathogen in the breeding programme may interfere with the breeding cycle. Multilocation and multiyear (national and international) testing of germplasm expands its exposure to different virulence spectra and to epidemics of different magnitudes under varying environmental profiles, which may, thus, increase the probability of identifying resistant genotypes with wider geographical adaptability and stability. In order to ensure reasonable selection pressure, 'hot spots', where the disease is frequent and severe and the virulence spectrum is wide, should be utilized for germplasm evaluation (e.g. for *S. tritici* Holletta, Ethiopia; Patzcuaro, Mexico). Screening for disease resistance at the seedling stage of fixed genotypes and of segregating populations in breeding schemes is not a routine procedure. Information on the interaction of host x pathogen with the environment, the effect of growth stage on disease response, the response to other plant pathogens, and the agronomic performance are essential determinants of any breeding programme.

Resistance to *Septoria tritici*

Resistance to *S. tritici* is based on quantitative assessment (percentage necrosis and/or pycnidia) or on indexing of symptoms on specific leaves with or without reference to the plant growth stage. An 'immune' response where no pycnidia are formed on a specific wheat cultivar within a *Triticum* spp. is rare. The categorization of resistance on the basis of a low level of symptoms has produced several thresholds to differentiate host response classes. One of the major difficulties in establishing a fixed differentiating response threshold is associated with variations in the expression of symptoms of the tested accessions under variable epidemics (Eyal and Talpaz, 1990). The level of symptoms of 21,000 highly susceptible or resistant wheat accessions over the 8 trial years remained fairly consistent. Highly susceptible accessions may express a high level of symptoms even under less favourable conditions, which may lead to an erroneous conclusion about the level of the epidemic within a trial, leading to underestimated severity on the screened germplasm. It was, therefore, suggested that comparative evaluations should be made with semi-dwarf/early-moderate maturing accessions expressing a low to moderate level of symptoms (±30% pycnidial coverage). The response of such accessions will fluctuate between years and locations according to the level of the epidemic established in the trial, and their stature/maturity profile will not be subjected to physiological bias. A less stringent approach may result in interpreting a low level of symptoms (poor epidemics, plant height/maturity interference) as genetically controlled resistance.

The incorporation of resistance from various sources (alien and cultivated wheat) into high-yielding cultivars lacking in resistance largely depends on appropriate screening under disease pressure at early phases of breeding programmes. The dependency on rainfall for spore splash and for the onset of infection introduces an unregulated major factor into screening of germplasm and breeding populations. This holds true particularly under semi-arid regimes where rainfall is irregular and epidemics are infrequent, and where pycnidiospores provide the major inoculum source. Any irregularity in selection pressure decreases the reliability of the screening and increases the amount of untested breeding material in consecutive generations.

Multilocation testing revealed differential symptom expression on genotypes classified as being resistant under certain conditions. It is likely that differences in virulence patterns (national and international) contribute to the differential host response to *S. tritici* (Eyal *et al.*, 1985; Eyal and Levy, 1987; Ballantyne, 1989; Perello *et al.*, 1989; Silfhout *et al.*, 1989; Kema *et al.*, 1996a,b). Resistance in the spring wheat accessions Bobwhite 'S' (CIMMYT, Mexico) and Kavkaz/K4500 L.6.A.4 (CIMMYT, Mexico) was reported to be controlled by several genes, with predominantly additive effects (Cordo *et al.*, 1994; Gilchrist and Velazquez, 1994; Jlibene *et al.*, 1994; Dubin and Rajaram, 1996). The South American and other wheat accessions containing cv. Frontana (Polyssu/Alfredo Chavez 6//Mentana) germplasm in their background (Bobwhite 'S', Colotana, Frontana, Enkoy, Fortaleza, IAS-20, Kavkaz/K4500 L.6.A.4, Klein Titan, Maringa, Lagoa Vermelha, Nova Prata, Toropi and Veranopolis) had continuously expressed low pycnidial coverage (seedling and adult) to the Israeli population of *S. tritici* (Eyal *et al.*, 1983; 1985; Eyal and Levy, 1987; Eyal and Talpaz, 1990; Eyal, 1995). Resistance to some of these Frontana-derived accessions was reported to be controlled by several recessive genes (Bobwhite 'S', Colotana and Klein Titan), whereas resistance in cv. Veranopolis (Trintecinco/Frontana) to Australian isolates of *S. tritici* was reported to be controlled by a single dominant gene (Wilson, 1979) (Table 23.1). The resistance of these accessions may not prove lasting if segments of the pathogen population adapt to this resistance following its incorporation into breeding programmes and the widening of its exposure to pathogen populations. Eyal (1995) suggested that the resistance in these accessions probably does not derive from the Russian parent cv. Aurora but rather from the Brazilian parent cv. Frontana, possibly in combination with other sources in such multiparental cultivars. Resistance of cvs Aurora, Bezostaya 1 and Kavkaz to Israeli isolate ISR398 is controlled by a partially dominant gene and probably another recessive gene and is not associated with the 1BL/1RS translocation from rye (Danon and Eyal, 1990). Somasco *et al.* (1996) reported that the semi-dwarf spring wheat cultivar Tadinia cross between European winter wheat and a spring cultivar (Tadorna/Inia 66) rendered immunity to the Californian population of *S. tritici* in both field and greenhouse conditions. Tadinia, Bulgaria 88 (Rillo and Caldwell, 1966) and Veranopolis (Wilson, 1979;

Table 23.1. Possible common Frontana background in spring wheat accessions resistant to *Septoria tritici.*

Wheat accessions	Pedigree[a]	Number of genes	Reference
Bet Lehem	**H574**/Lakhish 212	?	
Bobwhite 'S'	AU//KAL//BB/3/**WOP**	2 recessive	Gilchrist, 1994
Colotana	**Frontana**/Colonista	2 recessive	Danon and Eyal, 1990
Enkoy	Hbarand/4/WIS245/Sup51/3/**FR2**/**FR**//Yaqui	?	
Frontana	Polyssu/Alfredo Chavez 6//Mentana	?	
IAS 20 = IASSUL	Colonias//**Frontana**/Kenya 58	?	
Kavkaz/K4500 L.6.A.4	K4500 = **Enkoy**	several	
Milan 'S'	VS73.600/MRL/3/**BOW**/YR//TRF	?	Dubin and Rajaram, 1996
Veranopolis	Trintecinco/**Frontana**	1 dominant	Wilson, 1979

[a]Bold letters mean Frontana germplasm in background.

Ballantyne and Thomson, 1995) are the only accessions where a high level of resistance was reported to be controlled by single dominant genes. The gene for resistance in Tadinia was designated as *Stb4*, the one in Bulgaria 88 as *Stb1*, and the one in Veranopolis as *Stb2*. The authors stressed that some of the accessions reported to be resistant in the Californian study (Bulgaria 88, Cleo, Tadinia, Tadorna, UCD706-26 and UCD706-27) were moderately susceptible in Kansas, Chile and Israel. The single gene resistance to Septoria tritici blotch of the accessions Tadinia (resistance from Tadorna), Oasis (resistance from Bulgaria 88) and Veranopolis (resistance from Frontana ?) remained stable in California, Indiana and Australia, respectively. CIMMYT's wheat programme elevated resistance to Septoria tritici blotch in intermediate maturing, high-yielding semi-dwarfs by incorporating sources from China (in Catbird), France (in Milan), Brazil (in Corydon) and Russia (in Bobwhite 'S') (Dubin and Rajaram, 1996). There are few indications as to the lasting effectiveness of the recognized resistance sources following their incorporation into breeding material and upon increasing the area and distribution of the improved cultivars. Dubin and Rajaram (1996) stated that 'two to three genes are generally needed to impart acceptable levels of resistance'. The term 'acceptable' can be inferred to be the combined effect that several genes may have on reducing disease level.

Evidence for a decline in effectiveness was recently reported by Ahmed *et al.* (1996) for winter wheat cv. Gene which was highly resistant when released in 1992 in Oregon, and was found to harbour higher levels of Septoria tritici blotch in recent years. This cultivar and cv. Tadinia from California were highly resistant to *S. tritici* isolates from Texas, California and Oregon, whereas cvs Hill, Malcolm and Stephens were resistant to Texas and California isolates susceptible to isolates from Oregon. In the 1982-1984 national trials in the UK conducted under natural *S. tritici* infection, cv. Avalon expressed 6.4% infection and cv. Longbow was placed as the most susceptible (9.4%) (Bayles *et al.*, 1985). Cultivar Mercia, which was released several years after cvs Avalon and Longbow, expressed similar disease severity (17%) to these two cultivars (Bayles, 1991). The genetic basis of resistance of these and other highly resistant cultivars has not been studied, and it is, therefore, difficult to assess the mode of protection (specificity, non-specific, etc.) they possess and its durability. It is of interest to note that *M. graminicola* is operative in all three mentioned states, although its contribution to the virulence spectrum of *S. tritici* has not been investigated. The Dutch winter wheat cultivar Obelisk, that was considered resistant when released in 1985, is currently among the most susceptible cultivars (Kema *et al.*, 1996b).

The lengthy debate over specificity in the *S. tritici*-wheat pathosystem has postponed discussions on its implications, assuming that specificity does exist. The adaptability of segments of the pathogen population to resistant cultivars is evident from the study of Ahmed *et al.* (1996). This provided evidence for selection for cultivar-specific virulence, and indicated that resistance can be endangered by changes in the virulence of the pathogen population. Since monitoring of virulence in *S. tritici* is not a common practice, the strategy to be adopted for germplasm evaluation upon exposure to clonal or sexually introduced virulence and its interaction with resistance needs to be formulated.

The establishment of national and international virulence/resistance monitoring nurseries composed of identified resistance sources tested together with accessions of special interest (e.g. promising breeding lines) can provide a valuable tool for breeding, genetic and other purposes.

A longer incubation and latent period was reported by Jlibene and El Bouami (1995) for the bread wheat accessions VEE'S'/SNB'S' and Saada inoculated with two *S. tritici* isolates derived from bread wheat in Morocco. The incubation period in these two accessions was reported to be controlled by two complementary loci, and pycnidial density by one dominant independent locus. Leaf necrosis in VEE'S'/SNB'S' was controlled by one dominant locus but in Saada by two loci, one dominant and one recessive. The minimum number of independent loci involved in the inheritance of partial resistance components (incubation period, latent period, leaf necrosis and pycnidial density) would be five. The combined effect of these components in a breeding line may result in more effective and lasting protection, following exposure to a wider virulence spectrum and evaluation span (time and locations). The term 'slow-septoring' was suggested by Brönnimann (1982) to designate a reduced rate of Septoria nodorum blotch development. Genotypes known to be resistant to *S. tritici* in Western Australia (Millewa, CNT2, IASSUL, Corrigin and 77Z:893) exhibited low infection frequency, long latent period and low sporulation rate in a field trial conducted in 1990 (Loughman *et al.*, 1996). It is of interest to note that Wilson (1979) reported that resistance (level of symptoms) of the Brazilian accession IASSUL (IAS-20/Colonias//Frontana/Kenya 58) to *S. tritici* was determined by a single dominant gene. In the 1990 field trial the same accession expressed a low sporulation rate (58 x 10^5 spores g^{-1}), moderately high latent period (31 days) and low infection frequency (nine lesions 10 cm^{-2}). However, it exhibited a relatively low sporulation rate (21 x 10^5 spores g^{-1}), but moderate to high infection frequency and latent period in a glasshouse trial. It expressed low pycnidial coverage but rather tall stature and moderate to late maturity under Israeli conditions (Eyal *et al.*, 1983). It seems that the consistent low level of symptoms manifested by this accession can be assigned, in part, to the Frontana germplasm in its parentage (Eyal, 1995).

The use of tolerant cultivars where the yield of a susceptible cultivar is protected under an equivalent severe epidemic as compared to non-tolerant cultivars is not widely utilized (Zuckerman *et al.*, 1997). The phenomenon is poorly understood, although the advantages of reduced selection pressure on the pathogen increases its attractiveness as a protection measure. Eyal (1997) proposed a breeding strategy for the incorporation of tolerance into breeding schemes by combining resistance expressed as a low level of symptoms (tissue protection) and tolerance (yield protection) and stringently selecting for the latter by subjecting the breeding population to specific virulences. The strategy implies that when 'frontal' resistance is overcome by the pathogen, the 'background' tolerance may provide protection of yield.

Resistance to *Stagonospora nodorum*

Similarly to *S. tritici*, complete resistance has not been reported to *S. nodorum* in wheat. Specificity is much less distinct in the *S. nodorum* x wheat pathosystem. In moderately resistant cultivars, resistance may be controlled by the additive action of several genes, whereas in highly resistant cultivars, resistance may be governed by major resistance genes (Scharen and Eyal, 1983). The winter wheat accessions JCR-979, 81UWWMN2095 (Maris Huntsman//VPM/Moisson) and Red Chief and spring wheat cv. Frontana were highly resistant at the seedling stage to 33 isolates of *S. nodorum*. Nelson (1980) showed general combining ability effects to be highly significant for adult plant symptoms in a 6 x 6 diallel cross implying an additive gene effect. Certain crosses have shown significant specific combining ability which indicates non-additive gene action. Resistance in cvs Frondoso and Fronthatch was reported to be polygenic and could be explained by additive gene action (Mullaney *et al.*, 1982). Seedling resistance to *S. nodorum* was identified in the diploid wheats *Aegilops longissima* (genome S^lS^l), *A. speltoides* (genome SS) and *A. squarrosa* (genome DD) (Trottet and Dosba, 1983; Ecker *et al.*, 1990a,b).

Resistance to Stagonospora nodorum blotch on leaves was reported to be independent of resistance on the spikes and, therefore, seedling tests cannot replace field evaluations (Rosielle and Brown, 1980; Fried and Meister, 1987). The inheritance of both leaf and spike resistance was explained by Fried and Meister (1987) by additive-dominance mode. Bostwick *et al.* (1993) studied the inheritance of resistance to Stagonospora nodorum blotch in the Brazilian cultivar Cotipora (Veranopolis*2/Egypt NA101; Veranopolis = Trintecinco/Frontana) which expresses a low percentage of diseased tissue on both leaves and spikes. They suggested that combining flag leaf resistance, as in Cocker 84-27 with cv. Cotipora's leaf and spike resistance, may greatly improve the overall resistance level. Monosomic analysis showed that flag leaf resistance in cv. Cotipora is controlled by genes on chromosomes 3A, 4A and 3B whereas the spike resistance is controlled by genes on chromosomes 3A, 4A, 7A and 3B (Hu *et al.*, 1996). The data suggest that common genes are found on chromosomes 3A, 4A and 3B for flag leaf and spike resistance, but their magnitude may be different. Assuming one gene per chromosome, a minimum of four genes, on chromosomes 3A, 4A, 7A and 3B, are responsible for spike resistance. This gene number correlates well with the 3.16 gene number estimated from the genetic analysis by Bostwick *et al.* (1993).

King *et al.* (1983) stated that 'present evidence suggests that cultivar resistance to *S. nodorum* is non-specific and that isolates differ in aggressiveness but not in the range of cultivars attacked'. Partial resistance, characterized by a reduced rate of epidemic development, was evaluated based on the following components: infection frequency, latent period, size, shape and rate of lesion growth, spore production and its rate of increase (Jeger *et al.*, 1983; Lancashire and Jones, 1985). The combined effects of partial resistance are moderately heritable and generally appear to be expressed quantitatively. Jeger *et al.* (1981) suggested that cultivars with dissimilar combinations of resistance components may be used in mixtures, resulting in reduced levels of disease. This general conclusion was supported by Lancashire and Jones (1985) with the reservation that it holds true for cultivars differing in their latent period. The results they obtained demonstrate that within an apparently rather homogeneous group of resistant cultivars there is variability which may be used in cultivar mixtures. Significant variations in infection frequency, latent period and sporulation rate were observed in field and glasshouse trials among spring wheat genotypes in Western Australia (Loughman *et al.*, 1996). There was no correlation between the components of *S. nodorum* resistance. They, therefore, concluded that the lack of correlation between infection frequency and sporulation rate suggests that the genotypes were under independent genetic control, which thus may help to improve resistance by selection for low infection frequency, long latent period and low sporulation rate.

The resistance of winter spelt wheat (*Triticum spelta* L., genome AABBDD) cv. Oberkulmer to *S. nodorum* was studied using Quantitative Trait Loci (QTL) analysis (Messmer *et al.*, 1997). Two hundred and twenty-six recombinant F_5 inbred lines (RILs) derived from the cross between spelt wheat cv. Oberkulmer and winter wheat cv. Forno were analysed for disease severity (on leaf and spike), plant height, days to anthesis and ear morphology. The 121 markers (RFLP clones and microsatellite markers) tested on the 226 RILs yielded 156 segregated loci. Based on a preliminary map the investigators found up to seven QTL for resistance to leaf infection explaining 5-13% of the phenotypic variation, but only two were consistent over the 3 years the trail was conducted. Up to seven QTL were associated with spike resistance explaining 5-19% of the phenotypic variation. Only one QTL was common to leaf and spike resistance. QTL analysis is aiming towards developing tools that will be useful in marker-assisted selection for resistance to Stagonospora nodorum blotch in breeding programmes. Alignment of QTL with candidate genes (cloned genes or characterized sequences) based on anchored markers has great potential in identifying the actual gene underlying the QTL (Sorrells and Wilson, 1997). Placement of cloned genes and sequences on comparative maps, as suggested by Messmer *et al.* (1997) for *S. nodorum*, can facilitate the identification and selection of direct markers for QTL. The actual identity of the gene may be difficult to verify because of the possibility that the candidate is not the target gene but rather is tightly linked.

Concluding Remarks

Genetic protection, being economically and environmentally the favoured disease control measure, has not been readily exploited to provide measurable protection in wheat against the impact of Septoria tritici blotch and Stagonospora nodorum blotch. In only a few cases the wheat crop is protected by missionary breeding efforts of incorporated resistance against the two pathogens. Methodological difficulties hamper proper selection of suitable resistance sources, provision of frequent and continuous selection pressure, the question of virulence patterns, specificity, heritability, the effect of plant phenology on symptom expression and sporadic epidemics. All these, and other factors, contribute to discontinuity in providing a suitable level of genetic protection. Some of these difficulties are linked to the great variability in the pathogens, to unclear breeding strategies and to a lack of uniform methodologies.

The acceptance of specificity in the *Septoria tritici*-wheat pathosystem may contribute to an understanding of its magnitude and role in providing protection at the crop level rather than under controlled conditions. It is still not certain whether this type of resistance will provide the needed and lasting protection against diverse pathogen populations. The information gathered suggests that specific resistance provides protection under localized situations. The genetic basis, the mode of protection and the exposure to a wide spectrum of virulence of the recognized resistance sources (Bobwhite 'S', Bulgaria 88, IAS 20-IASSUL, Kavkaz/K4500 L.6.A.4, Veranopolis and others) have remained untested. The inheritance, incorporation and deployment of sources for quantitative resistance to *S. nodorum* need elucidation.

The International Monitoring *Septoria/Stagonospora* Nurseries philosophy executed by CIMMYT, Mexico should be adopted for both pathogens. It can provide information on cultivar x pathogen(s) interactions in regional, national and international domains. It will serve as a common basis for broad-base evaluation of germplasm to diverse pathogen populations under variable environmental conditions. Such an effort may identify basic elements related to the level of protection of old and new resistance sources, and may provide means for comparative evaluation of breeding material. It may further lay the foundation for better understanding of resistance souces, information which, in most cases, is scant or lacking.

With the implementation of new technologies to map host genomes, the genes associated with resistance to *S. tritici* and *S.nodorum* should be identified and placed on the maps. Once they are identified, their inheritance and mode of action will be easier to ascertain and their incorporation into breeding programmes and deployment into crop management systems will be more readily achieved.

References

Ahmed, H.U., Mundt, C.C., Hoffer, M.E. and Coakley, S.M. (1996) Selective influence of wheat cultivars on pathogenicity of *Mycospaeherella graminicola* (anamorph *Septoria tritici*). *Phytopathology* 86, 454-458.

Arseniuk, E., Fried, P.M., Winzler, H. and Czembor, H.J. (1991) Comparison of resistance of triticale, wheat and spelt in Septoria nodorum blotch at the seedling and adult plant stages. *Euphytica* 55, 43-48.

Ballantyne, B.J. (1989) Pathogenic variation in Australian cultures of *Mycosphaerella graminicola*. In: Fried, P.M. (ed.) *Proceedings of the 3rd International Septoria of Cereals Workshop*. Zurich, Switzerland, pp. 152-154.

Ballantyne, B. and Thomson, F. (1995) Pathogenic variation in Australian isolates of *Mycosphaerella graminicola. Australian Journal of Agricultural Research* 46, 921-934.

Bayles, R.A. (1991) Varietal resistance as a factor contributing to the increased importance of *Septoria tritici* Rob. and Desm. in the UK wheat crop. *Plant Varieties and Seeds* 4, 177-183.

Bayles, R.A., Parry, D.W. and Priestley, R.H. (1985) Resistance of winter wheat varieties to *Septoria tritici. Journal of the National Institute of Agricultural Botany* 17, 21-26.

Bostwick, D.E., Ohm, H.W. and Shaner, G. (1993) Inheritance of Septoria glume blotch resistance in wheat. *Crop Science* 33, 439-443.

Brokenshire, T. (1976) The reaction of wheat genotypes to *Septoria tritici. Annals of Applied Biology* 82, 415-423.

Brönnimann, A. (1982) Entwicklung der Kenntnisse über *Septoria nodorum* Berk. im Hinblick auf die Toleranz-oder Resistenzzüchtung bei Weizen. *Netherlands Journal of Agricultural Science* 30, 47-69.

Browning, J.A. (1979) Genetic protective mechanisms of plant-pathogen populations: Their coevolution and use in breeding for disease resistance. In: Harris, M.K. (ed.) *Biology and Breeding for Resistance*. Texas A & M University Press, Texas Publication MP-1451, pp. 52-75.

Cordo, C.A., Perello, A., Arriaga, H.O., Benedicto, G., Avila, V. and de Ziglino, I.R. (1994) Resistencia a la "mancha foliar" causada por *Septoria tritici* en el trigo pan (*Triticum aestivum* L.). *Revista de la Facultad de Agronomia* (La Plata) 70, 23-36.

Danon, T. and Eyal, Z. (1990) Inheritance of resistance to two *Septoria tritici* isolates in spring and winter bread wheat cultivars. *Euphytica* 47, 203-214.

Danon, T., Sacks, J.M. and Eyal, Z. (1982) The relationships among plant stature, maturity class, and susceptibility to Septoria leaf blotch of wheat. *Phytopathology* 72, 1037-1042.

Dubin, H.J. and Rajaram, S. (1996) Breeding disease-resistant wheats for tropical highlands and lowlands. *Annual Review of Phytopathology* 34, 503-526.

Ecker, R., Cahaner, A. and Dinoor, A. (1990a) The inheritance of resistance to Septoria glume blotch II. The wild wheat species *Aegilops speltoides*. *Plant Breeding* 104, 218-223.

Ecker, R., Cahaner, A. and Dinoor, A. (1990b) The inheritance of resistance to Septoria glume blotch III. The wild wheat species *Aegilops longissima*. *Plant Breeding* 104, 224-230.

Eyal, Z. (1995) Virulence in *Septoria tritici*, the causal agent of Septoria tritici blotch of wheat. In: Gilchrist, L., Van Ginkel, M., McNab, A. and Kema, G.H.J. (eds) *Proceedings of a Septoria tritici Workshop.* CIMMYT, Mexico D.F., pp. 27-33.

Eyal, Z. (1997) Genetic improvement for resistance. *Proceedings of the 10th Congress of the Mediterranean Phytopathological Union.* Montpellier, France, pp. 567-569.

Eyal, Z. and Levy, E. (1987) Variations in pathogenicity patterns of *Mycosphaerella graminicola* within *Triticum* spp. in Israel. *Euphytica* 36, 237-250.

Eyal, Z. and Talpaz, H. (1990) The combined effect of plant stature and maturity on the response of wheat and triticale accessions to *Septoria tritici*. *Euphytica* 46, 133-141.

Eyal, Z., Wahl, I. and Prescott, J.M. (1983) Evaluation of germplasm response to Septoria leaf blotch of wheat. *Euphytica* 32, 439-446.

Eyal, Z., Scharen, A.L., Huffman, M.D. and Prescott, J.M. (1985) Global insights into virulence frequencies of *Mycosphaerella graminicola*. *Phytopathology* 75, 1456-1462.

Fried, P.M. and Meister, E. (1987) Inheritance of leaf and head resistance of winter wheat to *Septoria nodorum* in a diallel cross. *Phytopathology* 77, 1371-1375.

Gilchrist, L. (1994) New *Septoria tritici* resistance sources in CIMMYT germplasm and its incorporation in the Septoria Monitoring Nursery. In: Arseniuk, E., Goral, T. and Czembor, P. (eds) *Proceedings of the 4th International Septoria of Cereals Workshop*. Radzikow, Poland, pp. 187-190.

Gilchrist, L. and Velazquez, C. (1994) Interaction to *Septoria tritici* isolate-wheat as adult plant under field conditions. In: Arseniuk, E., Goral, T. and Czembor, P. (eds) *Proceedings of the 4th International Septoria of Cereals Workshop.* Radzikow, Poland, pp. 111-114.

Hu, X., Bostwick, D., Sharma, H., Ohm, H. and Shaner, G. (1996) Chromosome and chromosomal arm locations of genes for resistance to Septoria glume blotch in wheat cultivar Cotipora. *Euphytica* 91, 251-257.

Jeger, M.J., Griffiths, E. and Jones, D.G. (1981) Influence of environmental conditions on spore dispersal and infection by *Septoria nodorum. Annals of Applied Biology* 99, 29-34.

Jeger, M.J., Jones, D.G. and Griffiths, E. (1983) Components of partial resistance of wheat seedlings to *Septoria nodorum*. *Euphytica* 32, 575-584.

Jlibene, M. and El Bouami, F. (1995) Inheritance of partial resistance to *Septoria tritici* in hexaploid wheat (*Triticum aestivum*). In: Gilchrist, L., Van Ginkel, M., McNab, A. and Kema, G.H.J. (eds) *Proceedings of a Septoria tritici Workshop*. CIMMYT, Mexico D.F., pp. 117-125.

Jlibene, M., Gustafson, J.P. and Rajaram, S. (1994) Inheritance of resistance to *Mycosphaerella graminicola* in hexaploid wheat. *Plant Breeding* 112, 301-310.

Kema, G.H.J. and van Silfhout, C.H. (1997) Genetic variation for virulence and resistance in the wheat-*Mycosphaerella graminicola* pathosystem III. Comparative seedling and adult plant experiments. *Phytopathology* 87, 266-272.

Kema, G.H.J., Annone, J.G., Sayoud, R., Van Silfhout, C.H., Van Ginkel, M. and de Bree, J. (1996a) Genetic variation for virulence and resistance in the wheat-*Mycosphaerella graminicola* pathosystem I. Interactions between pathogen isolates and host cultivars. *Phytopathology* 86, 200-212.

Kema, G.H.J., Sayoud, R., Annone, J.G. and Van Silfhout, C.H. (1996b) Genetic variation for virulence and resistance in the wheat-*Mycosphaerella graminicola* pathosystem II. Analysis of interactions between pathogen isolates and host cultivars. *Phytopathology* 86, 213-220.

King, J.E., Cook, R.J. and Melville, S.C. (1983) A review of *Septoria* diseases of wheat and barley. *Annals of Applied Biology* 103, 345-373.

Lancashire, P.D. and Jones, D.G. (1985) Components of partial resistance to *Septoria nodorum* in winter wheat. *Annals of Applied Biology* 106, 541-553.

Loughman, R., Wilson, R.E. and Thomas, G.J. (1996) Components of resistance to *Mycosphaerella graminicola* and *Phaesophaeria nodorum* in spring wheats. *Euphytica* 89, 377-385.

Messmer, M., Keller, M., Winzeler, M., Schachermayr, G., Feuillet, C., Gallego, F., Winzeler, H. and Keller, B. (1997) Identification of Quantitative Trait Loci (QTL) for *Septoria nodorum* resistance in a wheat by Spelt population. *Proceedings of International Triticeae Mapping Initiative (ITMI) Public Workshop*. Clermont-Ferrand, France, p. S12.

Mullaney, E.J., Martin, J.M. and Scharen, A.L. (1982) Generation mean analysis to identify and partition the components of genetic resistance to *Septoria nodorum* in wheat. *Euphytica* 31, 539-545.

Nelson, L.R. (1980) Inheritance of resistance to *Septoria nodorum* in wheat. *Crop Science* 20, 447-449.

Nelson, L.R. and Marshall, D. (1990) Breeding wheat for resistance to *Septoria nodorum* and *Septoria tritici*. *Advances in Agronomy* 44, 257-277.

Perello, A.E., Cordo, C.A., Arriaga, H.O. and Alippi, H.E. (1989) Variation in virulence in isolates of *Septoria tritici* Rob. ex. Desm. on wheat. In: Fried, P.M. (ed.) *Proceedings of the 3rd International Septoria of Cereals Workshop*. Zurich, Switzerland, pp. 42-46.

Rillo, A.O. and Caldwell, R.M. (1966) Inheritance of resistance to *Septoria tritici* in *Triticum aestivum* subsp. *vulgare*, Bulgaria 88. *Phytopathology* 56, 597.

Rosielle, A.A. and Brown, A.G.P. (1980) Selection for resistance to *Septoria nodorum* in wheat. *Euphytica* 29, 337-346.

Scharen, A.L. and Eyal, Z. (1983) Analysis of symptoms on spring and winter wheat cultivars inoculated with different isolates of *Septoria nodorum*. *Phytopathology* 73, 143-147.

Scott, P.R., Benedikz, P.W. and Cox, C.J. (1982) A genetic study of the relation between height, time of ear emergence and resistance to *Septoria nodorum* in wheat. *Plant Pathology* 31, 45-60.

Silfhout, C.H. Van, Arama, P.F. and Kema, G.H.J. (1989) International survey of factors of virulence of *Septoria tritici*. In: Fried, P.M. (ed.) *Proceedings of the 3rd International Septoria of Cereals Workshop*. Zurich, Switzerland, pp. 36-38.

Somasco, O.A., Qualset, C.O. and Gilchrist, D.G. (1996) Single-gene resistance to *Septoria tritici* blotch in the spring wheat cultivar Tadinia. *Plant Breeding* 115, 261-267.

Sorrells, M.E. and Wilson, W.A. (1997) Direct classification and selection of superior alleles for crop improvement. *Crop Science* 37, 691-697.

Tavella, C.M. (1978) Date of heading and plant height of wheat varieties as related to Septoria leaf blotch damage. *Euphytica* 27, 577-580.

Trottet, M. and Dosba, F. (1983) Analyse cytogénétique et comportement vis-à-vis de *Septoria nodorum* d'hybrides *Triticum* sp. x *Aegilops squarrosa* et de leurs descendances. *Agronomie* 3, 659-664.

Van Beuningen, L.T. and Kohli, M.M. (1990) Deviation from the regression of infection on heading and height as a measure of resistance to Septoria tritici blotch in wheat. *Plant Disease* 74, 488-493.

Wilson, R.E. (1979) Resistance to *Septoria tritici* in two wheat cultivars, determined by an independent, single dominant gene. *Australian Plant Pathology* 8, 16-18.

Zuckerman, E., Eshel, A. and Eyal, Z. (1997) Physiological aspects related to tolerance of spring wheat cultivars to Septoria tritici blotch. *Phytopathology* 87, 60-65.

Index

Figures in **bold** indicate major references.
Figures in *italic* refer to diagrams, photographs and tables.